Lecture Notes in Physics

For information about Vols. 1–67, please contact your bookseller or Springer-Verlag.

Lecture Notes
in Physics

Edited by J. Ehlers, München K. Hepp, Zürich
R. Kippenhahn, München H. A. Weidenmüller, Heidelberg
and J. Zittartz, Köln
Managing Editor: W. Beiglböck, Heidelberg

148

Advances in Fluid Mechanics
Proceedings of a Conference
Held at Aachen, March 26–28, 1980

Edited by E. Krause

Springer-Verlag
Berlin Heidelberg GmbH 1981

Editor

Egon Krause
Aerodynamisches Institut, RWTH Aachen
Wüllnerstr. zw. 5–7, D-5100 Aachen

ISBN 978-3-540-11162-7   ISBN 978-3-540-38635-3 (eBook)
DOI 10.1007/978-3-540-38635-3

2153/3140-543210

Contents

List of Speakers

Prof. Dr. F. Schultz-Grunow,   Institut für Allgemeine Mechanik der RWTH Aachen
Templergraben 55, 51 Aachen, BRD

Prof. Dr. H. W. Liepmann,   California Institute of Technology, Firestone Flight Sciences Lab.
Pasadena, California 91125, USA

Prof. Dr. H. O. Kreiss,   California Institute of Technology, Applied Mathematics,
Firestone Laboratory,
Pasadena, California 91125, USA

Prof. Dr. V. Rusanov,   Keldish Institute of Applied Mathematics,
Miusskaya sq. 4, 125047 Moscow, USSR

Prof. Dr. L. Ting,   New York University, Courant Institute of Mathematical Sciences
251 Mercer Street, New York, N.Y. 10012, USA

Dr. J. E. Green,   Royal Aircraft Establishment
Farnborough, Hants, Great Britain

M. Sirieix,
J. Délery,   O.N.E.R.A., 29 Avenue de la Division Leclerc,
92 Chatillon-sous-Bagneux (Seine), France

E. Stanewsky,   DFVLR, Institut für Experimentelle Strömungsmechanik
Bunsenstr. 10, 3400 Göttingen, BRD

Prof. Dr. W. J. Prosnak,   Polskiej Akademii Nauk
ul. Polna 54 m 27, 644 Warszawa, Poland

Prof. R. W. MacCormack,   NASA Ames Research Center, Moffett Field,
California 94035, USA

Prof. Dr. F. X. Wortmann,   Institut für Aerodynamik und Gasdynamik
Universität Stuttgart
Pfaffenwaldring 21, 7000 Stuttgart, BRD

Dr. H. Eckelmann,   Max-Planck-Institut für Strömungsforschung
Böttingerstr. 4-8, 3400 Göttingen, BRD

Prof. Y. C. Fung,   University of California, San Diego
Department of Applied Mechanics and Engineering Sciences
Bioengineering, M-005
La Jolla, California 92093, USA

Prof. C. G. Caro,   Imperial College of Science and Technology
Department of Aeronautics, Physiological Flow Studies Unit
Prince Consort Road, London SW7, Great Britain

Prof. R. Peyret,   Université de Paris VI, Mécanique Théorique
Tour 66-4, Place Jussieu, Paris 75230 Cedex 05, France

Editor's Preface

In 1929 the Aerodynamisches Institut of the Rheinisch-Westfälische Technische Hochschule Aachen inaugurated the first extension of its building with a scientific meeting to which the then director of the Institut, Theodore von Kármán, invited scientists from all over the world. About seventy accepted his invitation and over thirty papers were presented. The meeting, held in a rather informal manner, was one of the scientific events in aerodynamics of that time. It seems noteworthy that Dr. Julius Springer offered to print the contributions.

Almost 50 years later, the building of the Aerodynamisches Institut, bearing many scars from World War II and with parts of it close to collapse, was finally reconstructed during the period 1976 - 1980. It was felt that the new building should, like its first extension, be inaugurated with a scientific conference.

Since aerodynamics and fluid mechanics had undergone an enormous expansion in the intervening 50 years, it was clear from the very beginning of the planning of the conference that not the whole field but only those parts of fluid mechanics could be covered which relate to the work of the Institut. It was therefore decided to call on distinguished experts and ask them to describe the progress in the various branches of fluid mechanics in question. The following topics were chosen: biological flows, non-homogeneous flows, vortex motion, transition, turbulent shear flows, shock-wave boundary-layer interaction, solution of the conservation equations, wing theory, and aerodynamics of aircraft. Many approved of this concept. Approximately 300 scientists from 14 countries, among them scientists from the USA, the USSR, China, and the European countries, attended the conference, which was held March 26- 28, 1980 in the Kármán Auditorium of the RWTH Aachen (completed in 1977).

The opening address, the inaugural lecture, and the papers presented are published here in full, with a minimum of editorial changes.

The Deutsche Forschungsgemeinschaft generously gave financial support to the conference which without this help could not have come to pass. My colleagues Prof. em. Dr.phil. Dr.med.h.c. A. Naumann, Prof. Dr.-Ing. H. Zeller, and Dr. rer.nat. W. Limberg helped to arrange the programme and worked to make the conference a success. Special mention is made of Mrs. H. Rehfeld and Mr. H. Thal, who undertook the laborious task of preparing the manuscript. Grateful acknowledgement is also due to Mrs. D. Steinbach, Dipl.-Ing. U. Giese, and Dipl.-Ing. H. Henke, who assisted in reading the proofs.

Finally, I am indebted to Prof. W. Beiglböck of the Springer Verlag, Editor of the Lecture Notes in Physics, for arranging the publication of these Proceedings in this series.

Egon Krause
Aerodynamisches Institut
RWTH Aachen

Eröffnung [*]

F. Schultz-Grunow
Aachen

Meine Damen und Herren,

Magnifizenz hat mich gebeten, sie hier bezüglich der fachlichen Bedeutung dieser Einweihungs-
feier zu vertreten.

So kann ich im Namen der Hochschule dem Aerodynamischen Institut gratulieren zur Vollendung
seines Erweiterungsbaues und ihm danken, daß es ihm jede Mühe wert war, damit erweiterte
Lehr- und Forschungsmöglichkeiten auf einem zentralen Gebiet der Ingenieurwissenschaften zu
schaffen.

Der heutige Tag weckt Erinnerungen an die Anfänge der Aerodynamik in Aachen. Auch bei uns
entwickelte sie sich aus der Mechanik heraus. Schon der zweite Lehrstuhlinhaber für Mechanik,
kein geringerer als Arnold Sommerfeld (1900 - 1906), zeigte Interesse für Flüssigkeitsbewegun-
gen. Er schuf bei uns die Theorie der Lagerreibung, deren grundlegende Differentialgleichung
nach ihm benannt ist.

Ihm folgte bis 1913 Hans Reissner. Die sieben Jahre, die Hans Reissner bei uns verbrachte,
haben insbesondere dadurch einen nachhaltigen Einfluß ausgeübt, als er einer derjenigen war, die
an die Probleme der Flugtechnik mit wissenschaftlichen Methoden herangetreten sind. Der
Anregung und Pionierarbeit Hans Reissners verdankt die Hochschule die Entstehung des
Aerodynamischen Instituts, welches in Verbindung mit dem Lehrstuhl für Mechanik und
Aerodynamik 1912 - 1914 erbaut wurde (zitiert nach v. Kármán).

Bereits 1909 begann Reissner mit aerodynamischen Versuchen zunächst in dem von Hugo
Junkers geleiteten Maschinenlaboratorium, wo er eine Rundlaufvorrichtung und eine provisori-
sche Luftstromanlage betrieb, welche Geräte den Grundstock des Aerodynamischen Instituts
bildeten. Reissner lieferte neben dieser Versuchsarbeit grundlegende Beiträge zur Stabilität der
Flugzeuge, die die vielen Abstürze der Pionierzeit allen voran das Unglück Otto Lilienthals zu
klären verhalfen. Er war ein weitblickender Geist, denn er war es, der das erste Metallflugzeug,
die Reissnersche Ente, baute. Es hatte bereits eine selbsttragende Flügelhaut aus Wellblech.
Drei Bilder dieses Flugzeugs will ich zeigen, das letzte, weil es ein Datum trägt. Die
Entwicklung wurde von Reissner nicht weitergeführt, einmal aus finanziellen Gründen, und weil
es zum andern keine Flugzeugführer gab, um ein solch fortschrittliches Flugzeug steuern zu
können.

---

[*] The English translation is given on the subsequent pages.

Die Wahl des Nachfolgers von Reissner war ein Glücksfall. Theodor von Kármán, ein Doktorand und Mitarbeiter Ludwig Prandtls, der bereits in jungen Jahren durch die Kármánsche Wirbelstraße hochberühmt geworden war, erhält nicht den erhofften Ruf nach München zur Nachfolge August Föppls und war dadurch für Aachen frei. Er kam 1913 mit dem mächtigen Rüstzeug der Prandtlschen Aerodynamik nach Aachen. Mit einem seltenen Schwung, einer ausgeprägten Persönlichkeit, die mit ihrem Witz und Humor ausgezeichnet in die Aachener Landschaft paßte, und mit einer phänomenalen Schöpferkraft wurde das Institut und das Lehrgebiet nach modernen Gesichtspunkten aufgebaut. 1928 konnte ein Windkanal nach Göttinger Muster mit einer Drei-Komponenten-Waage nach Wieselsberger in Betrieb genommen werden. Er hatte einen ungewöhnlichen Platz, nämlich auf dem Dach des Instituts und wurde so zum Wahrzeichen der T.H. Aachen. Ich zeige ihn im Bild. Der Efeubewuchs des Gebäudes hat aber nichts mit der Tätigkeit in dessen Innerem zu tun. Der Kanal überlebte den 2. Weltkrieg, wurde von den Alliierten abgerissen, dann nochmals aufgebaut und 1946 endgültig abgebrochen.

Es war damals in Aachen eine berühmte Zeit, denn Aachen verfügte auch über eine reiche Zahl hervorragender Mathematiker, die sich für Aerodynamik interessierten. 1911 stellte Kutta hier seine berühmte Auftriebsformel auf, nachdem er schon vorher in München durch seine Dissertation, die das heute in der Numerik nicht mehr wegzudenkende Runge-Kutta-Verfahren brachte, von sich reden gemacht hatte. Blumenthal gab erste Methoden zur Berechnung von Tragflügelströmungen. Hamel leitete zähe Strömungen ab, deren Stromlinien auch Stromlinien einer Potentialströmung sein können, es sind die Spiralströmungen. Trefftz half v. Kármán die äußerst wichtige Familie der Kármán-Trefftz-Profile theoretisch herzuleiten.

Und damit komme ich zu den Mitarbeitern. Unser Ehrendoktor und jetziger Ehrenprofessor am Courant-Institute in New-York, K. Friedrichs, den v. Kármán als Mathematiker aus Göttingen an sein Institut holte, schrieb eine bekannte Arbeit über den freitragenden Flügel als Rahmentragwerk. Unser Ehrendoktor, Professor Gabrielli, der zu unserer großen Freude heute mit seiner Gattin unter uns weilt, promovierte über die Torsionssteifigkeit freitragender Flügel in der Rekordzeit von einem Jahr. Der Amerikaner Wattendorf promovierte über den Landestoß von Wasserflugzeugen. Er war ein guter Trompetenbläser und trug dadurch zur fröhlichen Geselligkeit im Institut bei. Andere Namen sind Scheubel und Hermann, der den Großteil der Windkanäle für Peenemünde konzipierte.

Auf dem Gebiet des Unterrichts wurde neben der Kármánschen Vorlesung eine Vorlesung über Flugzeugbau eingerichtet, die dem damaligen ersten Assistenten W. Klemperer übertragen wurde. Gleichzeitig sammelte sich eine Gruppe Studenten zur Flugwissenschaftlichen Vereinigung Aachen, die in enger Zusammenarbeit mit dem Aerodynamischen Institut Segelflugzeuge baute. V. Kármán setzte sich dafür ein, daß in der Rhön Segelflug betrieben werden durfte. Dort gewann die Vereinigung 1920 den Ersten Preis und 1922 brach W. Klemperer mit 13 Minuten Flugzeit den von Orville Wright gehaltenen Weltrekord von neun Minuten Flugdauer. Heute ist es nur noch die Nacht, die die Flugzeit begrenzt.

Die sogenannten Aachener Vorträge, eine 1929 abgehaltene Konferenz über Aerodynamik und verwandte Gebiete stellt ähnlich wie die Innsbrucker Vorträge einen Merkstein in der Entwicklung der Aerodynamik dar. Was Rang und Namen hatte traf sich hier. Ich zeige ein Bild der Teilnehmer. Vorne rechts v. Kármán, links vorne seine Schwester Pepe, die über ihn mit Strenge wachte, in der Mitte natürlich Prandtl mit einem ungewöhnlich strengen Gesicht, daneben Reissner, in der näheren Umgebung v. Mises, Ackeret, Tollmien, weiter hinten Busemann mit zerwühltem Haarschopf, dann links Miss Swain, mehr in der Mitte Goldstein, Rosenhead, hinten rechts Lerbs und links Friedrichs und der lange Trefftz, um niemand zu verdecken.

1934 nahm die große Epoche v. Kármán ihr Ende. Nach kurzer Zwischenzeit folgte C. Wieselsberger. Er führte die Hochgeschwindigkeits-Aerodynamik ein, verstarb aber leider viel zu früh. Sein Nachfolger Seewald richtete einen heute viel in Anspruch genommenen Ventilprüfstand ein, sein Nachfolger, Herr Naumann, baute das Institut in seiner jetzigen Form mit Windkanälen aus und errichtete ein von den Medizinern viel gefragtes biomedizinisches Labor. Sein Nachfolger, der jetzige Leiter, Professor Krause, führt die numerische Behandlung von Strömungen ein, welche Tätigkeit kürzlich mit der Berechnung der Taylor-Wirbel im Kugelspalt einen beachtenswerten Erfolg erzielt hat. Die neuere Zeit kommt kurz, weil sie sich im Institut manifestiert, das man nachher besichtigen kann.

Alle Anstrengungen in Lehre und Forschung bleiben erfolglos, wenn wie jetzt, die Vorbildung von der Schule her völlig unzureichend ist. Die Studienreformen sind ein Schlag ins Wasser, das Päppeln der Jugend hat sich nicht bewährt. Ähnlich wie früher für Medizin und Philologie das große Latinum Vorbedingung war, müssen wir auf ein fachbezogenes Abitur pochen, wenn schon das bewährte alte Abitur heute durch eine Vielfalt von verschiedenen Fach-Abiturs ersetzt worden ist. Es geht nicht, daß wir in mühevoller Paukarbeit in zwei Semestern nachzuholen versuchen müssen, was auf der Schule von Anfang an versäumt wird. Nicht nur soll es dabei auf das Wissen ankommen, sondern auf das Training des Denkapparates. Wie ein Vater eines Aachener Freundes von mir zu seinem Sohn sagte: Die Hauptsache ist, daß das Gehirn turnen lernt. Ich will hier nur auf Tatsachen den Finger legen und keineswegs in Belehrungen ausufern. Ich muß es Ihnen überlassen, sehr ernst darüber nachzudenken, wie das geändert werden kann und, wo es auch sei, an einer Besserung kräftig mitzuarbeiten.

Ich schließe, indem ich dem Aerodynamischen Institut eine erfolgreiche Zukunft wünsche, eingebettet in seine große Vergangenheit.

Opening Address

Ladies and Gentlemen,

Our vice-chancellor asked me to speak here on his behalf about the scientific meaning of this inaugural ceremony.

First of all I congratulate the Aerodynamisches Institut in the name of the Hochschule on the completion of its reconstruction and thank the institute for making every effort to create additional teaching and research facilities in a central field of the engineering sciences.

This day wakens rememberances of the beginnings of Aerodynamics in Aachen. As at other universities also here it evolved out of Mechanics. The second holder of the professorial chair for Mechanics, no one less than Arnold Sommerfeld (1900 - 1906), showed interest in fluid motion. Here in Aachen he proposed the theory of lubrication, the fundamental differential equation which is named after him.

He was followed by Hans Reissner until 1913. The seven years which Hans Reissner worked here brought lasting influence, since he was one of those, who approached the problems of aviation with scientific methods. The Hochschule owes the foundation of the Aerodynamisches Institut to Hans Reissner's initiative and pioneering work. It was built in conjunction with the chair for Mechanics and Aerodynamics 1912 to 1914.

1909 Reissner had already began with aerodynamic experiments, first in the Mechanical Laboratory, directed by Hugo Junkers, where he urged on a whirling arm and on a provisional air stream apparatus. These devices constituted the experimental foundation of the Aerodynamisches Institut. In addition to the experimental work, Reissner wrote fundamental papers on the stability of airplanes, whereby he helped to clarify the numerous crashes of the pioneering time in particular the crash of O. Lilienthal. He was a man of farsightedness and imagination and it was he who built the first all-metal plane, the Reissner Canard. It already had a self-supporting wing fabric out of corrugated metal. I will show three pictures of this airplane, the last as it bears a date. The development was not continued by Reissner, mainly for financial reasons, but also, because no airplane pilot existed, who was able to pilot such a sophisticated airplane.

The choice of Reissner's successor was a stroke of luck. Theodore von Kármán, a doctoral candidate and Ludwig Prandtl's assistant, who had already become famous at a young age through the Kármán vortex street, was not offered the chair at Munich he had hoped for as successor of August Föppl, and therefore, was free to come to Aachen. He arrived in Aachen 1913 equipped with an immense knowledge of Prandtl's aerodynamics. The Institut and the curriculum were built up in accordance with modern viewpoints with the rare enthusiasm of a strong personality, whose wit and humor fit excellently into the Aachen landscape, and by a phenomenal creative power. In 1928 a windtunnel of the Göttingen type with a three-component balance as developed by Wieselsberger could be set in operation. The tunnel had an unusual place, namely on the roof of the Institut and thereby it became a distinctive mark of the T.H. Aachen. Here is a picture of it. The tunnel survived World War II, was turn down by the Allies , then reconstructed , and was turn down completely in 1946.

Aachen was in its golden age since at that time it had an impressive number of excellent mathematicians who were interested in aerodynamics. Here in 1911 Kutta developed his famous formula for the lift, after he had already previously caused a great stir in Munich through his dissertation, in which he derived the Runge-Kutta-method, an indispensible part of modern numerics. Blumenthal proposed some of the earliest methods for calculating flows about wings. Hamel derived a solution for viscous flows, the streamlines of which can also be identified with the streamlines of potential flows; they are the spiral flows. Trefftz helped v. Kármán to derive theoretically the extremely important family of the Kármán-Trefftz-Profiles.

And with this I come to the assistants. Our honorary doctor and present honorary professor at the Courant-Institute in New York, K. Friedrichs, whom v. Kármán engaged as a mathematician for his institute from Göttingen, wrote a renowned work about the cantilever wing as a frame system. Our honorary doctor, Professor Gabrielli, who together with his wife give us the great honour of their presence, wrote his doctoral dissertation about the torsional strength of cantilever wings in the record time of less than one year. The American Wattendorf wrote a dissertation about the impact of landing of seaplanes. He was a good trumpet player and contributed to the spirited social life of the Institut. Other names are Scheubel and Hermann, who drew up, the major plans for the windtunnels in Peenemünde.

In the teaching field, in addition to Kármán's course, a course in airplane design was initiated, which was given by the first assistant at that time, W. Klemperer. At the same time a group of students founded the student aeronautics society Aachen (FVA) which in close cooperation with the Aerodynamisches Institut built gliders. V. Kármán used all his influence in seeing to it that the group could participate in the gliding contests in the Rhön hills. There the society won the first price in 1920 and in 1922 W. Klemperer broke the world record of 9 minutes flight time, held by Orville Wright, with a flight time of 13 minutes. Today it is only the night which limits the flying time.

The so-called Aachen Lectures, a conference on aerodynamics and related subjects, held in 1929, was, like the "Innsbruck Lectures" a milestone in the development of aerodynamics. It was the gathering of the most prominent of the scientific elite. I would like to show a picture of the participants. In front on the right von Kármán, on the left his sister Pepe, who kept a stern eye on him, in the middle, of course, Prandtl with an unusual serious expression, next to him Reissner, in his immediate vicinity v. Mises, Ackeret, Tollmien, farther back Busemann with dishevelled hair, then on the left side Miss Swain, more to the center Goldstein, Rosenhead, in the back to the right Lerbs and to the left Friedrichs, and the tall Trefftzs, in the back in order not to stand in front of anybody.

In 1934 the great epoch of v. Kármán came to an end. A short time thereafter C. Wieselsberger became his successor. He introduced high-speed aerodynamics, but most unfortunately, passed away much too early. His successor Seewald set up a test stand for safety valves which is in great demand today and his successor, Mr. Naumann, equipped the Institut with windtunnels and gave it its present form. His successor, the present director, Professor Krause, is introducing numerical analysis of flows; not long ago this activity achieved noteworthy success with the calculation of Taylor vortices in the spherical gap. The more recent developments will not be dealt with in depth at this time in as much they can be seen later on in the Institut.

All efforts in teaching and research are fruitless, if as at the present time the basic secondary education is completely inadequate. All reforms of our educational system have been in vain, the coddling of the youth has proven to be a failure. Similar to earlier times when the Matriculation Latin was a prerequisite for medicine and philology we must insist on a subject-orientated school-leaving examination, even if the old time-tested school-leaving examination has nowadays to be replaced by a variety of different specialized school-leaving examinations. It is impossible for us to try to make up in two terms through irksome cramming what was neglected in school from the very beginning. We are not only concerned about factual knowledge but even more about training of the brain muscles. As a father of one of my Aachen friends said to his son: "The main thing is that the brain learns how to turn a somersault." I only want to point out facts and by no means do I want to become pedantic. I must leave it up to you to think very seriously about how this could be changed and, wherever it may be, to vigorously help to improve things.

I close by wishing the Aerodynamisches Institut a successful future, rooted in its great past.

The first windtunnel of the Göttingen type located on the roof of the Aerodynamisches Institute; completed under v. Kármán in 1928.

The Reissner Canard. First all-metal airplane with cantilever wing built around 1912.

9

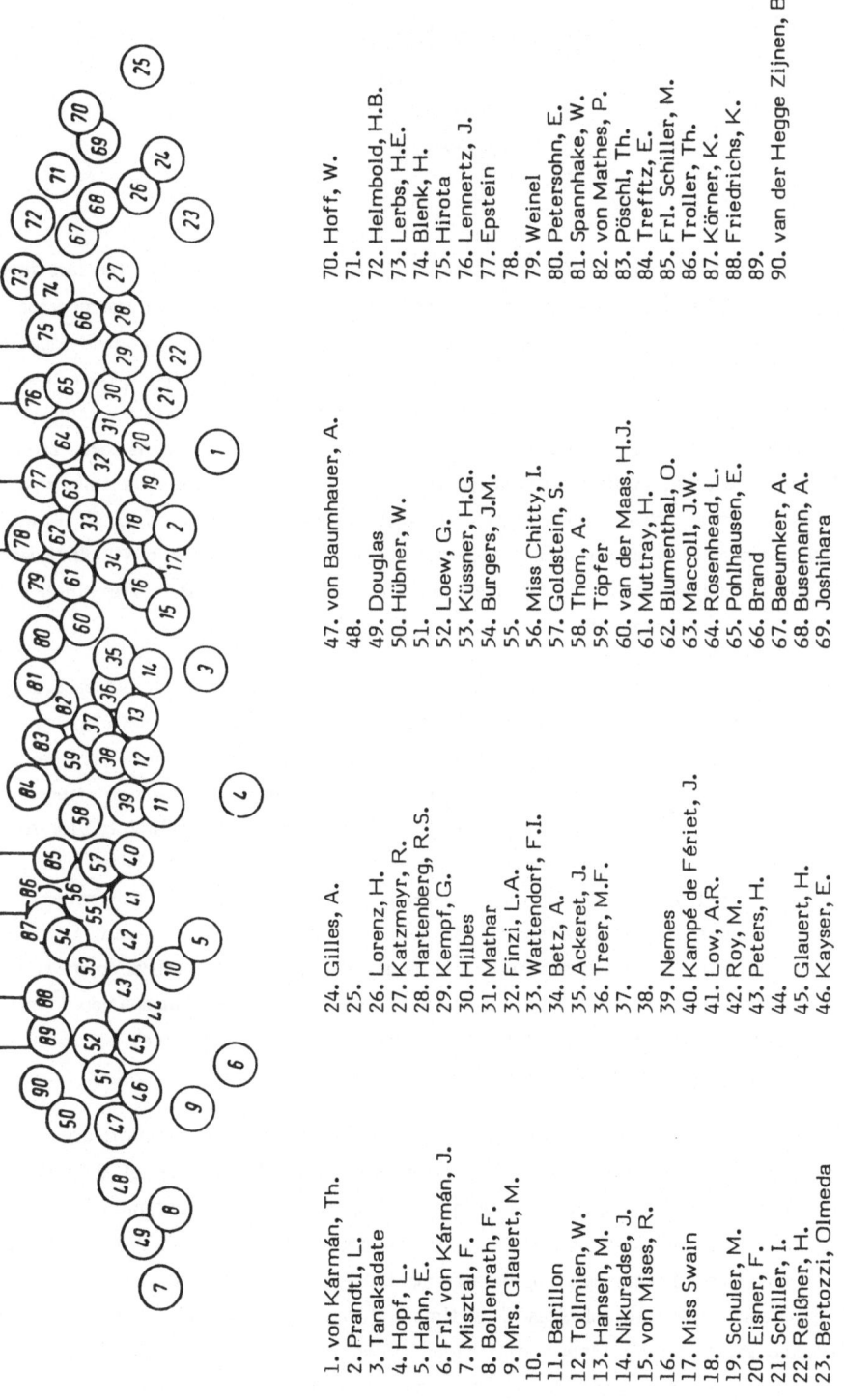

1. von Kármán, Th.
2. Prandtl, L.
3. Tanakadate
4. Hopf, L.
5. Hahn, E.
6. Frl. von Kármán, J.
7. Misztal, F.
8. Bollenrath, F.
9. Mrs. Glauert, M.
10.
11. Barillon
12. Tollmien, W.
13. Hansen, M.
14. Nikuradse, J.
15. von Mises, R.
16.
17. Miss Swain
18.
19. Schuler, M.
20. Eisner, F.
21. Schiller, I.
22. Reißner, H.
23. Bertozzi, Olmeda

24. Gilles, A.
25.
26. Lorenz, H.
27. Katzmayr, R.
28. Hartenberg, R.S.
29. Kempf, G.
30. Hilbes
31. Mathar
32. Finzi, L.A.
33. Wattendorf, F.I.
34. Betz, A.
35. Ackeret, J.
36. Treer, M.F.
37.
38.
39. Nemes
40. Kampé de Fériet, J.
41. Low, A.R.
42. Roy, M.
43. Peters, H.
44.
45. Glauert, H.
46. Kayser, E.

47. von Baumhauer, A.
48.
49. Douglas
50. Hübner, W.
51.
52. Loew, G.
53. Küssner, H.G.
54. Burgers, J.M.
55.
56. Miss Chitty, I.
57. Goldstein, S.
58. Thom, A.
59. Töpfer
60. van der Maas, H.J.
61. Muttray, H.
62. Blumenthal, O.
63. Maccoll, J.W.
64. Rosenhead, L.
65. Pohlhausen, E.
66. Brand
67. Baeumker, A.
68. Busemann, A.
69. Joshihara

70. Hoff, W.
71.
72. Helmbold, H.B.
73. Lerbs, H.E.
74. Blenk, H.
75. Hirota
76. Lennertz, J.
77. Epstein
78.
79. Weinel
80. Petersohn, E.
81. Spannhake, W.
82. von Mathes, P.
83. Pöschl, Th.
84. Trefftz, E.
85. Frl. Schiller, M.
86. Troller, Th.
87. Körner, K.
88. Friedrichs, K.
89.
90. van der Hegge Zijnen, B. G.

Participants of the conference on aerodynamics held at the Aerodynamisches Institut in 1929 on the occasion of the inauguration of the first extension building of the Institut.

Education, Training, and Research in the Engineering Sciences

H. W. Liepmann
Pasadena

After struggling with the wording for a lecture in German - only my third in the last forty years - I have now the unenviable task of writing a reasonable facsimile of the address in English. It is not easy to translate the three words used in the German title into equally concise English while keeping the same meaning. The word "Bildung" is closely akin to the term "liberal arts education" "Ausbildung" to "vocational training". But neither conveys quite the same ideas as the corresponding German, at least not to me. "Forschung" into "research" is fortunately a conformal translation. Worse than that, ideas which are easily discussed in a talk, in the presence of an audience with feedback, tend to sound stuffy and pompous when written down. I lose somehow the ease of presentation and feel that I might sound like Nolte in Wilhelm Busch: "Drum soll ein Kind die weisen Lehren der alten Leute hoch verehren, die haben alles hinter sich und sind, gottlob, recht tugendlich!"

Any university has to contribute to the culture of the time through research and pass on to younger minds the heritage of the past and the challenge of the future. The selection of the proper mix of general education, professional training, and research is an eternal problem for which no general solution can be found. Schools of technology in particular face an often self-contradictory criticism. They are accused of being too theoretical or too applied, too far from industry or governed by industry. Technology has become the whipping boy of modern times, so much so that an unhealthy and unproductive bad conscience pervades the technical professions. Natural sciences and technology are sometimes portrayed as anti-culture or anti-humanity. These attacks are to me ridiculous expressions of frustration. I consider natural sciences and technology as much a part of the general culture as, say, the arts. I admit that one can do more damage by mishandling an engineering project than by mishandling a symphony but to quote from one of may favorite aphorisms: Never mistake impotence for virtue!

The dictionary defines the culture of a group as its shared customs, ideas, and attitudes. This definition leaves me somewhat cold since it does not emphasize the heritage of accomplishments from which the ideas and attitudes have grown. A true education should make it at least possible to appreciate this past heritage and to contribute to the future. It is commonly expected that an educated human will appreciate music, literature, and the arts and that he will be aware of the social and political ideas of his time and their historical roots. Appreciation of the natural sciences, let alone engineering, is seldom if ever expected and these fields of human endeavor are hardly considered today as part of the culture of our times.

This lack of appreciation is indeed a strange regression from former times. The inscription above the entrance of Plato's academy reads: "Let no one ignorant of geometry enter". The cultural elite in ancient Athens or during the height of Islamic splendor in Damascus, Cordova or Samarkand would have ostrasized anybody lacking in appreciation of mathematics and astronomy. In the quadrivium of the middle ages, mathematics and music were combined in the masters' curriculum.

Our century will probably be less remembered for its contributions to the arts than for its revolutionary ideas in physics and biology, the rise of the computer, and the creative engineering which allowed man to leave the ground to travel the air and finally outer space. This feat, together with the subsequent burst of new understanding of the universe, extends the horizon of mankind more than even the terrestrial discoveries of the fifteenth century and the realization that the earth is round and finite. These contributions of science and technology to an understanding of the world we live in and our own makeup will rank as a permanent part of our culture.

The loud and persistent clamor for increased education of technologists in the social sciences and humanities can be and should be answered by a demand for increased appreciation of the natural sciences and technology by social scientists and humanists, and indeed by anybody who subscribes to a true liberal education. There are more technologists today who appreciate Shakespeare, Mozart, and van Gogh than humanists who appreciate the beauty of Maxwell's equations, the lines of an aircraft, or the marvelous intricacy of an integrated circuit.

The lack of real understanding of contemporary technology among social scientists and lawyers who, after all, make decisions on questions ranging from the uses of atomic energy to problems of pollution is, I think, far more dangerous than any lack of appreciation of social sciences among engineers. To educate an electorate, or at least a legislature, capable of making decisions on the basis of understanding of technology rather than fear and mystique is probably the most important task of education today, and very likely the crucial test for the survival of a democratic society.

The term "engineering education" implies exposure to subjects beyond purely professional techniques. Part of such exposure is, in fact, mandatory from a purely professional point of view: In a time of rapidly changing technology, the professional techniques of the day are rapidly outmoded, new fields of engineering endeavor appear suddenly, and the background in fundamentals must be sufficiently broad to make respecialization possible. Beyond this strictly professional point of view, intellectual pleasure and curiosity plus the possiblity of fruitful interaction with a greater part of the human society are the driving forces for extended education. Obiously, only part of such an extension can be provided by a formal study program, but the foundation can be laid and interest stimulated by an early opportunity to sample other fields. In spite of pressure for more courses in a narrow engineering discipline, enough time must be left for options in cultural subjects, where culture is interpreted in a broad sense and where much is left to individual taste. Subjects such as economics and psychology, music and literature are obvious candidates for such cultural exposure but cosmology, genetics and, say, number theory qualify as well. After all, an extension of consciousness beyond a narrow professional outlook is intended. It matters little whether the extended fields of study are contiguous to the professional field or not, provided only that the overall coverage is sufficiently broad. Indeed, the choice of cultural interest is an extremely personal one; one of my own noncontiguous interests, in Islamic history, stems from my early study in Istanbul.

Professional options for an engineer range from detailed design to advanced research and development and top management positions. Any technical university has either to come to grips with these wide variations by providing an equally wide choice of curricula, or else has to deliberately choose educational policies directed at only part of this wide spectrum. My home base, the California Institute of Technology is a very small Institute with a very high faculty to student ratio and a highly selective student body of which less than one thousand are in engineering and applied science. Here schooling of large numbers of line engineers is clearly impossible and training is directed toward top R & D and managerial positions. My opinions and consequently my remarks here are evidently colored by this background.

In a time of rapidly changing technology, a broad education of engineers in the basic sciences becomes a necessity since specialization in a narrow field cannot be expected to last a professional lifetime. The capacity to become familiar with a new and different engineering task depends on the ability to penetrate an unfamiliar discipline and this ability requires a solid background in mathematics and physics. Today an engineer faced at times with designs involving lasers or superconducting magnets or Josephson junctions has to have some feel for quantum effects. The need for new and better materials brings the structural engineer into contact with solid state physicists and again a certain understanding of foreign fields becomes important. I lived through the development of supersonic and space flight and witnessed at close range the frustration of engineers with too limited a background in mathematics and physics. The need for an increased background in natural science, especially mathematics, not only for engineers but for general education as well, puts increased emphasis on pre-university schooling - a chapter by itself. Unfortunately, attempts to inject exaggerated rigor into mathematical curricula have been counter productive, scaring away students with a native intuitive feel for mathematics and the physical sciences.

The often repeated accusation that natural scientists in general and engineers in particular cannot communicate and do not understand humanity, is, I believe, baloney - to use a very descriptive American slang expression. Failure to communicate is a common shortcoming often but not always connected with professional narrowness and by no means restricted to only one part of the professional community. During at discussion an the Second International Congress of Engineering Education, the impossible language in the instruction manuals for household appliances was cited as an example of the total failure of engineers to communicate. The argument was countered by pointing to the language of tax forms. Mathematical symbols and, today, Fortran form a large part of an engineer's language. Conformal translation into colloquial language is not always easy. Verbal aptitude obviously is the mainstay of a poet or playwright. These differences exist. Communication, however, implies the transfer of ideas and for this purpose a certain verbal aptitude is necessary but not sufficient. Clarity of thought and precision in the formulation of ideas is even more important and here schooling in basic mathematics helps a great deal. Even outstanding verbal aptitude cannot overcome a lack of precise ideas. Advice due to, I believe, Mort Sahl makes this point in a lighter vein: "If you cannot communicate, the least you can do is shut up!"

In the professional training of engineers, the difficult choice between basic science, specialized engineering courses, and research is further complicated by the problem of appropriate interaction with industry. Both faculty and students have to have industrial experience of some form or other in order to appreciate the requirements and specific problems of industry. Conversely, through this interaction, industry has to learn to appreciate the specific problems and aims of an educational institution. Here I favor a loose interaction somewhat in the sense the word is used in physics for two systems that retain their individuality. Faculty consulting in industry, and participation in government committees and boards, are necessary, but ownership or complete management of firms by faculty is detrimental since it leads sooner or later to conflict of interest. The opposite situation of industrial engineers consulting with a school through a temporary appointment or part-time presence on the campus is equally important and beneficial. Appropriate student exposure to industry problems is more difficult. Temporary jobs in industry seldom supply the intended overall view of the aims and methods of industry. Occasional seminars and colloquia given by members of industrial staff help. The best way we have found so far is a seminar course called "Case Studies in Engineering". The design and construction of a specific product is discussed by the very engineers responsible for each phase of the product. For example, a Hughes communication satellite project was presented in a set of about 20 seminars given by the Hughes engineering and management staff. At different times the Lockheed L-1011, the Douglas DC-10, the Mariner and Voyager Spacecraft projects were presented. These presentations covered the whole process, including financing and governmental regulations, in order to demonstrate the exacting constraints under which design and construction of a real engineering project proceeds. As a fringe benefit for the industry involved, these seminar courses are one of the best means for advanced recruiting.

Joint research with industry can be very helpful as long as both partner are fully aware of their different strengths, weaknesses, and aims. A university laboratory has access to a number of young, fresh, but inexperienced minds and its principal output is new knowledge. Industry has access to experienced engineers and its output is hardware. Industry usually wants fast answers to specific problems; university research tends toward slower progress in the understanding of a field. Both interest and capabilities overlap but are by no means identical; a failure to realize some of the inherent differences in outlook and approach have led to some rather silly accusations from both sides.

In the administration of any laboratory or university, an increased urge for "documentation" of performance and "democratization" of the decision making is being felt. I can appreciate the reason for this urge but deplore the consequences: It is impossible to evaluate a young faculty member in such a way that the pertinent information can be codified on an IBM card. A few members of the administration and faculty have a particularly good "hunch" or "feel" for potential that cannot be documented in standard form. Committees and committee reports necessarily tend to overemphasize documentable standards in promotion and appointments. Academic committees are invaluable for advice but seldom for decisions. The well-known American joke that "a camel is a horse designed by a committee" is sometimes close to the unfortunate truth. There does not exist a fail-safe system for running an educational establishment. The one definable task of an administration is to protect the ones who do most from the ones who do least.

Finally, I hold strong ideas on teaching and research and their interrelation at the university level. Research and teaching are the raison d'etre of a faculty. Creative teaching is nearly impossible without a deep involvement in research. Without the continuous exasperation and occasional euphoria of the struggle to do research at the limit of the state of the art, teaching tends to become stale.True, there are on any faculty a few excellent researchers who fail to be inspiring teachers, at least in the classroom. Similarly, there are a few inspired teachers who contribute little to research. These are the exceptions. Very few research results per se can compare with the lasting impact of forming the next generation through teaching and example. Still an overriding interest in personal research coupled with disdain for teaching is not at all uncommon. This, to me, is a plain misjudgement of relative importance and a preference for Macbeth over Banquo, which I do not share.

Numerical Solution of Conservation Laws

H. O. Kreiss
Pasadena

1. Introduction

During a course on the numerical solution of partial differential equations we discussed a paper by G. Sod [8] where different methods to calculate the solution of a shock tube problem were compared. The problem was the following: A membrane separating a gas in two different states was suddenly removed. Then a moving shock, a rarefaction wave and a contact discontinuity appeared. The methods he compared were those of Hyman, Godunov, Lax-Wendroff, McCormack, Rusanov, Glimm and Shastra. The results of the first five methods were suspiciously similar. Also the number of grid points was very large.

To understand this, we did our own experiments for simplified problems. Using the first five methods we solved

$$\partial u/\partial t + \frac{1}{2} \partial u^2/\partial x = 0 \qquad (1.1)$$

and

$$\partial u/\partial t + \partial u/\partial x = 0 \qquad (1.2)$$

with periodic boundary conditions and initial values

$$u(x,0) = \begin{cases} 1 & \text{for } \frac{1}{3} \le x \le \frac{2}{3}, \\ 0 & \text{elsewhere.} \end{cases}$$

For (1.1) we obtain a shock and a rarefaction wave and for (1.2) we have travelling (contact) discontinuities.

The results were rather disappointing. In order to obtain solutions which did not oscillate too much "artificial viscosity" had to be added. The amount needed was so large that the results were rather alike, i.e. all approximations were just first order accurate. (Already G. Sod [8] remarked that one had to add artificial viscosity). The number of points needed to obtain acceptable results was too large.

It turned out that Hyman's method [1] performed best. It is fourth order accurate in space and third order accurate in time and can be considered as an improved Leap-frog method. Let

$$D_0 v_\nu = \frac{v_{\nu+1} - v_{\nu-1}}{2 \, \Delta x} \;, \quad D_+ v_\nu = \frac{v_{\nu+1} - v_\nu}{\Delta x} \;, \quad D_- v_\nu = \frac{v_\nu - v_{\nu-1}}{\Delta x}$$

denote the usual centered, forward and backward difference operators and

$$Q\ v_\nu = \frac{4}{3} \frac{v_{\nu+1} - v_{\nu-1}}{2\Delta x} - \frac{1}{3} \frac{v_{\nu+2} - v_{\nu-2}}{4\Delta x}$$

the standard fourth order accurate approximation of $\partial v/\partial x$.

For the equation

$$\partial u/\partial t + \partial f(u)/\partial x = 0$$

Hyman's method can be written as

$$v_\nu^*(t_{n+1}) = v_\nu(t_{n-1}) - 2\Delta t\ Qf_\nu(t_n)$$

$$v_\nu(t_{n+1}) = \frac{1}{5}(4v_\nu(t_n) + v_\nu(t_{n-1})) - 2\Delta tQ(2f_\nu(t_n) + f_\nu^*(t_{n+1})) +$$

$$\frac{6}{5}\ \epsilon\Delta t\Delta x D_+ D_- v_\nu(t)\ ,$$

$$f_\nu^*(t_{n+1}) = f(v_\nu^*(t_{n+1})).$$

Here ϵ denotes the coefficient of the artificial viscosity. Typical results (due to J. Barker) after one revolution (t = 1) are given in the next eight figures. It shows that one needs a large number of points to describe the discontinuities well. Also the "best" is different for the shock and the contact discontinuities.

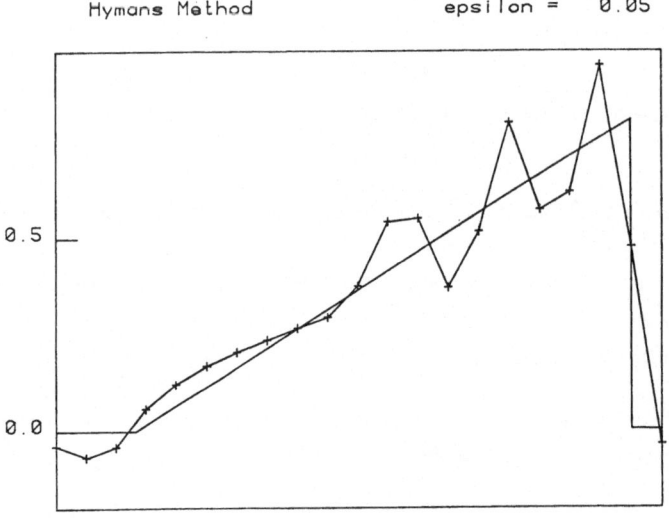

Hymans Method   epsilon =   0.05

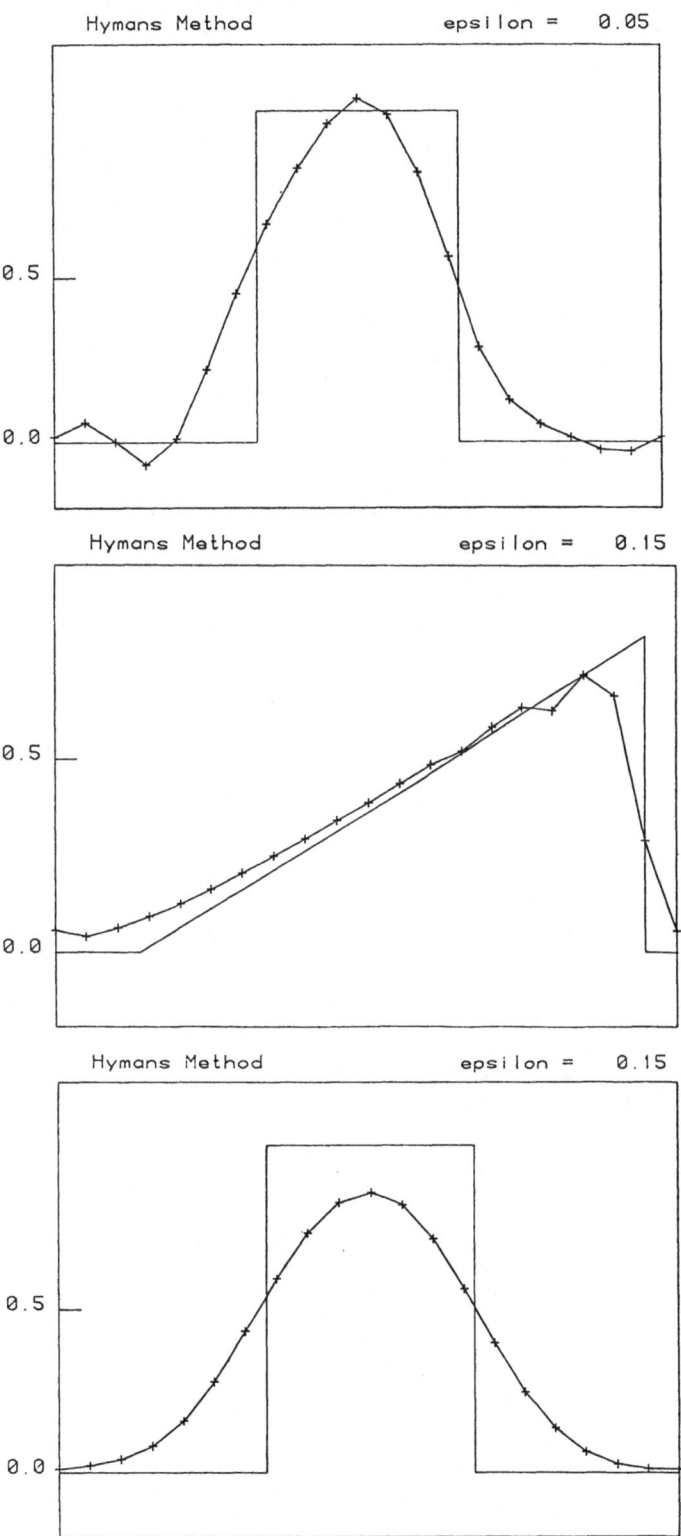

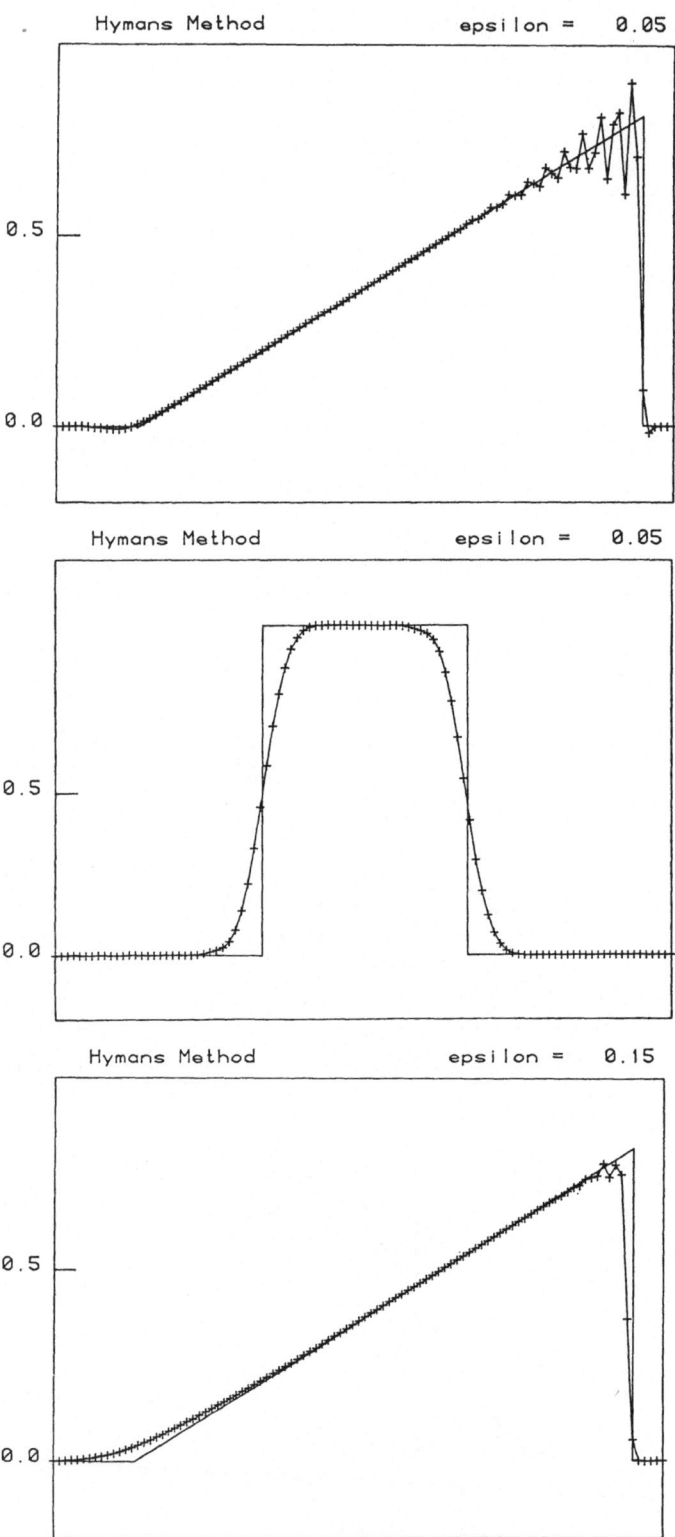

Hymans Method   epsilon =   0.15

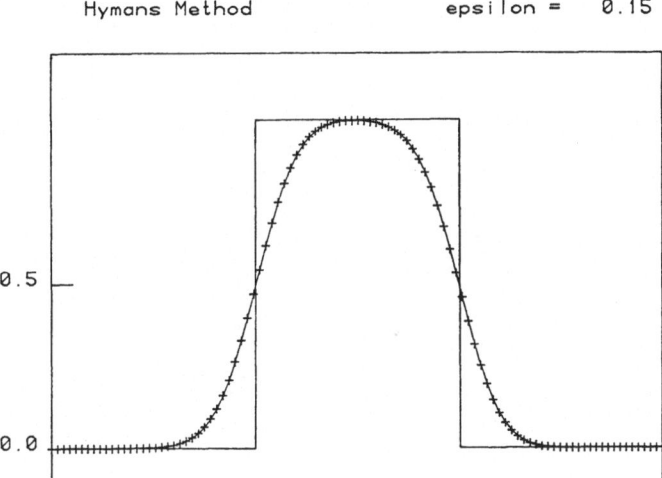

Recently numerical methods for singular pertubation problems have been developed and they will be discussed in the following sections. These methods can also be used for shock calculations. Consider the conservation law

$$\partial u/\partial t = \tfrac{1}{2} \partial(u^2)/\partial x + \varepsilon \partial^2 u/\partial x^2 \ , \quad 0 < x < 1 \qquad (1.3)$$

with periodic boundary conditions and approximate it by the backward Euler method

$$\varepsilon \partial^2 u/\partial x^2(x,t+\Delta t) + \tfrac{1}{2}\partial(u^2(x,t,+\Delta t))/\partial x - \tfrac{1}{\Delta t} u(x,t+\Delta t) = -\tfrac{1}{\Delta t} u(x,t)$$
$$(1.4)$$

Thus at every time step we have to solve a singular pertubation problem. The next two plots show calculations (due to D. Brown). In the first we start with initial data

$$u(x,0) = \begin{cases} \tfrac{1}{2} \text{ for } 1/3 \le x \le 2/3 \\[2mm] -\tfrac{1}{2} \text{ elsewhere} \end{cases}$$

which leads to a steady shock. In the second we start with initial data

$$u(x,0) = \begin{cases} 4/5 \text{ for } 1/3 \le x \le 2/3 \\[2mm] 1/5 \text{ elsewhere} \end{cases}$$

which gives us a moving shock.

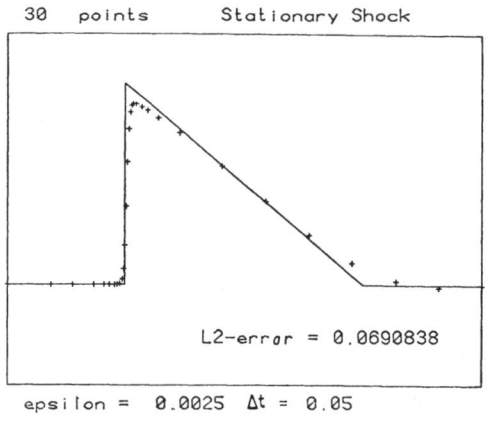

30   points   Stationary Shock

L2-error = 0.0690838

epsilon =   0.0025   Δt = 0.05

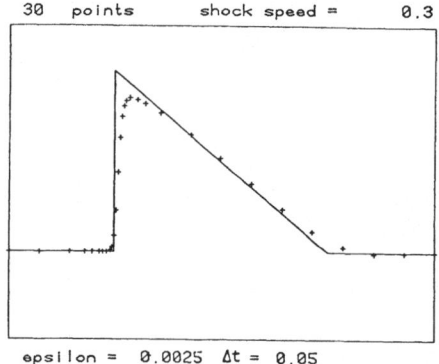

30   points   shock speed =   0.3

epsilon =   0.0025   Δt = 0.05

The trouble with the above approximation is that for moving shocks the "backward Euler method" requires small time steps. This can be remedied in the following way:

1) The above method need only to be applied in a neighbourhood of the shock.
2) Then we can introduce local "moving" coordinates and need only to calculate steady shock profiles.

D. Brown has already written a program using these principles and exept for minor "bugs" the method works. A preliminary plot follows. (We believe that the slight oscillations behind the shock are due to the fact that the local shock region was too small).

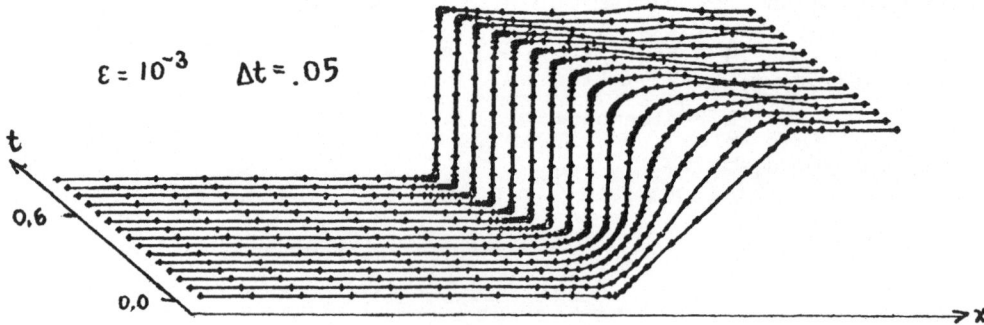

$\varepsilon = 10^{-3}$ $\Delta t = .05$

We expect that the procedure can be generalized to more than one space dimension by using the method of splitting. The first step is to solve the difference equations on every x-line. In particular we obtain the center of every shock profile (i.e. the point where in the moving coordinate system the appropriate characteristic changes sign). Connecting these points we obtain the shock centers on the y-lines. Then we can construct the shock profiles on the y-lines and solve the difference equations in the y-direction.

2. Singular Perturbation Problems

Consider a system of ordinary differential equations

$$dy/dx = A_0(x)y + F(x), \quad 0 \le x \le 1, \tag{2.1}$$

with n linearly independent boundary conditions

$$R_0 y(0) + R_1 y(1) = g$$

Here $y' = (y^{(1)}..., y^{(n)})$, [1] is a vector function with n components, and R_0, R_1 and $A(x)$ $\in C^1$ [2] are n x n matrices.

We want to solve the above problem by difference approximations. For that reason we divide the x-axis into subintervals of variable length h_j with grid points $x_0 = 0$, $x_v = \sum_{j=0}^{v-1} h_j$, $v = 1,2,....N$, $x_N = 1$, and denote by $u_v = u(x_v)$ vector functions defined on the grid:

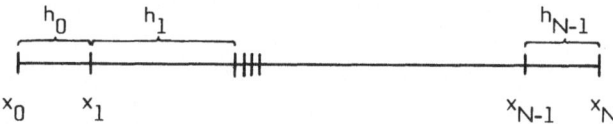

Let $h = \max h_j$. The case that we can choose h so small that $h|A| \ll 1$ has been treated many times before. Our aim is to treat the case $h|A| \gg 1$, i.e. we want to discuss methods for stiff equations. There are essentially two difficulties. 1) The matrix A has large eigenvalues of both signs. 2) There are turning points, i.e. these large eigenvalues are changing sign.

We start with a simple example. Consider the system

---

[1] If y is a vector then y' denotes its transpose and y* its adjoint. The vector norm is defined by $|y| = \max|y^{(i)}|$. Similar notations hold for matrices, for example
$|A| = \sup_y |Ay|/|y|$. Furthermore for vector functions
$\|y(x)\| = \max_{0 < x < 1} |y(x)|$ denotes the maximum norm.

[2] $A(x) \in C^j$ if the elements of A are j times continuously differentiable.

$$\varepsilon \ dy/dx = \begin{bmatrix} -1 & 0 & 0 \\ 0 & 0 & 0 \\ 0 & 0 & +1 \end{bmatrix} y = Ay, \ 0 \le x \le 1, \ \varepsilon > 0 , \tag{2.3}$$

with boundary conditions

$$y^{(1)}(0) + \alpha_0 y^{(2)}(0) + \beta_0 y^{(3)}(0) = g_1,$$

$$y^{(2)}(0) + \gamma_0 y^{(3)}(0) = g_2 , \tag{2.4}$$

$$y^{(3)}(1) + \alpha_1 y^{(2)}(0) + \beta_1 y^{(1)}(1) = g_3.$$

Here $\varepsilon > 0$ is a small positive constant. The general solution of (2.3) is given by

$$y^{(1)}(x) = \exp[-x/\varepsilon] y^{(1)}(0), \ y^{(2)}(x) \equiv y^{(2)}(0),$$

$$y^{(3)}(x) = \exp[(x-1)/\varepsilon]] y^{(3)}(1) .$$

Introducing it into the boundary conditions gives us

$$\begin{bmatrix} 1 & \alpha_0 & \beta_0 & e^{-1/\varepsilon} \\ 0 & 1 & \gamma_0 & e^{-1/\varepsilon} \\ \beta_1 e^{-1/\varepsilon} & \alpha_1 & 1 \end{bmatrix} \begin{bmatrix} y^{(1)}(0) \\ y^{(2)}(0) \\ y^{(3)}(1) \end{bmatrix} = \begin{bmatrix} g_1 \\ g_2 \\ g_3 \end{bmatrix}$$

Therefore, neglecting terms which are exponentially small,

$$y^{(2)}(0) \approx g_2, \ y^{(1)}(0) \approx g_1 - \alpha_0 g_2, \ y^{(3)}(1) \approx g_3 - \alpha_1 g_2$$

This shows that away from the boundary layers the solution of (2.3), (2.4) is smooth.

We approximate the above problem by a standard difference approximation, namely the trapezoidal rule

$$\frac{u_{\nu+1} - u_\nu}{h} = A(\frac{u_{\nu+1} + u_\nu}{2}), \quad \nu = 0,1,2,\ldots \ N-1; \tag{2.5}$$

on a uniform mesh, i.e. $h_j = h$. The desired solution shall satisfy the boundary conditions (2.4). The general solution of (2.5) is given by

$$u_\nu^{(1)} = \kappa^\nu u_0^{(1)}, \ u_\nu^{(2)} = u_0^{(2)}, \ u_{N-\nu}^{(3)} = \kappa^\nu u_N^{(3)} \tag{2.6}$$

where

$$u = \frac{1 - h/2\varepsilon}{1 + h/2\varepsilon}$$

Introducing (2.6) into the boundary conditions (2.4) gives us

$$
\begin{bmatrix}
1 & \alpha_0 & \beta_0\,\kappa^N \\
0 & 1 & \gamma_0\,\kappa^N \\
\beta_1\kappa^N & \alpha_1 & 1
\end{bmatrix}
\begin{bmatrix}
u_0^{(1)} \\
u_0^{(2)} \\
u_N^{(3)}
\end{bmatrix}
=
\begin{bmatrix}
g_1 \\
g_2 \\
g_3
\end{bmatrix}
\tag{2.7}
$$

If h is so small that $h/2\epsilon \ll 1$ then $\kappa \sim e^{-h/\epsilon}$ and it is obvious that the solution of (2.5) behaves like the solution of the differential equations. However, if $h/2\epsilon \gg 1$ then $\kappa \sim -1$ and the solution of (2.5) does not approximate the solution of the differential equations at all, because $u_\nu^{(1)}, u_\nu^{(3)}$ oscillate wildly. In particular, if $g_1 = \alpha_0\,g_2$, $g_3 = \alpha_1 g_2$ then $y^{(1)}(x)$, $y^{(3)}(x)$ are exponentially small and the solution of the differential equation is smooth up to the boundary. The corresponding solution of the difference approximation is still wildly oscillating. There are two ways to overcome this difficulty.

1) We use the trapezoidal rule for all components but introduce in the boundary layers $0 \le x \le \eta$, $1 - \eta \le x \le 1$, $\eta = (\epsilon|\log\epsilon|)$, new "stretched" independent variables, such that the boundary layer solutions are smooth functions of these new variables. Then we use a uniform grid in the new variables. In the old variables we get a nonuniform mesh, which is so fine that the boundary layer solutions change slowly from one point to the next. Away from the boundary layers we use a uniform mesh, i.e. $h_\nu = h$ for $\eta \le x \le 1-\eta$. This technique was used extensively by G. Pearson [7].

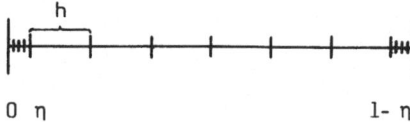

$$
\begin{array}{cc}
0\ \eta & \qquad\qquad\qquad 1-\eta
\end{array}
$$

In the boundary layer regions the differential equations are well approximated by the difference equations. Therefore, for $h \ll \epsilon$, we get essentially

$$
u^{(1)} \approx
\begin{cases}
\exp[-x_\nu/\epsilon]\,u_0^{(1)} & \text{for } 0 \le x_\nu \le \eta \\[4pt]
\pm\,(-1)^\nu u^{(1)}(\eta) & \text{for } \eta \le x_\nu \le 1-\eta, \quad u_\nu^{(2)} \equiv u_0^{(2)} \\[4pt]
\exp[-(x_\nu-1+\eta)/\epsilon]\,u^{(1)}(1-\eta) & \text{for } 1-\eta \le x_\nu \le 1,
\end{cases}
$$

and correspondingly for $u_\nu^{(3)}$. Observe that $u_\nu^{(1)}$, $u_\nu^{(3)}$ oscillate wildly in the interval $\eta \le x \le 1-\eta$. However, choosing η sufficiently large, the amplitude is so small that is has no effect. It is clear that now the solution of the difference equations approximates the solution of the differential equations well.

The drawback of this method is that we have to use the refinement even if the boundary conditions are such that the actual solution of the differential equation is smooth up to the boundary. The same is true if turning points are present.

We have to construct the mesh such that all solutions through the turning point become smooth. For nonlinear problems one often does not know the position of the turning point and the behavior of the solution. This makes the construction of the mesh rather difficult.

2) Instead of using the trapezoidal rule for all components we could use one-sided schemes for the first and last component. For example,

$$\varepsilon \frac{u_{\nu+1}^{(1)} - u_{\nu}^{(1)}}{h} = -u_{\nu+1}^{(1)} \ , \quad \varepsilon \frac{u_{\nu+1}^{(3)} - u_{\nu}^{(3)}}{h} = u_{\nu}^{(3)} \ , \tag{2.8}$$

$$\varepsilon \frac{u_{\nu+1}^{(2)} - u_{\nu}^{(2)}}{h} = 0 .$$

The general solution of (2.8) has the form (2.6) and $u_0^{(j)}$, $j = 1,2,3$ are given by (2.7) but now

$$\kappa = \frac{1}{1 + h/\varepsilon} .$$

if now $\varepsilon \ll h$ we still have $|\varkappa| \ll 1$ and it is clear that the solution of the difference equations resembles the solution of the differential equations. This technique can be used for rather general systems. However, the boundary layers are not resolved adequately. In general this will result in an additional error of order $O(\varepsilon)$. To obtain accurate results one has to resolve the boundary layers, (see for example [7]), or a combination of the asymptotic expansion of the boundary layer solution and the approximation (2.8) (see for example [6]).

One can refine the scheme (2.8) considerably and we shall do this in the next section.

The main problem for the numerical solution of singular pertubation problems is to find the mesh on which the solution varies slowly. There are two possibilities to do this.

1) One can use the behavior of the coefficients of the differential equation to determine the variation of the solution. This approach has been discussed in [2, 4]. An extended version of [5] is under preparation.

2) In this paper we want to refine the mesh adaptively. Starting with some mesh one computes the numerical solution and adds or deletes mesh points according to the variation of the numerical solution. More details are given in [3] and [5].

3. Difference Approximation for a Scalar Equation

Consider a scalar equation

$$dy/dx = a(x)y + f(x), \ 0 \le x \tag{3.1}$$

with initial data

$$y(0) = y_0$$

Here $a(x)$, $f(x)$ are complex-valued functions, which can be large. We are not interested in the oscillatory stiff case. Therefore we assume that there are constants ρ, c of "moderate size" and that

$$\text{Re}(a) \leq -\rho^{-1}(|a(x)| - c). \tag{3.2}$$

We approximate (3.1) by

$$\frac{u_{\nu+1} - u_{\nu}}{h_{\nu}} = \alpha_{\nu}\, a_{\nu} u_{\nu} + (1-\alpha_{\nu})\, a_{\nu+1} u_{\nu+1} + \alpha_{\nu} f_{\nu} + (1-\alpha_{\nu})\, f_{\nu+1} \tag{3.3}$$

which can be written as

$$u_{\nu+1} = A_{\nu} u_{\nu} + h_{\nu} F_{\nu} \tag{3.4}$$

where

$$A_{\nu} = \frac{1 + h_{\nu}\, \alpha_{\nu}\, a_{\nu}}{1 - h_{\nu}(1-\alpha_{\nu})\, a_{\nu+1}} \quad , \quad F_{\nu} = \frac{\alpha_{\nu} f_{\nu} + (1-\alpha_{\nu})\, f_{\nu+1}}{1 - h_{\nu}(1-\alpha_{\nu})\, a_{\nu+1}}$$

We want to choose the α_{ν} in such a way that the following conditions are satisfied:

1) The method is second-order accurate for $|h_{\nu} a_{\nu}| \ll 1$.

2) If $\text{Re}(a_{\nu}) \ll -1$ then the solutions of the homogeneous equation $v_{\nu+1} = A_{\nu} v_{\nu}$ decay rapidly. This we express by

$$|A_{\nu}| \leq (1 + h_{\nu} \gamma_{\nu})^{-1}$$

where $\gamma_{\nu} = \sigma \max(|\text{Re}(a_{\nu})|, |\text{Re}(a_{\nu+1})|)$, $\sigma = \text{const} > 0$.

3) If $\text{Re}(a_{\nu}) \gg 1$ then the solutions of $v_{\nu+1} = A_{\nu} v_{\nu}$ increase rapidly. This we express by

$$|A_{\nu}| \gtrless 1 + h_{\nu} \gamma_{\nu}$$

4) The α_{ν} are Lipschitz continous functions of $a_{\nu} h_{\nu}$.

An easy calculation shows that these conditions are satisfied if we choose α_{ν} in the following way:

I.   If $\text{Re}(h_{\nu} a_{\nu}) \leq 0$ and $\text{Re}(h_{\nu} a_{\nu+1}) \leq 0$ then

$$\alpha_{\nu} = \begin{cases} 1/2 & \text{if } |\text{Re}(h_{\nu} a_{\nu})| \leq 1 \\[2mm] \dfrac{1}{2|\text{Re}(h_{\nu} a_{\nu})|} & \text{if } |\text{Re}(h_{\nu} a_{\nu})| > 1 \end{cases}$$

II.   If $\text{Re}(h_{\nu} a_{\nu}) \geq 0$ and $\text{Re}(h_{\nu} a_{\nu+1}) \geq 0$ then

$$\alpha_\nu = \begin{cases} 1/2 & \text{if } |\text{Re } (h_\nu a_{\nu+1})| \leq 1 \\ 1 - \dfrac{1}{2|\text{Re }(h_\nu a_{\nu+1})|} & \text{if } |\text{Re } (h_\nu a_{\nu+1})| > 1 \end{cases}$$

III.   If $\text{Re}(h_\nu a_\nu) > 0$ and $\text{Re}(h_\nu a_{\nu+1}) < 0$ then $a_\nu = 1/2$.

IV. If $\text{Re}(h_\nu a_\nu) < 0$ and $\text{Re}(h_\nu a_{\nu+1}) > 0$ then introduce a new point x_ν^* with $x_\nu \leq x_\nu^* < x_{\nu+1}$ where $\text{Re}(h_\nu a(x_\nu^*)) = 0$. The I or II is applicable.

REMARK: For linear equations the conditions that α_ν is a Lipschitz continous function of $h_\nu a_\nu$ is not necessary. One could use the standard procedure

$$\alpha_\nu = \begin{cases} 1 & \text{if } \text{Re}(h_\nu a_\nu) > 1 \\ 1/2 & \text{if } |\text{Re}(h_\nu a_\nu)| \leq 1 \\ 0 & \text{if } \text{Re}(h_\nu a_\nu) < -1 \end{cases}$$

This is the procedure we proposed in [4]. However, for nonlinear equations we use "Newton" and the discontinous change of the formulae can cause convergence problems. Also, if one wants to use Richardson extrapolation one needs an even smoother transition.

We shall now describe the procedure to refine the mesh. Assume we have computed the solution of (3.3) on a mesh $0 = x_1 < x_2 < ... < x_N = 1$. Let

$$J_{\nu,1} = \frac{u_\nu - u_{\nu-1}}{h_\nu}, \qquad J_{\nu,2} = \frac{J_{\nu+1,1} - J_{\nu,1}}{h_\nu + h_{\nu+1}}$$

denote the first and second devided diffferences respectively. Under reasonable assumptions (see [5]) one can prove that the error $\max_\nu |y(x_\nu) - u_\nu|$ can be estimated by $\max_\nu M[x_{\nu-1}, x_\nu, x_{\nu+1}]$ where

$$M[x_{\nu-1}, x_\nu, x_{\nu+1}] = (h_\nu^2 + h_{\nu+1}^2)(|J_{\nu,2}| + |J_{\nu,1}| + |u_\nu|).$$

Therefore the strategy is to add points if $M[x_{\nu-1}, x_\nu, x_{\nu+1}]$ is too large and to delate points if $M[x_{\nu-1}, x_\nu, x_{\nu+1}]$ is very small. In detail we procede as follows. Let Δ denote a threshold constant, and assume we have constructed the mesh for $x \leq x_{\nu-1}$.

1) If $M[x_{\nu-1}, x_\nu, x_{\nu+1}] > \Delta$ then we add the points $x_{\nu-1} + 1/2\, h_{\nu-1}, x_\nu + 1/2\, h_\nu$.

2) If $M[x_{\nu-1}, x_\nu, x_{\nu+1}] < 1/3\,\Delta$ then we investigate $M[x_{\nu-1}, x_{\nu+1}, x_{\nu+2}]$. If also $M[x_{\nu-1}, x_{\nu+1}, x_{\nu+2}] < 1/3\,\Delta$ then we delate the point x_ν and investigate $M[x_{\nu-1}, x_{\nu+2}, x_{\nu+3}]$. If also $M[x_{\nu-1}, x_{\nu+2}, x_{\nu+3}] < 1/3\,\Delta$ then we delate $x_{\nu+1}$, etc.

This procedure gives us a new mesh. However, numerical experience has shown that one should not change the mesh size too fast. Therefore we add more points such that

$$\tfrac{1}{3} \leq |h_{\nu}/h_{\nu+1}| \leq 3 \tag{3.5}$$

This is done by the following procedure.

If $|h_{\nu}/h_{\nu+1}| < 1/3$ then we add the point $x_{\nu+1} + 1/2\, h_{\nu+1}$.

If $|h_{\nu}/h_{\nu+1}| > 3$ then we add the point $x_{\nu} + 1/2\, h_{\nu}$.

This process is repeated until (3.5) is satisfied everywhere.

The next step is to calculate the solution of (3.3) on the new mesh.

4. Difference Approximations for Systems

In applications the systems are often of the form

$$dy/dx = (\tfrac{1}{\varepsilon} A(x) + B(x))y + F(x) \tag{4.1}$$

Here $\varepsilon > 0$ is a small constant and

$$A(x) = \begin{bmatrix} a_{11}(x)\ a_{12}(x)\ \dots\ a_{1n}(x) \\ a_{21}(x)\ \dots\ \dots\ \dots\ a_{2n}(x) \\ \dots\dots\dots\dots\dots\dots\dots \\ a_{n1}(x)\ \dots\ \dots\ \dots\ a_{nn}(x) \end{bmatrix} \qquad B(x) = \begin{bmatrix} b_{11}(x)\ b_{12}(x)\ \dots\ b_{1n}(x) \\ b_{21}(x)\ \dots\ \dots\ \dots\ b_{2n}(x) \\ \dots\dots\dots\dots\dots\dots\dots \\ b_{n1}(x)\ \dots\ \dots\ \dots\ b_{nn}(x) \end{bmatrix}$$

are smooth functions of x.

If $A(x)$ is upper triangular, i.e. $a_{ij} = 0$ for $i > j$, then we can write (4.1) formally as n scalar equations

$$dy^{(i)}/dx = (\tfrac{1}{\varepsilon} a_{ii} + b_{ii})y^{(i)} + G^{(i)} \tag{4.2}$$

where

$$G^{(i)} = \frac{1}{\varepsilon} \sum_{j=i+1}^{n} a_{ij}\, y^{(j)} + \sum_{\substack{j=1 \\ j \neq i}}^{n} b_{ij}\, y^{(j)} + F^{(i)}$$

Thus we can use the scheme and the refinement procedure of the last section for every equation (4.2).

If $A(x)$ is not upper triangular then we have to transform $A(x)$ to upper triangular form. This can be done analytically or by the Q-R method which is economical. Assume that we want to calculate the solution of the difference equation on a mesh $0 = x_1 < x_2 < \ldots < x_N = 1$.

Then we construct unitary matrices U_i such that

$$U_i^* A(x_i) U_i = \begin{bmatrix} \tilde{a}_{11}(x_i) \cdots\cdots\cdots \tilde{a}_{1n}(x_i) \\ 0 \quad \tilde{a}_{22}(x_i) \cdots\cdots \tilde{a}_{2n}(x_i) \\ \cdots\cdots\cdots\cdots\cdots\cdots \\ 0 \cdots\cdots\cdots\cdots 0 \; \tilde{a}_{nn}(x_i) \end{bmatrix}$$

In every interval $x_i \le x \le x_{i+1}$ we introduce a new variable by

$$y = U\tilde{y} \, , \qquad U(x) = U_i + (U_{i+1} - U_i) \frac{x - x_i}{x_{i+1} - x_i}$$

and obtain from (4.1)

$$d\tilde{y}/dx = U^*(x) (\tfrac{1}{\varepsilon} A(x) + B(x)) U(x)\tilde{y} - U^*(dU/dx)\tilde{y} + U^* F.$$

Now $U^* A\, U$ is upper triangular in the mesh points and we can apply the previous method.

5. Numerical Examples

In this section we consider second-order equations

$$\varepsilon d^2 y/dx^2 = d(a(x)y)/dx + b(x)y \, , \qquad -c \le x \le d$$

$$y(-c) = \alpha \, , \qquad\qquad y(d) = \beta \qquad\qquad (5.1)$$

We write them as a first-order system by introducing a new variable

$$dv/dx = b(x)y$$

Then we can integrate (5.1) and obtain

$$dy/dx = (a(x)/\varepsilon)y + f_1, \qquad f_1 = \varepsilon^{-1} v$$

$$\qquad\qquad\qquad\qquad\qquad\qquad\qquad\qquad (5.2)$$

$$dv/dx = f_2 \, , \qquad\qquad f_2 = by.$$

We think of the system as two scalar equations and apply the method developed earlier. In particular, the second equation will always be approximated by

$$\frac{v_{\nu+1} - v_\nu}{h_\nu} = \tfrac{1}{2}(b_{\nu+1} y_{\nu+1} + b_\nu y_\nu) \, . \qquad\qquad (5.3)$$

We present now a number of computer printouts and explain the notations on the first graph (compare with [3]).

$$EPS*D2Y/DX2 = D((-X\overset{\wedge}{3} + X/2)*Y)/DX + (3*X\overset{\wedge}{2} + 0.2)*Y = 0.$$

stands for

$$\varepsilon d^2y/dx^2 = d((-x^3 + \tfrac{1}{2}x)y)/dx + (3x^2 + 0.2)y$$

EPS = 1.0E-5 means $\varepsilon = 10^{-5}$.

TOL = 0.1. It is of no consequence for linear equations. For nonlinear equations it tells the machine to stop the Newton iteration when the residue is smaller than the tolerance.

DELTA = 0.03 denotes the value of the threshold constant in the mesh refinement procedure.

ITERATIONS. It is of no consequence for linear equations. For nonlinear problems it counts the number of Newton iterations.

REFINEMENTS. It denotes the number of mesh refinements.

EPS*D2Y/DX2=D((-X^3+X/2)*Y)/DX + (3*X^2+0.2)*Y

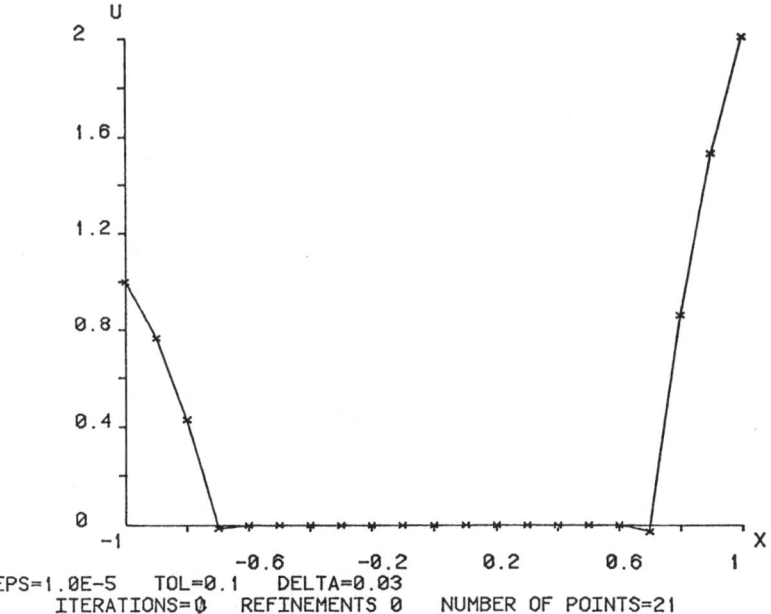

EPS=1.0E-5   TOL=0.1   DELTA=0.03
        ITERATIONS=0   REFINEMENTS 0   NUMBER OF POINTS=21

EPS*D2Y/DX2=D((-X^3+X/2)*Y)/DX + (3*X^2+0.2)*Y

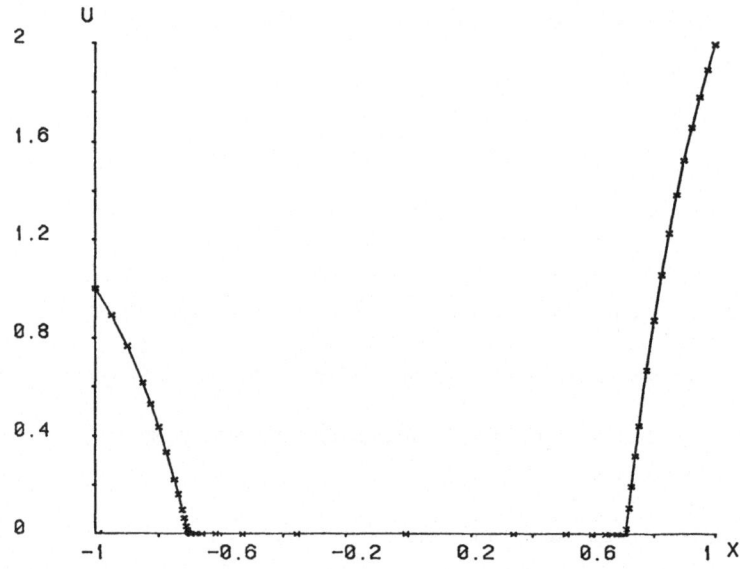

EPS=1.0E-5   TOL=0.1   DELTA=0.03
ITERATIONS= 0   REFINEMENTS 9   NUMBER OF POINTS=46

EPS*D2Y/DX2=D(-(SIN(PI*X))^2*Y)/DX + (-X+1+PI*SIN(2*PI*X))*Y

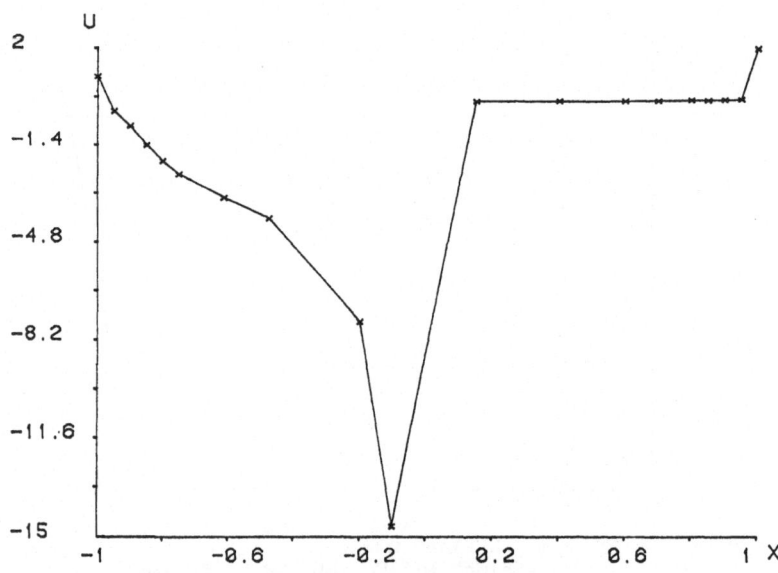

EPS=1.0E-5   TOL=0.1   DELTA=0.08
ITERATIONS=0   REFINEMENTS 1   NUMBER OF POINTS=19

EPS*D2Y/DX2=D(-(SIN(PI*X))^2*Y)/DX + (-X+1+PI*SIN(2*PI*X))*Y

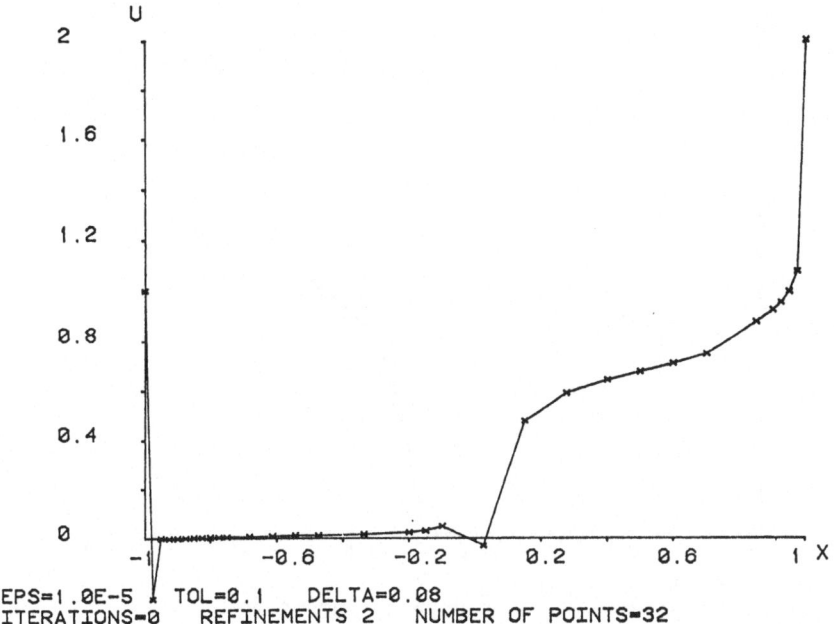

EPS=1.0E-5   TOL=0.1   DELTA=0.08
ITERATIONS=0   REFINEMENTS 2   NUMBER OF POINTS=32

EPS*D2Y/DX2=D(-(SIN(PI*X))^2*Y)/DX + (-X+1+PI*SIN(2*PI*X))*Y

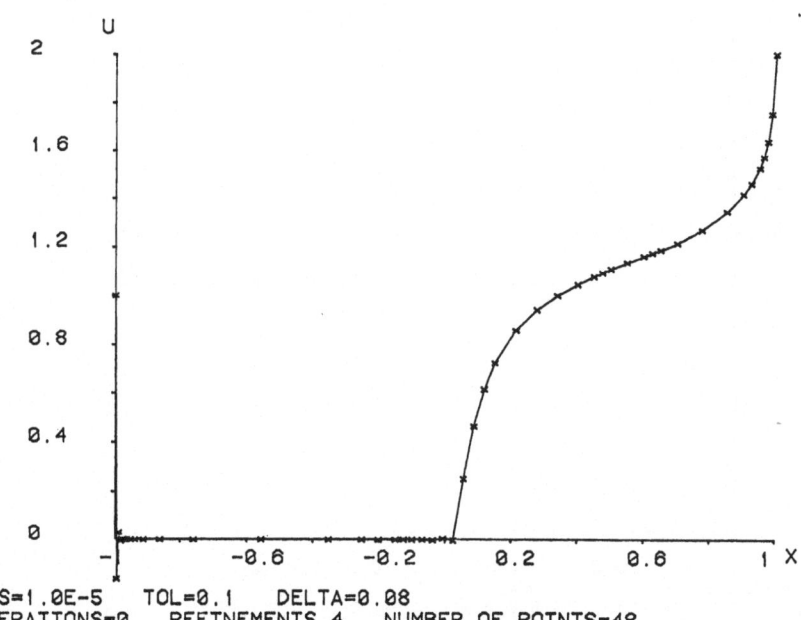

EPS=1.0E-5   TOL=0.1   DELTA=0.08
ITERATIONS=0   REFINEMENTS 4   NUMBER OF POINTS=48

EPS*D2Y/DX2=D(-(SIN(PI*X))^2*Y)/DX + (-X+1+PI*SIN(2*PI*X))*Y

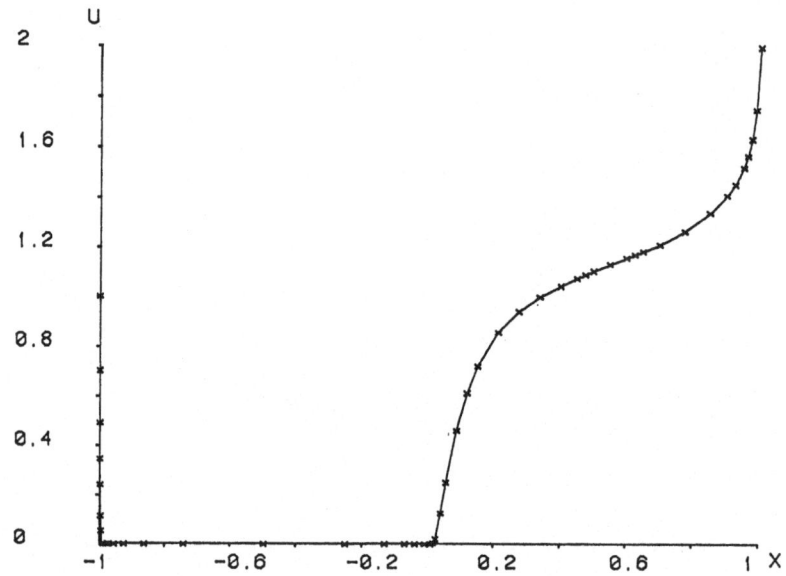

EPS=1.0E-5   TOL=0.1   DELTA=0.08
ITERATIONS=0   REFINEMENTS 8   NUMBER OF POINTS=50

EPS*D2Y/DX2=D(-X*X*Y)/DX+(2X-1)*Y

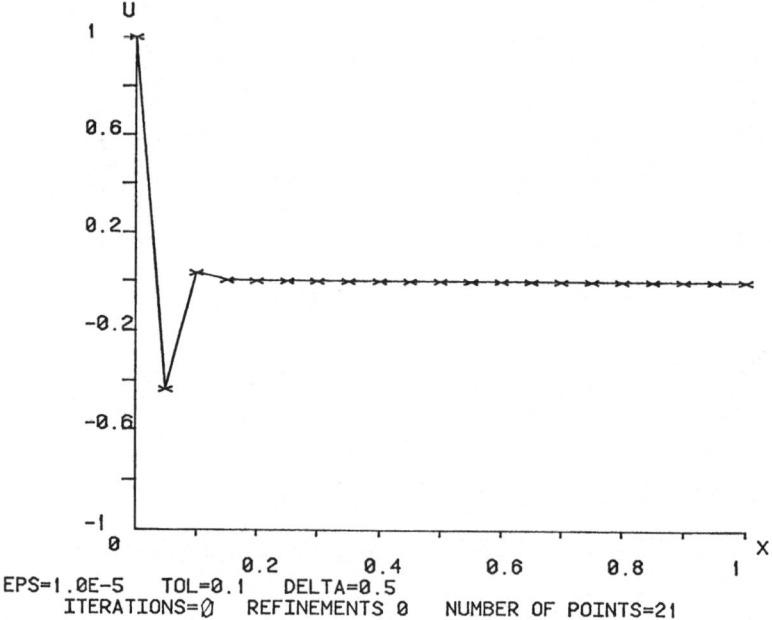

EPS=1.0E-5   TOL=0.1   DELTA=0.5
    ITERATIONS=0   REFINEMENTS 0   NUMBER OF POINTS=21

EPS*D2Y/DX2=D(-X*X*Y)/DX+(2X-1)*Y

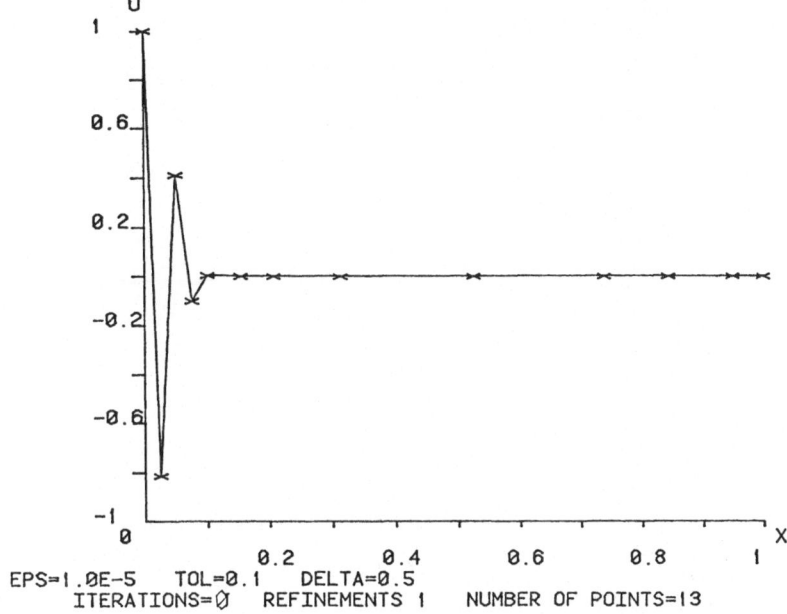

EPS=1.0E-5   TOL=0.1   DELTA=0.5
      ITERATIONS=0   REFINEMENTS 1   NUMBER OF POINTS=13

EPS*D2Y/DX2=D(-X*X*Y)/DX+(2X-1)*Y

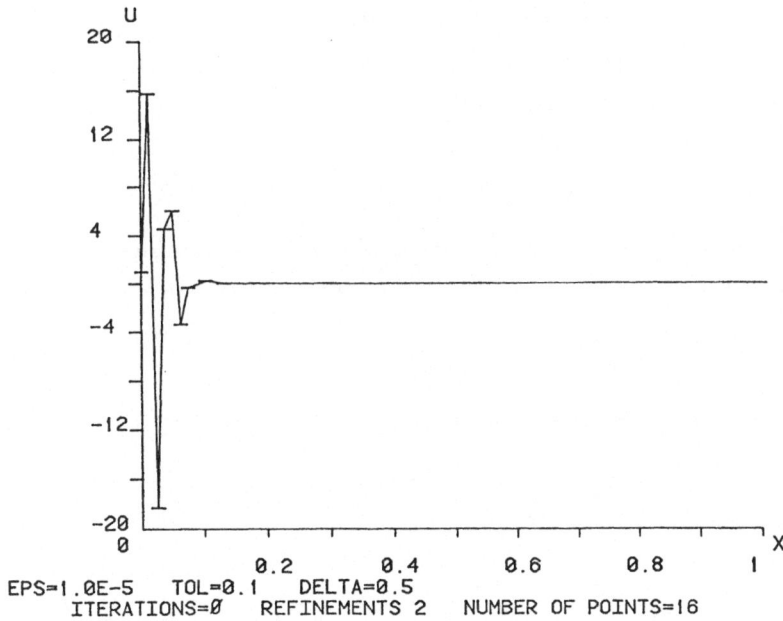

EPS=1.0E-5   TOL=0.1   DELTA=0.5
      ITERATIONS=0   REFINEMENTS 2   NUMBER OF POINTS=16

EPS*D2Y/DX2=D(-X*X*Y)/DX+(2X-1)*Y

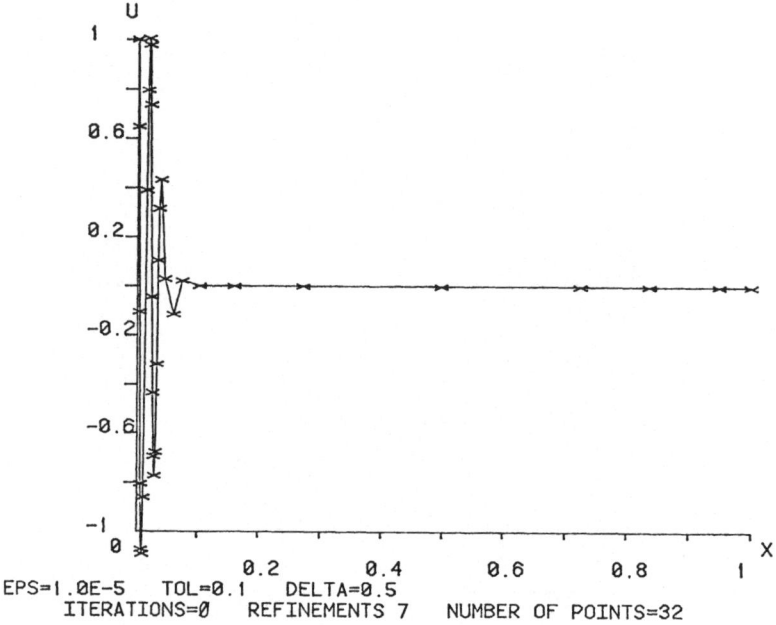

EPS=1.0E-5   TOL=0.1   DELTA=0.5
ITERATIONS=0   REFINEMENTS 7   NUMBER OF POINTS=32

6. Nonlinear Equations

We consider now nonlinear equations

$$\varepsilon d^2 y/dx^2 = d(a(x,y))/dx + b(x,y), \qquad -c \leq x \leq d,$$

$$y(-c) = \alpha, \quad y(d) = \beta \qquad\qquad (6.1)$$

and rewrite them as first-order systems

$$\varepsilon dy/dx = a(x,y) + v$$

$$dv/dx = b(x,y) \qquad\qquad (6.2)$$

We use Newton's method to solve the system. Let $y^{(n)}$, $v^{(n)}$ be an approximation to the solution of (6.2). Then we linearize (6.2) around $y^{(n)}$, $v^{(n)}$ to obtain new approximations

$$y^{(n+1)} = y^{(n)} + \tilde{y}, \qquad\qquad v^{(n+1)} = v^{(n)} + \tilde{v},$$

where u, v are solutions of the linearized system

$$\varepsilon d\tilde{y}/dx = a_y(x,y^{(n)})\tilde{y} + \tilde{v} + f_1, \qquad\qquad a_y = \partial a/\partial y,$$

$$d\tilde{v}/dx = b_y(x,y^{(n)}) \tilde{y} + f_2, \qquad\qquad (6.3)$$

with

$$f_1 = -\varepsilon d(y^{(n)})/dx + a(x,y^{(n)}) + v^{(n)}, \quad f_2 = -v^{(n)} + b(x,y^{(n)})$$

The linear system (6.3) is solved by our method.

As an example we have considered the equation

$$\varepsilon d^2 y/dx^2 = -\frac{1}{2} d(y^2)/dx + y, \quad -1 \le x \le 1$$

$$y(-1) = -1, \quad y(+1) = 2.$$

and solved it for

$$\varepsilon = 0.1, \ 0.05, \ 0.02, \ 0.01, \ 0.005, \ 0.002, \ 0.001 \ .$$

The initial guess was a straight line between $y = -1$ and $y = 2$. We used "the method of continuation", i.e. we used the computed solution as an initial guess to solve the problem for the next ε. The printouts are only given for $\varepsilon = 10^{-1}, 10^{-2}, 10^{-3}$.

EPS*D2Y/DX2=D(-Y*Y/2)/DX+Y

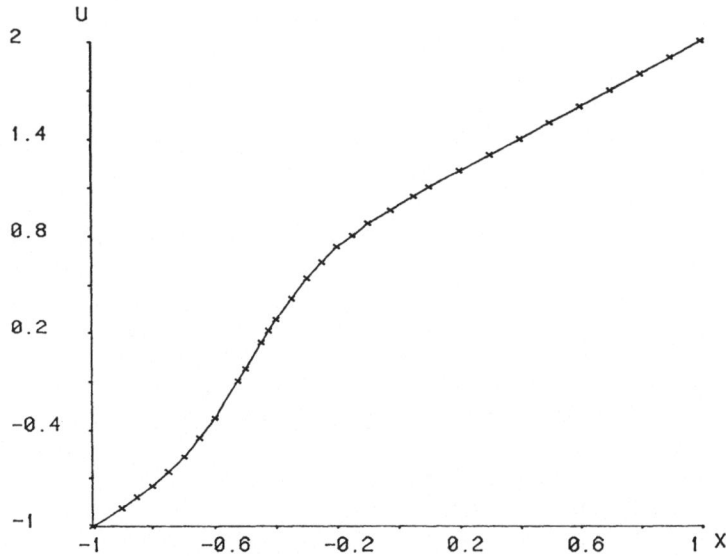

EPS=0.1   TOL=0.01   DELTA=0.05
ITERATIONS=2   REFINEMENTS 2   NUMBER OF POINTS=31

EPS*D2Y/DX2=D(-Y*Y/2)/DX+Y

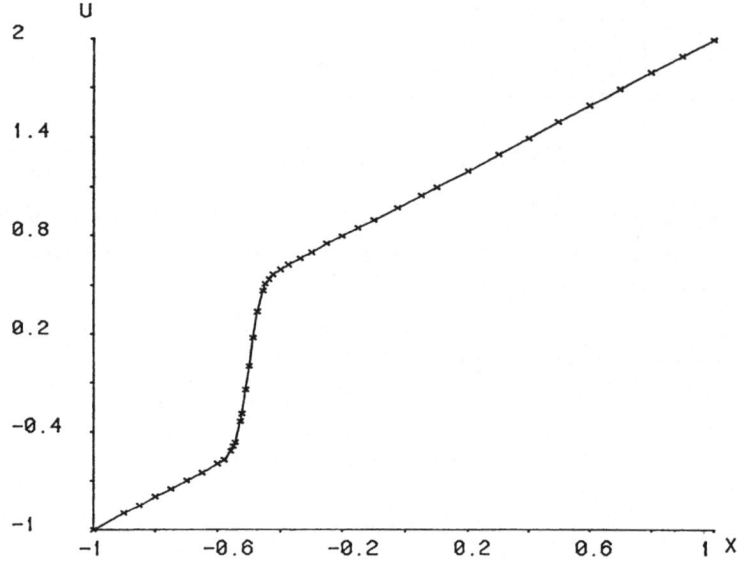

EPS=0.01   TOL=0.01   DELTA=0.05
ITERATIONS=2   REFINEMENTS 2   NUMBER OF POINTS=42

EPS*D2Y/DX2=D(-Y*Y/2)/DX+Y

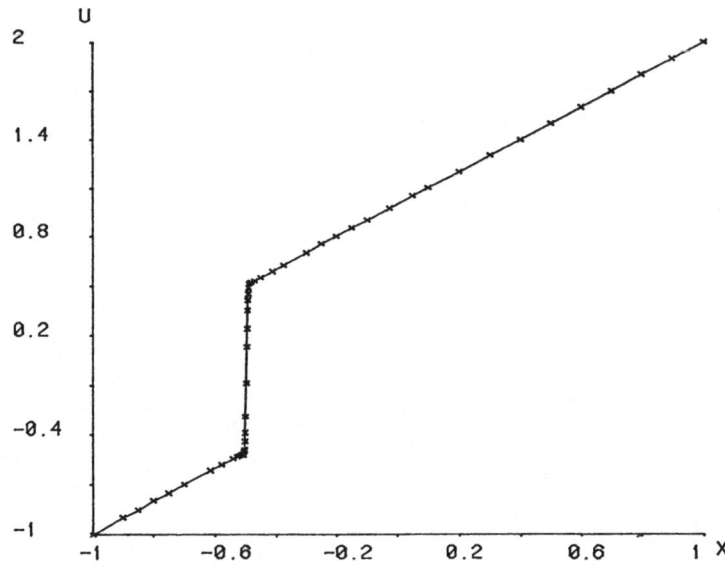

EPS=1.0E-3   TOL=0.01   DELTA=0.05
ITERATIONS=1   REFINEMENTS 4   NUMBER OF POINTS=52

EPS*D2Y/DX2=D(-Y*Y/2)/DX + Y

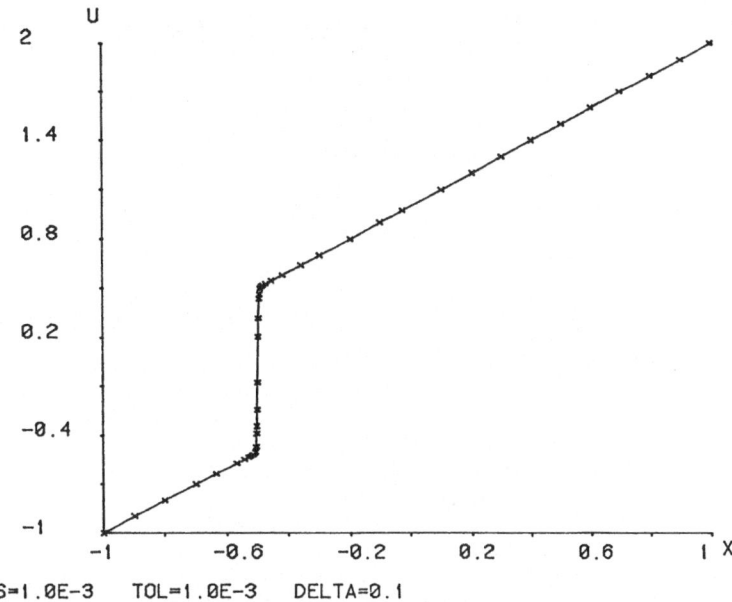

EPS=1.0E-3   TOL=1.0E-3   DELTA=0.1
ITERATIONS=8   REFINEMENTS 1   NUMBER OF POINTS=44

7. References

[1] H y m a n, J.M.: A method of lines approach to the numerical solution of Conservation Laws. Preprint Los Alamos Scientific Laboratory, LA-UR-79-837 (1979).

[2] K r e i s s, H.-O.: Numerical methods for singular pertubation problems, SIAM-AMS Proceedings, 10, 1976.

[3] K r e i s s, B. and K r e i s s, H.-O., to appear

[4] K r e i s s, H.-O., and N i c h o l s, N.: Numerical methods for singular pertubation problems. Dept. of Computer Sciences, Uppsala University, Report No. 57 (1975).

[5] K r e i s s, H.-O. and N i c h o l s, N., to appear.

[6] M i r a n k e r, W.L. and M o r r e e u w, J.P.: Semianalytic numerical studies of turning points arising in stiff boundary value problems. Math. Comp. 28 (1974), pp. 1017-1034.

[7] P e a r s o n, C.E.: On a differential equation of boundary layer type. J. Math. Phys. 47 (1968), pp. 134-154.

[8] S o d, G.: A survey of numerical methods for compressible fluids. Courant Institute Report COO-3077-145 (1977).

Numerical Methods for Computation of Multi-Dimensional Discontinuous Gas Flow

V. Rusanov
Moscow

1. Introduction

In this paper some questions connected with the calculation of inviscid discontinuous gas flows are considered. The paper is not a review, and our presentation is by no means a complete study of the field. Despite the recent advances in the development of the numerical methods for multi-dimensional discontinuous gas flows the number of unsolved problems is rather great. Here we discuss some basic questions as well as the difficulties of their solution and the prospects for overcoming these difficulties.

2. Statement of the Problem

An inviscid gas flow in a three-dimensional space is described by the density ρ, the pressure p and the velocity components u_x, u_y, u_z depending on the space coordinates x, y, z and the time t. We shall assume further that ρ, p, u_x, u_y, u_z are the components of a formal vector w. The components of w satisfy the wellknown gas dynamic differential equations which contain also thermodynamic quantities (internal energy and entropy) being the functions of p and ρ.

In the gas filled region the surfaces could exist on which the components of w are non-continuous. The relationship between the components of w on the both sides of the discontinuity surface S at some point P is local and depends on the normal v_p to S and on the velocity D_p of S along this normal.

There are two kinds of strong discontinuities in gas flow: the shock waves and the tangential or contact discontinuities. The former are named "nonlinear", and the latter are named "linear". Roughly speaking, the difference between them is that on the "linear" discontinuity the velocity D does not depend on the characterstic function mostly varying at the discontinuity, while at the "nonlinear" discontinuity this dependence is rather strong. An exact definition of this difference may be found in [1]. The existence of the two kinds of discontinuities and the difference between them are of great importance for the theory of gas dynamic equations as well as for the development of the numerical methods. In the gas flow field the surfaces could also exist on which not the values of w but its derivatives with respect to (x, y, z) and t are non-continuous. These are called the weak discontinuity surfaces.

In practice, the boundaries of the region, where the solution must be found, almost always consist of the discontinuity surfaces. Some other boundaries may be introduced into the computations, but, as a rule, they usually are specified by the numerical technique used, and not by the physics. Hence, there is every reason to consider the vector w to be the function differentiable over the whole domain except for the finite (though possibly great) number of the strong and weak discontinuity surfaces.

Though such representation of the solution makes the theoretical analysis more complicated as compared with the introduction of the generalized functions, it is more appropriate from the physical and practical viewpoints. For the practical purposes, the method yielding the result which, in the "weak" sense, is close to the desired solution is not so efficient as the method using the smoothness of the functions (in the domains where it exists) to achieve the higher accuracy.

While developing such a method, however, one has to overcome a number of difficulties. The basic of them is that the boundaries of the smoothness domains are not known a priori. Moreover, their number and the topology of the flow region division into the smoothness domains may also be unknown. Since the gas dynamic equations are nonlinear, new discontinuities may arise with time, while the old ones may disappear. The total number of the arising and interacting discontinuities may be so great that the consideration of each of them is out of question (especially, in the multi-dimensional case). Due to this, the methods of computation "through" the discontinuities were developed, which do not require any information about the presence or the absence of discontinuities. An essential and so far not eliminated flaw of the through computation methods is that the value of w near the discontinuity surface is computed with more or less distortion, and there is some difficulty in finding the actual location of the discontinuity in the flow field.

Sometimes it is possible to determine the topology of the discontinuities beforehand and to select the discontinuities most important in the problem under consideration. Some of them may be the boundaries of the flow region, the others may be considered as the boundaries of the subregions with the corresponding boundary conditions. Then the method has to be developed for the computation of a flow in the region with unknown boundaries whose location must be computed simultaneously with flow. This method is free from the flaws of the through computation technique but its applicability is limited by the number of boundaries.

All the existing methods of computation of the gas flows with discontinuities are based on one of the two approaches discussed above, or on their combination. In this paper we consider some aspects of each of these approaches based on the finite-difference approximations. We restrict ourselves only to the computations in the Eulerian coordinates and do not consider the methods based on the introduction of the Lagrangian or the Eulerian-Lagrangian coordinates.

3. The Through Computation Methods

Neumann and Richtmyer were the first to propose a through computation method for the shock wave in their well known work [2] published in 1950. The main idea was to introduce an artificial viscosity that would prevent the "tilting" of the wave profile. Now it is used, in one or other form, in all the through computation methods.

The viscosity may be either introduced in the differential equations beforehand or appear as the "scheme" viscosity due to the dissipative terms in the difference scheme. Due to its presence a stretched discontinuity profile of w with the large (but not infinite) gradients is generated instead of the abrupt "jump" in the gas dynamic functions. The parasitic oscillations may occur near the discontinuity. Their amplitude may reach 20-30 percent of the jump and they may rather slowly diminish with the moving of the discontinuity. The main task in constructing the through computation schemes is to prevent the profile stretching and oscillation.

Aside from the involvement of some or other form of viscosity, the difference scheme for the through computations must be conservative. The differential equations usually could be written in conservative form due to the physical sense of the problem. This form is usually used for the construction of the difference schemes for through computations.

The locality of the conditions at the discontinuity and their independence of the wave curvature allow one to study as the first approximation the properties of the difference schemes for the plane discontinuity with the constant values of w on the both sides of it (the step-type discontinuity). It proves to be possible to get theoretical solution to some basic questions while studying the one-dimensional problem. The numerical investigation shows that the results obtained are basically valid for the multidimensional problems, too. We shall briefly consider these results, which allow us to evaluate the through computation methods from the general point of view, to understand the causes of the scheme flaws and to show some ways for their elimination.

3.1 The Structure of the Discontinuity Profile in the Numerical Solution (Linear Discontinuity)

Let us first consider the simple one-dimensional linear problem:

$$u_t + au_x = 0, \quad u(x,0) = u^0(x) = \begin{cases} 0, & x < 0 \\ 1, & x > 0 \end{cases} \tag{3.1}$$

Let us introduce the steps $\Delta x = h$ and $\Delta t = \tau$, and write down the explicit scheme on the floating grid:

$$u^{n+1}(x) = \sum_{\nu=-\nu_1}^{\nu_2} a_\nu u^n(x + \nu h); \quad -\infty < x < +\infty \tag{3.2}$$

Let us study the asymptotes of the fuction $u^n(x)$ at $n \to \infty$ and τ, h fixed. Going over to the Fourier transform we get the amplification factor

$$G(kh) = \sum_{\nu=-\nu_1}^{\nu_2} a_\nu \, e^{i\nu kh} \tag{3.3}$$

where k is the wave number.

Let the scheme (3.2) have the order p. Then [3]

$$G(kh) = e^{-i\sigma kh} \{1 - \frac{\varepsilon b_{p+1}}{(p+1)} (kh)^{p+1} + 0((kh)^{p+2})\} \tag{3.4}$$

Here $\sigma = a\tau/h$ is the Courant number, b_{p+1} is real, and $\varepsilon = 1$ when p is odd and $\varepsilon = -i$ when p is even. From the stability of scheme (3.2) it follows that if p is odd then $b_{p+1} > 0$. If p is even b_{p+1} can always be made positive by changing the direction of the axis x. So we shall assume that in all cases $b_{p+1} > 0$.

If the scheme (3.2) is stable the theorem [3, 4] is valid that for any fixed ξ and $n \to \infty$ the value $u^n(n\sigma\tau + \xi (b_{p+1}n)^{1/p+1}$ approaches the function

$$F_p(\xi) = \int_{-\infty}^{\xi} f_p(z)\,dz, \quad F_p(\infty) = 1 \tag{3.5}$$

where

$$f_p(z) = \begin{cases} \dfrac{1}{\pi} \displaystyle\int_0^\infty e^{-\frac{s^{p+1}}{p+1}} \cos(zs)\,ds, & p \text{ is odd,} \\[4mm] \dfrac{1}{\pi} \displaystyle\int_0^\infty \cos(\frac{s^{p+1}}{p+1} + zs)\,ds, & p \text{ is even.} \end{cases} \tag{3.6}$$

Or, in other way, in any finite interval $[-X, X]$ the function $u^n(x)$ is close to the function

$$u^n(x) \approx F_p(\frac{x - n\sigma\tau}{(b_{p+1}n)^{1/p+1}h}) \tag{3.7}$$

if n is great enough.

Figures 1-3 show the profiles of the functions $F_p(\xi)$ for $p = 1, 2, 3$.

From formulae (3.5) and (3.6) we get the asymptotes of the functions $F_p(\xi)$ as follows:

1) At $\xi \to -\infty$:

$$F_1(\xi) \sim |\xi|^{-1} e^{-\xi^2/2}$$

$$F_2(\xi) \sim |\xi|^{-3/4} \cos(\frac{2}{3}|\xi|^{3/2} - \frac{\pi}{4})$$

$$F_3(\xi) \sim |\xi|^{-2/3} \; e^{-3/8|\xi|^{4/3}} \cos(\frac{3\sqrt{3}}{8} \, |\xi|^{4/3} - \frac{\pi}{6}) \qquad\qquad (3.8)$$

2) At $\xi \to +\infty$:

$$1 - F_1(\xi) \sim \; \xi^{-1} \; e^{-\xi 2/2}$$

$$1 - F_2(\xi) \sim \; \xi^{-3/4} \; e^{-2/3\xi^{3/2}} \qquad\qquad (3.9)$$

$$1 - F_3(\xi) \sim \xi^{-2/3} \; e^{-3/8\xi^{4/3}} \cos(\frac{3\sqrt{3}}{8} \, \xi^{4/3} - \frac{\pi}{6})$$

The consideration of figures 1-3 and formulae (3.7)-(3.9) allows to make the following conclusions:

1)   When calculated with the scheme of order p, the discontinuity profile not only shifts by the value n a τ in accordance with the equation (3.1), but also stretches with the velocity proportional to $n^{1/p+1}$.

2)   The qualitative nature of the "stretched" profile near the discontinuity and at $\xi \pm \infty$ essentially depends on the scheme parity. At $\xi < 0$ the even-order schemes have much stronger oscillations damping very slowly at $\xi \to -\infty$. The schemes of order above one generate nonmonotonous profiles. In other words, all the schemes of order above one are asymptotically nonmonotonous.

These conclusions drawn for the scheme (3.2) and the one-dimensional equation are general enough. They seem to be valid for the wider classes of linear schemes including the implicit ones.

Thus, in the framework of the linear schemes uniform at all points of the grid, the only way to diminish the "stretching" of the linear discontinuity is to increase the order of the scheme accuracy. Since the even-order schemes yield much stronger oscillations and the schemes of order five and above can not be implemented in practice, the third-order schemes are optimal. Their various versions are constructed now by many authors.

The use of the homogeneous linear schemes, however, does not solve the problem of calculation of the linear (contact) discontinuities. Even with the third-order scheme, the profile stretching with the velocity $n^{1/4} \sim t^{1/4}$ becomes unacceptable when n is great enough. It proves to be impossible to establish not only the location of the tangential discontinuity but even the fact of its existence.

Many tried to improve the computation of the linear discontinuity. The greatest success seemed to be achieved in the works [5, 6]. The method proposed there has a clear theoretical justification and yields very good results in the one-dimensional case. Its generalization for the multi-dimensional cases, however, meets some difficulties which are inevitable for such complex problems. It seems to be interesting to study the capabilities of the method [5] in application to the third-order schemes in order to increase the computation accuracy in the regions of smoothness.

3.2 The Structure of the Nonlinear Discontinuity Profile in the Numerical Solution

The investigation of the profile structure of the nonlinear discontinuity generated by the difference scheme is a much more complicated problem, so we restrict ourselves only to some preliminary results which are, however, of fundamental significance. We consider first the case of a nonlinear equation

$$\frac{\partial u}{\partial t} + \frac{\partial F(u)}{\partial x} = 0 \tag{3.10}$$

with the initial data corresponding to the shock wave moving with the velocity D, such as

$$u(x,0) = u^0(x) = \begin{cases} u_1, & x < 0 \\ u_2, & x > 0 \end{cases} \qquad D = \frac{F(u_1) - F(u_2)}{u_1 - u_2} \tag{3.11}$$

The exact solution has the form $u(x, t) = u^0(x - Dt)$.

To solve the problem (3.10)-(3.11) let us use the difference scheme

$$u^{n+1}(x) = \Phi\{u^n(x + vh)\}_{v=-v_1}^{v_2}, \tag{3.12}$$

where Φ is the nonlinear function of the quantities

$$u^n(x + vh), \quad -v_1 \le v \le v_2$$

Now we raise the following questions:

1. Has equation (3.12) the stationary (to the accuracy of the shift by nD) solution v ( )
    satisfying the equation

$$v(\xi - D\tau) = \Phi\{v(\xi + vh)\}_{v=-v_1}^{v_2} \tag{3.13}$$

and the boundary conditions $v(-\infty) = u_1$, $v(+\infty) = u_2$

2. What are the properties of the function $v(\xi)$ if it exists?

3. Does there exist $\lim_{n \to \infty} u^n(\xi + nD\tau)$ equal to $v(\xi)$?

    If it exists, then in what sense?

These questions were implicitly formulated by Lax [7] for the equation with $F(u) = u^2/2$ and for the first-order scheme proposed by him. In [8] the existence conditions for $v(\xi)$ were clarified for an arbitrary three-point difference scheme. Recently I.V. Besmenov has proved the existence of limit of $u^n(\xi + nD\tau)$ for the monotonic schemes.

The numerical analysis shows that the function $v(\xi)$ not only is continuous but also it has the high degree of smoothness. In contrast to the linear discontinuity the shock profile stretches not infinitely but for a finite length depending on the discontinuity and difference scheme parameters. The functions $v(\xi)$ computed for the equation (3.10), $F(u) = 0.5u - 0.1u^2$, with schemes of accuracy $p = 1, 2, 3$ are plotted in Figs. 4-6.

Note that though $v(\xi)$ is a smooth function on the floating grid, it does not look the same in the calculations on the fixed grid even if n is large enough. The fixed grid is shifted at each step by the value $- D\tau$ with respect to the coordinate ξ. The points with the coordinates $x_m - nD\tau$ are chosen on the continuous smooth curve to form the set of values of the grid function $u_m^n = v(x_m - nD\tau)$. In Fig. 7 are given the distributions of u_m^n at three successive time steps $p = 2$, $n = 200, 201, 202$. It is difficult to notice the connection between these distributions and the continuous smooth function $v(\xi)$ in Fig. 5. However, by matching the grid functions for several time-steps on one plot, shifting each of them by $- nD\tau$, we shall have all the points u_m^n on the curve $v(\xi)$.

The analysis of the numerical results shows that the scheme order basically has the same influence on the shock profile as in the linear case. The higher the scheme order the less stretching of the profile is in the vicinity of the shock. The oscillations near the shock are much stronger for the even-order schemes, but the first-order scheme does not always generate the monotonous function $v(\xi)$. Thus, for the nonlinear equation the first-order scheme is not necessarily asymptotically monotonous.

3.3 On the Discontinuity Profile Structure for the System of Equations

Let us consider the Cauchy problem for the system of hyperbolic equations

$$\frac{\partial f(w)}{\partial t} + \frac{\partial F(w)}{\partial x} = 0 \qquad (3.14)$$

with the discontinuous initial conditions

$$w(x,0) = \begin{cases} w_1, & x < 0 \\ w_2, & x > 0 \end{cases}$$

w_1 and w_2 being such as to satisfy the conditions at the discontinuity

$$F(w_2) - F(w_1) = D[f(w_2) - f(w_1)] \qquad (3.15)$$

Then the Cauchy problem has the exact solution

$$w(x,t) = \begin{cases} w_1, & x - Dt < 0 \\ \\ w_2, & x - Dt > 0 \end{cases}$$

First we consider the linear (contact) discontinuity case. If (3.14) is a system of gas dynamic equations, then in the contact discontinuity case, when u = const and p = const (3.14) reduces to an equation with respect to ρ of the form $\rho_t + u\,\rho_x = 0$. Therefore all the conclusions in section 3.1 are valid.

In a general case of the linear system we have $F = Aw$, where A is a constant matrix. Reducing A to a diagonal form, we get a split system of equations of the form $W_t^{(i)} + a^{(i)} W_x^{(i)} = 0$, $i = 1,2,...k$. The conclusions of section 3.1 are valid for each of these equations. The value of D must be equal to one of $a^{(i)}$, and only the corresponding component of $W^{(i)}$ is noncontinuous.

A quite different situation takes place in the case of the nonlinear shock discontinuity. The detailed numerical investigation shows that under certain conditions the difference scheme has the stationary (to within the shift) solution of the form $w = w(x-Dt)$ [8] . However, the profiles of the components of w have rather complicated structure due to their interaction. It is possible to investigate and describe this structure if instead of the distribution of the $w(\xi)$ components to consider the functions corresponding to the Riemann invariants J_+, J_- and entropy S calculated for the stationary profile.

Namely, let $X(w)$ be a matrix reducing the matrix $A(w) = (df/dw)^{-1}(dF/dw)$ to the diagonal form, i.e. $X(w)AX^{-1}(w) = \Lambda(w)$, where $\Lambda(w)$ is a diagonal matrix with the elements $\lambda_+ = u \pm c$ and $\lambda_0 = u$. Let $v(\xi)$ be a stationary profile. Now we define the vector-function $\Psi(\xi)$ depending on $v(\xi)$ by the formula

$$\Psi(\xi) = \Psi_1 + \int_{-\infty}^{\xi} X(w(\eta))\,\frac{dv}{d\eta}\,d\eta \tag{3.16}$$

The components of $\Psi(\xi)$ are $J_+(\xi)$, $J_-(\xi)$ and $S(\xi)$.

For the three-point difference scheme the structure of the $\Psi(\xi)$ components has been studied both analytically and numerically. The results show that only one of the invariants J_+ or J_-, which has the maximal jump across the shock, behaves in the same way as the solution of the single equation (3.10). Its profile can be made monotonous by the appropriate choice of the scheme parameters. The profile of the other invariant and the entropy, however, always has more or less strong oscillations whose amplitude depends only on the Courant number and/or the wave intensity. These oscillations may be noticeable in the profile of the $w(x)$ components.

Fig. 8 shows the distributions of the pressure p_m^n and u_m^n in the shock profile, where the small oscillations can be seen. The p_m^n distribution at a single step and the stationary profile $p(\xi)$ obtained numerically is given in enlarged scale in Fig. 9. An apparent irregular nature of oscillations at a single step is explained by a random distribution of the p_m^n values on the regular oscillating curve $p(\xi)$.

The distributions of the $\psi(\xi)$ components obtained by the integration over the stationary profile are given on an enlarged scale in Fig. 10. The invariant J_+ with near a monotonous profile has the maximal jump on the shock moving to the right. The oscillations are observed in the profiles of the entropy S and of the second invariant J_-. They are especially strong in J_-, and generate oscillations in p_m^n and u_m^n shown in Figs. 8 and 9. Note that the damping factors and the oscillation frequencies evaluated theoretically for J_- and S differ no more than by 0,5% from those calculated with the difference scheme.

The above results were obtained for the family of the first-order three-point schemes in the case of the nonlinear system of equations. The investigation of second-order schemes shows that there are strong oscillations in all the components as in the case of one equation. Therefore the detailed study of the second-order schemes are of no interest. At the same time it is very important for the theory of numerical methods to investigate the shock profile structure in the nonlinear third-order scheme; the sufficiently comprehensive results in this direction have not been obtained yet.

3.4 On the Shock Profile Structure in the Multi-Dimensional Problems

In the multi-dimensional problems two additional factors arise that influence the discontinuity profile structure. These are a curvature of the discontinuity surface and its orientation with respect to the mesh. If the mesh is fine enough one may consider the discontinuity surface to be plane in the first approximation. Then we get a problem on calculation of the plane discontinuity whose movement direction is oriented arbitrarily with respect to the mesh. Let us consider the two-dimensional case. Let h_1, h_2 be the mesh size in space and the discontinuity moves in the direction making the angle Θ to the axis x (Fig. 11).

Let us first consider one equation

$$\frac{\partial u}{\partial f} + \frac{\partial F(u)}{\partial x} + \frac{\partial G(u)}{\partial y} = 0 \tag{3.17}$$

and the simple first-order difference scheme

$$u_{m,\ell}^{n+1} = u_{m,\ell}^n + \Phi_{m+1/2,\ell}^x - \Phi_{m-1/2,\ell}^x + \Phi_{m,\ell+1/2}^y - \Phi_{m,\ell-1/2}^y$$

$$\Phi_{m+1/2,\ell}^x = -\frac{q_1}{2}(F_{m+1,\ell}^n + F_{m,\ell}^n) + \frac{\omega}{2}(u_{m+1,\ell}^n - u_{m,\ell}^n) \tag{3.18}$$

$$\Phi_{m,\ell+1/2}^y = -\frac{q_2}{2}(G_{m,\ell+1}^n + G_{m,\ell}^n) + \frac{\omega}{2}(u_{m,\ell+1}^n - u_{m,\ell}^n)$$

where $u^n_{m,\ell} = u(x_m, y_\ell, t^n)$, $F^n_{m,\ell} = F(u^n_{m,\ell})$

$$G^n_{m,\ell} = G(u^n_{m,\ell}); \quad q_j = \tau/h_j$$

Assuming that the solution of equation (3.17) depends only on the combination $\xi = x \cos\Theta + y \sin\Theta$ let us introduce the orthogonal coordinates ξ and $\eta = -x \sin\Theta + y \cos\Theta$. Changing the variables in (3.17), we get the one-dimensional equation.

$$\frac{\partial u}{\partial t} + \frac{\partial}{\partial \xi}\{F_\Theta(u)\} = 0 \qquad\qquad (3.19)$$

with the function $F_\Theta(u) = F(u)\cos\Theta + G(u)\sin\Theta$

Let us write the difference scheme (3.18) on the floating mesh

$$u^n_{m+\mu,\ell+\lambda} \rightarrow u^n(x+\mu h_1, y+\lambda h_2); \quad \mu,\lambda = -1,0,1$$

We shall search for the solution depending only on the coordinate ξ, denoting it by $U^n_\Theta(\xi)$.

Then $u^n(x,y) = U^n_\Theta(x\cos\theta + y\sin\theta)$

$$u^n_{m\pm 1,\ell} \rightarrow u^n(x\pm h_1)\cos\theta + y\sin\theta) = U^n_\Theta(\xi\pm h_1\cos\theta)$$

$$u^n_{m,\ell\pm 1} \rightarrow U^n_\Theta(\xi\pm h_2\sin\theta)$$

For the function $U^n_\Theta(\xi)$ we get the one-dimensional five-point difference scheme with the points ξ, $\xi\pm h_1\cos\Theta$, $\xi\pm h_2\sin\Theta$. This scheme principally does not differ from that considered above and the study of $U^n_\Theta(\xi)$ as $n\rightarrow\infty$ could be carried out in the same way both for the scheme (3.18) and for the more complex schemes.

The numerical experiments show that for the linear discontinuity the speed of the profile stretching is also of the order $O(n^{1/p+1})$, where p is the scheme order and the difference between the discontinuity profiles for different Θ is very small. Let us present some examples for the third-order scheme described below in section 3.5. The profile stretching speed was evaluated by the bahaviour of tg α^n, as a function of n, where α^n is the maximum angle between the tangent to the profile $U^n_\Theta(\xi)$ and the axis ξ. The calculations showed that at $50 \le n \le 500$ the formula tg $\alpha^n \sim n^{-1/4}$ is valid to high accuracy. From Fig. 12 one can see that the corresponding values in the logarithmic coordinates lie on the straight line lg tg $\alpha^n = -0.25$ lgn+const. The shifted profiles of the functions $U^n_\Theta(\xi)$, $\Theta = \pi/4$ are shown in Fig. 12 for n = 200 and n = 500, with the normalization $\overline{\xi} = \xi/n^{1/4}$ taken into account.

The influence of the angle Θ upon the profile $U^n_\Theta(\xi)$ is illustrated in Fig. 13 where the profiles for different angles are plotted for the same value of n. It can be seen that the said influence is very small.

Similar results were obtained for the nonlinear equation, but in this case the dependence of the profile on the angle Θ can be stronger. Indeed, for different Θ the function $F(u)\cos\Theta + G(u)\sin\Theta$ can be quite different depending on the choice of $F(u)$ and $G(u)$.

For the system of equations, the question about the discontinuity profile is much more complicated. The system of the gas dynamic equations is invariant with respect to the axes rotation. The introduction of the coordinates (ξ,η) leads to the one-dimensional problem if the velocity components u_x, u_y are transformed into U_ξ, U_η and $\partial U_\eta/\partial\eta = 0$ as well as $\partial p/\partial\eta = 0$, $\partial\rho/\partial\eta = 0$. We get the one-dimensional problem in which U = const (not necessarily zero). However, the changing (x, y) to (ξ,η) in the difference scheme does not always lead to the simple one-dimensional scheme. Due to this, when the discontinuous solution of the system of equations is calculated the distortions can appear not only in the transversal direction (stretching and oscillations) but along the shock, too (deflection from rectilinearity). The last distortions are usually periodic due to the periodic structure of the mesh. It is an interesting and important problem to construct the difference schemes free from this fault, i.e. "invariant" with respect to the rotation of the coordinate axes.

3.5 The Third-Order Schemes

As mentioned above the third-order schemes have some advantages compared to the first- and second-order schemes. Today we have a number of methods of construction of the third-order schemes, mainly for the one-dimensional problems. The development of the third-order schemes and their application to the multi-dimensional problems meet some difficulties, which are common for all methods when the number of the problem dimensions increases. In this section we shall consider the third-order difference scheme for the two-dimensional problem and discuss some of its properties [9, 10] . The scheme is based on the iteration procedure similar to the Runge-Kutta method for the ordinary equations.

We consider the system

$$\frac{\partial w}{\partial t} = \frac{\partial F}{\partial x} + \frac{\partial G}{\partial y} + f \tag{3.20}$$

where $F = F(w, x, t)$, $G = G(w, x, t)$, $f = f(w, x\, t)$ and introduce the mesh with the steps h_1, h_2, τ . Let $w_{m,l}^n$ be the mesh function at the point (x_m, y_l, t^n), $x_m = mh_1$, $y_l = lh_2$, $t^n = n\tau$. Let the values of $w_{m,l}^n$ be given and necessary to find $w_{m,l}^{n+1}$.

The following sequence of iterations is defined (Fig. 14):

$$w_{m+1/2,\,l\;1/2}^{n+(1)} = \Phi^{(1)}\,(w_{m,l}^n,\,w_{m+1,l}^n,\,w_{m,l+1}^n,\,w_{m+1,l+1}^n) \tag{3.21}$$

$$w_{m,l}^{n+(2)} = \Phi^{(2)}\,(w_{m+\mu,\,l+\lambda}^n,\;w_{m+\frac{\mu}{2},\;l+\frac{\lambda}{2}}^{n+(1)})$$

$$w_{m,\ell}^{n+(3)} = \phi^{(3)} (w_{m+\mu,\ell+\lambda}^{n}, w_{m+2\mu,\ell+2\lambda}^{n}, w_{m+\frac{\mu}{2},\ell+\frac{\lambda}{2}}^{n+(1)}, w_{m+\mu,\ell+\lambda}^{n+(2)})$$

where λ, μ independently take the values of -1, 0, 1. The functions $\phi^{(i)}$ are the composite functions of their arguments through the functions F, G and f. They contain also a number of scalar coefficients as the scheme parameters.

Letting $w_{m,l}^{n+1} = w_{m,l}^{(2)}$ one can obtain the first- or second-order scheme, depending on the choice of parameters, and letting $w_{m,l}^{n+1} = w_{m,l}^{n+(3)}$ the third-order scheme is obtained for some choice of parameters. The $\phi^{(3)}$ contains also the coefficient γ which does not influence the approximation but ensures the stability of the scheme.

The stability of the third-order scheme can be investigated only numerically, but the result can be presented in a rather simple form. As an example, let us consider the case of one equation with the constant coefficients

$$\frac{\partial u}{\partial t} = a\frac{\partial u}{\partial x} + b\frac{\partial u}{\partial y} \tag{3.22}$$

Denote $\sigma = \tau \sqrt{(a/h_1)^2 + (b/h_2)^2}$ and $tgx = (bh_1)/(ah_2)$. Fig. 15 shows the stability boundaries in the plane (σ, γ) for different x. The lines $\gamma = 1$ and the line $\gamma = 2(7-2\sigma^2) \sigma^{2/3}$ corresponding to $x = \pi/4$ are limiting. The limit value of σ is 0.479, and the maximum time step is given by

$$\tau_{max} \leq \sigma_{max} [(a/h_1)^2 + (b/h_2)^2]^{-1/2}$$

The curve $\gamma = 2(4-\sigma^2) \sigma^2/3$ corresponding to the one-dimensional scheme similar to (3.21), for which $\sigma_{max} = 1$ at $\gamma = 2$, is shown for comparison in Fig. 15.

The computations with the use of the scheme (3.21), some of which are mentioned above, show that in both the qualitative and quantitative aspects the discontinuity structure and stretching differ only by little from those calculated with the one-dimensional third-order scheme. Let us discuss two main flows of the third-order schemes and the possible ways of their remedy.

The first one is that the discontinuity profile is non-monotonuous in any third-order scheme. From the general theory it follows that in order to eliminate the oscillations it is necessary to change the order of approximation to the first one in the neighbourhood of the discontinuity. Note that we say "the neighbourhood", since at the point of discontinuity the approximation order is always zero. In order to change somehow the scheme in the vicinity of the discontinuity one must know its location, i.e. check all the mesh points for the discontinuity. Then the problem splits into two ones: to determine whether the discontinuity exists, to "detect" its location and simultaneously (or subsequently) to change the scheme.

In the one-dimensional case, there are the identification criteria for the shock wave (or the strong compression wave), and the contact discontinuity; after they are found the recalculation in the discontinuity neighbourhood can be performed [*]. Note that is it essential in the multi-dimensional cases to preserve the high order of approximation in the smooth domain. Coming back to the scheme (3.21), we should note that the change of its order of accuracy is of no difficulty and does not require additional computational time.

Thus the main problem, which is interesting by itself consists in the detection of the discontinuity. This problem which arises in any through computation method has not yet been satisfactorily solved, especially in the multi-dimensional cases. Below we consider some approach to this problem. It should be noted, however, that the exact localization of the discontinuity in the form of a surface or a line actually contradicts the idea of the through computation. In some cases, it is not so important to determine the discontinuity location as to indicate the regions where the discontinuity is absent and to ensure the maximum accuracy which the solution smoothness allows.

The necessity to have a regular orthogonal mesh is the second inconvenience of the third-order scheme. It is practically impossible to construct even the second-order scheme on an irregular mesh, not saying of the third-order one. At the same time, the domains with curvilinear or even unknown boundaries occur in abundance in gas dynamic problems. To find his way out one should transform the real flow domain onto another so that the curvilinear mesh will be transformed into the rectangular one. If such a transformation can be implemented with the aid of the functions of desired smoothness the problem is solved.

As an example, we consider equation (3.20) which is to be solved in the domain ABCD having the form of a curvilinear quadrangle (Fig. 16). Let the transformation

$$x = X(\xi,\eta), \quad y = Y(\xi,\eta); \quad \xi = \phi(x,y), \quad \eta = \Psi(x,y) \tag{3.23}$$

transfer ABCD to the rectangle A'B'C'D' in the plane (ξ,η).

By substituting the variables in (3.20) we get

$$\frac{\partial w}{\partial t} = (\frac{\partial F}{\partial \xi}\xi_x + \frac{\partial G}{\partial \xi}\xi_y) + (\frac{\partial F}{\partial \eta}\eta_x + \frac{\partial G}{\partial \eta}\eta_y) + f \tag{3.24}$$

Due to (3.23) $\xi_x, \xi_y, \eta_x, \eta_y$ depend actually on ξ,η so the system (3.24) can be reduced, without difficulty, to the divergent form similar to (3.20), as

$$\frac{\partial w}{\partial t} = \frac{\partial}{\partial \xi}(\xi_x F + \xi_y G) + \frac{\partial}{\partial \eta}(\eta_x F + \eta_y G) + \tilde{f} \tag{3.25}$$

---

[*] The method proposed in [5] seems to be promising, when the checking is made simultaneously with the scheme correction. However the extension of this method to the two-dimensional problems meets some difficulties also connected with the detection of the discontinuities.

where

$$\tilde{f} = -(\frac{\partial \xi_x}{\partial \xi} + \frac{\partial \eta_x}{\partial \eta}) F - (\frac{\partial \xi_y}{\partial \xi} + \frac{\partial \eta_y}{\partial \eta}) G + f$$

If the transformation (3.23) cannot be written in an explicit form or in the form of simple analytical expressions, one may use local splines of appropriate smoothness. It means that first the grid functions $x_{m,l}$ and $y_{m,l}$ are given on a set of points (ξ_m, η_l), and then for these functions the local splines of desired smoothness $X(\xi,\eta)$ and $Y(\xi,\eta)$ are constructed, allowing the (numerical) inversion. We are speaking of local splines because the construction of nonlocal ones is practically impossible in the two- and multi-dimensional cases. Also the local splines could be applied to the wider class of domains as compared with the curvilinear quadrangles.

We have a further complication when the functions X and Y depend on time, i.e. the calculation domain is not a cylinder in the coordinates x, y, z. As mentioned above, X and Y can depend on some unknown functions calculated during the computation. In this case the formulae for the variables transformation becomes much more complicated and the possibility of practical implementation of this method depends on a specific problem.

As the first step in the scheme construction with transformation of variables we consider the case of a stationary supersonic axisymmetric flow about a body (Fig. 17). The initial coordinates are τ and z; one of the new variables (ξ, η)namely, $\eta = z$ plays the role of time. As a matter of fact, the problem turns out to be one-dimensional. The initial data are given for $\eta = \eta_o$. It is required to find the solution in the domain $0 \le \xi \le 1$, $\eta > \eta_o$. The derivatives σ_x, σ_y etc. depend on an unknown function determining the configuration of the bow shock wave. The calculation of points in the bow wave and near it was performed by using the method described in section 4. This method preserves the third-order approximation up to the boundary. The interior shock forming in the flow is calculated with the through computation technique. The computation results are given in Figs. 18 and 19, where the field of characteristics with the shock (see 3.5) and the gas dynamic functions of ξ for different η are plotted. For comparison in Fig. 19 the gas dynamic function are plotted with the use of the second-order scheme. In this case the oscillations are strongly revealed even in the domain where the shock has not yet formed.

3.6 Identification of Discontinuities in the Through Computation Method

In the through computation method the neighborhoods of discontinuities appear as zones of large gradients of functions representing the solution. The problem of the detection of discontinuities is not only to find such zones but to establish the fact that the discontinuity really exists in this zone, as well as to determine its type and quantitative characteristics. As far as we know, the problem of detection in such formulation has not yet recieved its complete solution even in the one-dimensional case.

The widely used practice of detecting the discontinuities consists in the computation of the isolines of gasdynamic functions after which the zones of the large gradients can be determined and, hence, the zones with the virtual discontinuities identified. The accuracy in the determination of the discontinuity location by this method, as well as by others, depends on the degree of the profile stretching.

In the work [11] a method was proposed to detect the shock waves. The method employs the properties of the shock profile generated by a difference scheme. The concept used may be illustrated by considering the plane shock wave case. Let the trajectory of the shock be described by the equation $x = Dt$ in the exact solution. The equations for the characterstics which intersect on the shock wave, are

$$x - x_- = \lambda_- t, \quad x - x_+ = \lambda_+ t$$

where $x_- < 0$, $x_+ > 0$ and $\lambda_- > D > \lambda_+$ (Fig. 20).

Let us consider the shock profile generated by the floating mesh difference scheme and moving with the velocity D. Accordingly, we can obtain the profile of the characteristic velocity distribution, $\lambda(\xi)$, $\xi = x-Dt$, moving with the velocity D so that for $|\xi| > K$

$$\lambda(\xi) \approx \lambda_- \quad \text{or} \quad \lambda(\xi) \approx \lambda_+$$

depending on the sign of ξ.

The profile $\lambda(\xi)$ is monotonuous or it is not, depending on the difference scheme. Since $\lambda(\xi)$ is a continuous function, there exists a point $\hat{\xi}$ such that $\lambda(\hat{\xi}) = D$. If the $\lambda(\xi)$ is monotonuous the point $\hat{\xi}$ is unique. Now we consider the field of characteristics constructed by using the function $\lambda(\xi)$ i.e. a family of solutions of the equation

$$dx/dt = \lambda(x - Dt)$$

It is easy to show that the characteristics constructed with the function $\lambda(\xi)$ will asymptotically and from both sides approach a line described by the equation (Fig. 20)

$$x = Dt + \hat{\xi}$$

This line represents a trajectory of the exact wave, shifted by $\hat{\xi}$ being a distance between the exact position of the wave and the point on the profile $\lambda(\xi)$, at which $\lambda = D$.

The numerical calculations show that in all cases the value of $\hat{\xi}$ has an order of the mesh size, and if the function $\lambda(\hat{\xi})$ is nonmonotonuous the point $\hat{\xi}$ is also unique. The nonmonotonuous nature of $\lambda(\xi)$ can be seen only in the fact that the approach of the characterstics to the line $x = Dt + \hat{\xi}$ is oscillatory at its early stage. By applying this method to the pracitical computations, even rather weak waves can be detected. Fig.18 gives a clear view of the approaching characteristics and the formation of the shock from the compression wave.

4. Exact Calculation of Discontinuities as the Flow Region Boundaries

The proper physical formulation of a gas dynamic problem with initial and boundary conditions provides the conditions of its local correctness in a differential formulation, which can be easily verified if necessary. The main problems arising in numerical calculations refer to the choice of a proper coordinate system and to a writing of the difference scheme near the boundary. The first of these problems was mentioned above when the difference schemes for an irregular mesh were discussed. Now we shall consider only the questions in connection with the difference scheme, assuming the boundary to be a coordinate surface.

When a difference scheme is constructed near the boundary it is necessary to provide first the same order of approximation as at the interior points, and then the stability of the boundary value problem for the difference equations. The higher the order of approximation is the more difficult it is to meet both requirements.

We consider first the one-dimensional case. Let us have for the hyperbolic system

$$\frac{\partial w}{\partial t} = \frac{\partial F}{\partial x} + f \tag{4.1}$$

in the domain $x \geq 0$, $t \geq 0$ the mixed problem:

$$w(x,o) = w^o(x) \quad (4.2) \qquad \qquad \psi(w(o,t),t) = 0 \tag{4.3}$$

Here $F = F(w, x, t)$, $f = f(w, x, t)$ is the vector-function with the r components, and $\psi(w, t)$ is the vector-function with the r_1 components.

Let $A(w,x,t) = \frac{\partial F}{\partial w}$ and $(B(w,t) = \frac{\partial \psi}{\partial t}$ be the matrices of the orders $r \times r$ and $r_1 \times r$ respectively. We make the following assumptions:

1. Among the eigenvalues $\lambda(w, x, t)$ of the matrix there exist $r^+ = r - r_1^-$ positive ones, $\lambda_j^+ > 0$, $j = 1,...,r$ and $r^- = r_1$ negative ones, $\lambda_j^- < 0$, $j = 1,..., r^-$. There exist the matrices $P(w, x, t)$ and $Q = p^{-1}$ which locally reduce A to a diagonal form. The matrices P and Q can be divided into blocks corresponding to the positive and negative

$$P = \genfrac{\{}{\}}{0pt}{}{P^+}{P^-} , \quad Q = \{Q^+, Q^-\} \tag{4.4}$$

2. For all $t \geq 0$ the rank B equals to r^-, i.e.

$$\det(BQ^-)\big|_{x=o} \neq 0, \qquad \det\genfrac{\{}{\}}{0pt}{}{P^+}{B}\big|_{x=o} \neq 0 \tag{4.5}$$

The conditions 1. and 2. provide the local correctness of the mixed problem (4.1) - (4.2).

Let us describe the principle of the difference scheme construction for the mixed problem under consideration. Let $x_m = mh$, $t^n = n\tau$, and w_m^n be a mesh function. First we write down the difference scheme for the Cauchy problem (4.1) - (4.2), as:

$$(w_m^{n+1} - w_m^n)/\tau = \{\mathcal{L}_\tau(w^n)\}_m \tag{4.6}$$

where $\mathcal{L}_\tau$ is the finite-difference operator providing the approximation of order p.

Let the $\mathcal{L}_\tau$ depend on the values $w_{m+\nu}^n$, $-\nu_1 \le \nu \le \nu_2$, and ν_1, $\nu_2 \ge 0$. Then we can find the values of w_m^{n+1} for $m \ge \nu_1$ with the aid of (4.6). The points with $m \ge \nu_1$ are named interior points and the points with $m = \ell = 0,1,\dots, \nu_1 - 1$ are named boundary points. At the boundary points the difference schemes of the order p are written as

$$(w_\ell^{n+1} - w_\ell^n)/\tau = \{\mathcal{L}_\tau^{(\ell)}(w^n)\}_\ell \tag{4.7}$$

We have $\nu_1^{(\ell)} \le \ell$ for $\mathcal{L}_\tau^{(\ell)}$.

The values of w_ℓ^{n+1} at the points $\ell = 1,\dots, \nu_1 - 1$ are obtained through (4.7). At the point $\ell = 0$ on the boundary, the boundary conditions (4.3) must be combined with the difference equations for the characteristic combinations corresponding λ_j^+ and approaching the boundary. To do this we multiply the left hand side of (4.7) by P^+. Then we get

$$P^+ w_0^{n+1} = P^+ [w_0^n + \tau\{\mathcal{L}_\tau^{(0)}(w^n)\}_0]$$

$$\Psi(w_0^{n+1}, t^{n+1}) = 0 \tag{4.8}$$

Due to (4.5), this system has the solution w_0^{n+1}.

The matrix P is evaluated at some point (w^*, x^*, t^*). If $w(0, t)$ changes with time not too fast, one may set $w^* = w_0^n$, $x^* = x_0$, $t^* = t^{n+1}$.

Next we consider the system

$$\frac{\partial w}{\partial t} = \frac{\partial F}{\partial x} + \sum_{j=1}^n \frac{\partial G_j}{\partial y_j} + f \tag{4.9}$$

where F, G_j, and f are the vector-functions depending on w, x, y_j with r components. For (4.7) the mixed problem is set in the region $t \ge 0$, $x \ge 0$, $|y_j| < +\infty$. It is formulated as follows:

$$w(x,y,0) = w^0(x,y), \quad y = (y_1,\dots,y_n) \tag{4.10}$$

$$\Psi(w(0,y,t),t) = 0 \tag{4.11}$$

where Ψ is the vector-function with r_1 components.

We shall assume that the problem (4.9) - (4.11) is locally correct and, hence, conditions 1. and 2. are satisfied. Introducing the mesh with steps $\Delta x = h$, $\Delta y_j = h_j$, $\Delta t = \tau$ and assuming $x_m = mh$, $y_{j,\ell j} = \ell_j h_j$, $t^n = n\tau$, we introduce the mesh function $w_{m,\ell}^n$, $\ell = (\ell_1,\dots,\ell_n)$. Now, replacing the index m in formulae (4.6) - (4.8) with the index $(m,\ell) = m,\ell_1,\dots,\ell_n$ we get the general form of the difference scheme approximating the mixed problem (4.9) - (4.11).

There are no fundamental difficulties in the implementation of the method proposed, and its complexity depends on the desired order of approximation.

The situation is somewhat worse with the stability. Its investigation is usually carried out on linearized equations with "frozen" coefficients. In this case one may apply the theory of stability of the mixed problem, developed in [12, 13] for the difference equations. But its practical use requires a rather complex numerical analysis even for one dimension. In constructing the operators $\mathscr{L}_\tau^{(\ell)}$, it is expedient to introduce free parameters not affecting approximation, and then to determine their values from the stability condition.

In the case of a one-dimensional problem and a third-order scheme, the scheme near the boundary was constructed and its stability fully investigated [14]. This scheme was used in the above example of the flow problem (see Figs. 17-18).

5. About the Testing of the Computation Methods for Discontinuous Flows

The "step" type discontinuity is usually applied for testing the computational methods for non-continuous flows. It has an advantage of simple and convenient comparison with the exact solution, but unfortunately it does not allow to estimate the influence of the discontinuity upon accuracy of calculation in the smooth domain. Because the "smooth" region in this test is a region of constant functions all the schemes behave similarly in this region and yield absolute accuracy.

For actual estimation of accuracy in the region where the functions are smooth but not constant, special tests are needed. In the paper [15] the principles for constructing such tests were proposed for one-dimensional problems. By using these principles, tests with the arising of shocks, their interaction, reflection, etc. can be developed. In all cases, for computing the exact solution it is sufficient to solve numerically a few (depending on the test complexity) ordinary differential equations with lagging arguments, which is of no problem for to-day's computers.

The results of the schemes investigations by means of these tests were given in [14, 16]. An important fact was established that as the distance from the discontinuity increases an error of the computation having the zeroth order near the discontinuity approaches the error determined by the order of approximation.

6. Conclusion

The theoretical and computational results presented here allow to conclude the following: In calculating the discontinuous solutions the third-order difference schemes have some advantages as compared with the first- and second-order schemes. The third-order schemes provide the higher accuracy in the smooth region or, when yielding the same accuracy, they allow to increase the mesh size, which is very important in the multi-dimensional case. Besides, they produce a less stretched profile in the neighborhood of the discontinuity, and much weaker oscillations as compared with the second-order schemes.

However, some technical and fundamental difficulties prevent the wide use of the third-order schemes in computation of multi-dimensional problems. The main technical difficulties are associated with the development of irregular-mesh schemes as well as the schemes near the boundaries with the order of approximation and stability preserved. The fundamental difficulties common for all the through computation methods are associated with an infinite stretching of the linear discontinuity profile, and with the occurrence of parasitic oscillations near the discontinuity.

Some considerations have been mentioned above as to overcoming the technical difficulties. The use of computers in the development and the investigation of difference scheme seem to be of great help in solving this problem.

In order to overcome the fundamental difficulties new ideas are needed. From the results presented it follows that to improve considerably the quality of the discontinuity calculation without violating the accuracy in the smooth regions, adjacent to the discontinuity zones one has to pass over from uniform schemes to non-uniform ones using the detection of the discontinuity location. A great number of existing papers and articles dealing with the improvement of the quality of discontinuity calculations imply the same conclusion.

The main problem here consists in detecting the discontinuity location. The problem grows much more complex in the two-dimensional case, but even then it does not appear to be hopeless. Once the solution is obtained, we can attain the goal by modifying the scheme in the discontinuity zone and by passing over to the standard scheme in the smooth domain.

7. References

[1] L a x, P.D.: Hyperbolic systems of conservation laws. Comm. Pure Appl. Math., 10, (1957), pp. 537-566.

[2] v o n N e u m a n n, J., R i c h t m y e r, R.D.: A method for numerical calculation of hydrodynamic shocks. J. Appl. Phys., 21, (1950), pp. 232-237.

[3] Z h u k o v, A.I.: The limiting theorem for difference operators. (Russian) Uspekhi Matematicheskikh Nauk, 14, (1959), pp. 129-136.

[4] K u z n e t z o v, N.N.: The asymptotic of the solutions of finite-difference problem. (Russian) Zhurnal Vychislit. Matem. i Matemat. Phiziki, 12, (1972), pp. 334-351.

[5] B o r i s, J.P., B r o o k, D.L.: Flux-corrected transport. I. Shasta, A Fluid Transport Algorithm that Works. J. Comp. Phys., 11, (1973), pp. 38-69.

[6] B r o o k, D.L., B o r i s, J.P., H a i n, K.: Flux-corrected transport II. Generalization of the Method. J. Comp. Phys., 18, (1975), pp. 248-283.

[7] L a x, P.D.: Weak solutions of non-linear hyperbolic equations of hydrodynamics. Comm. Pure Appl. Math., 7, (1954), pp. 159-193.

[8] R u s a n o v, V.V.: Non-linear analysis of the shock profile in difference schemes. Proc. Sec. Int. Conf. on Numer. Meth. in Fluid Dynamics. In "Lecture Notes in Physics", Springer Verlag, 8, (1970).

[9] R u s a n o v, V.V.: On difference schemes of third order accuracy for non-linear hyperbolic systems. J. Comp. Phys., 5, (1970), pp. 507-516.

[10] I z v o l s k i i, V.A., R u s a n o v, V.V.: The construction and investigation of the third-order difference schemes. (Russian) Preprint of the Keldysh Inst. Priklad. Matemat. Akad. Nauk N 3 (1979).

[11] R u s a n o v, V.V.: Processing and analysis of computational results for multidimensional problems of aerodynamics. Proc. Third Int. Conf. on Numer. Meth. in Fluid Dynamics. In "Lecture Notes in Physics", (1970).

[12] K r e i s s, H.-O.: Stability theory for difference approximations of mixed initial boundary value problems. Math. Comp., 22, (1968), pp. 703-714.

[13] K r e i s s, H.-O.: Theory of difference approximations for mixed initial boundary value problems. II. Math. Comp., 26, (1972), pp. 649-686.

[14] R u s a n o v, V.V., N a z h e s t k i n a, E.I.: Boundary conditions in difference schemes for hyperbolic equations. Proc. Third GAMM-Conference on Numerical Method in Fluid Mechanics. Köln (1979).

[15] R u s a n o v, V.V.: A test case for checking computational Methods for gas flows with discontinuities. In GAMM-Workshop "Boundary Algorithms for Multidimensional Inviscid Hyperbolic Flows". K. Förster (Ed.), Verl. F. Vieweg & Sohn, Braunschweig, Wiesbaden, (1978), pp. 100-125.

[16] R u s a n o v, V.V.: Advanced techniques for computation supersonic flows. Proc. 15th AIAA Aero. Space Sci. Meeting AIAA paper, (1977), pp. 77-173.

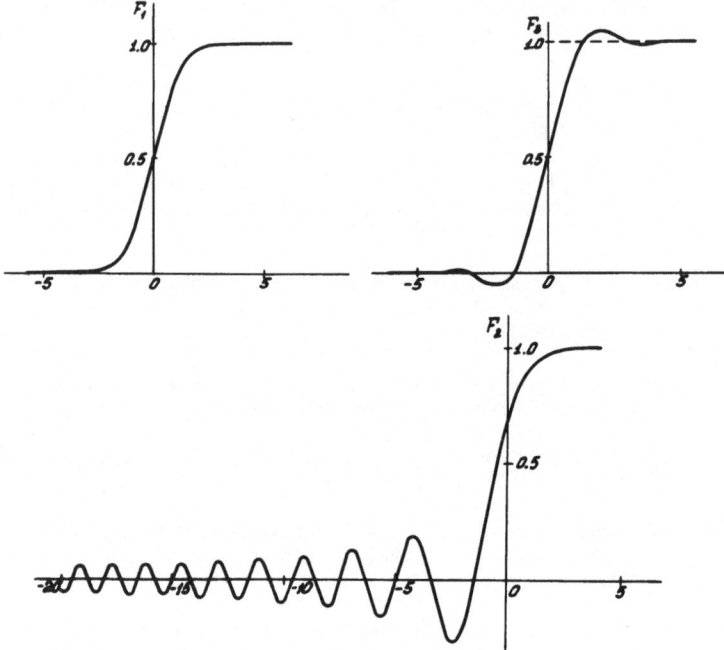

Fig. 1 - 3

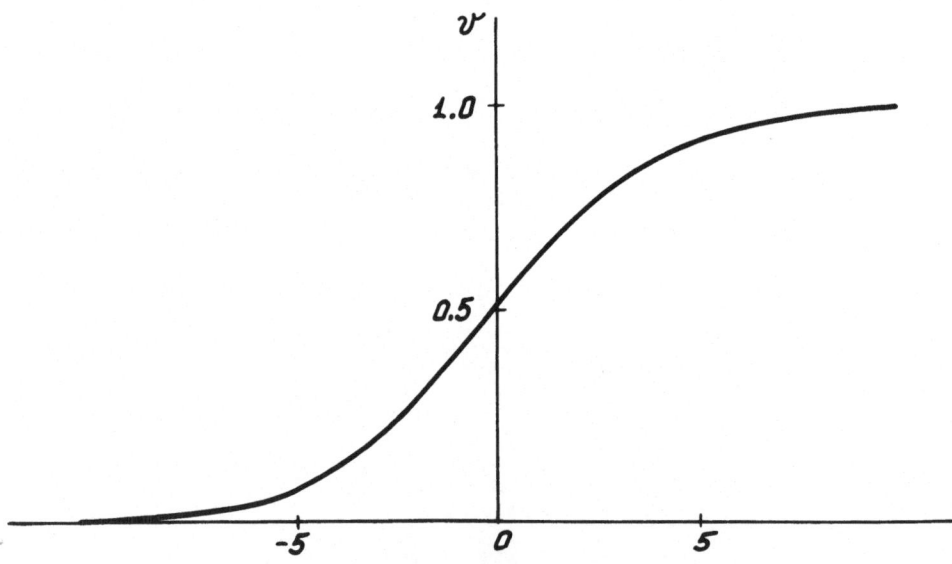

Fig. 4

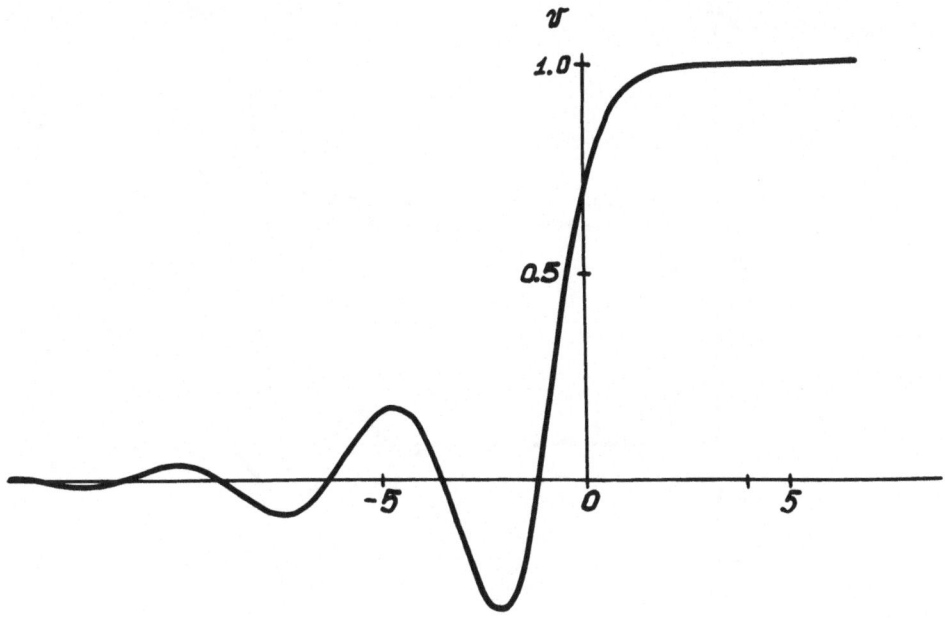

Fig. 5

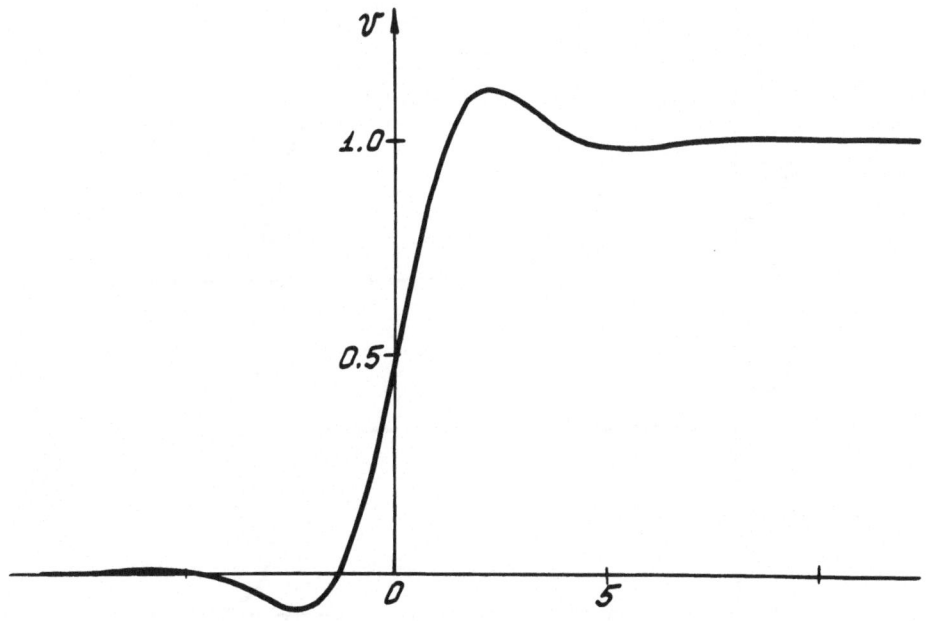

Fig. 6

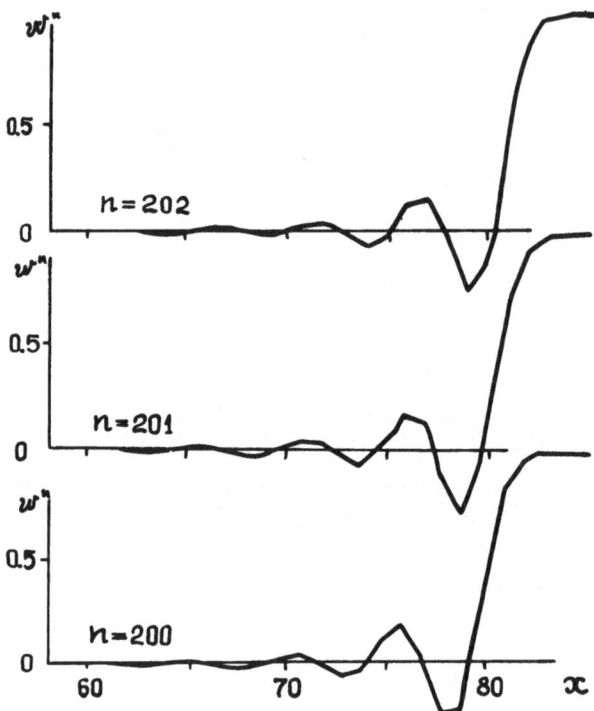

Fig. 7

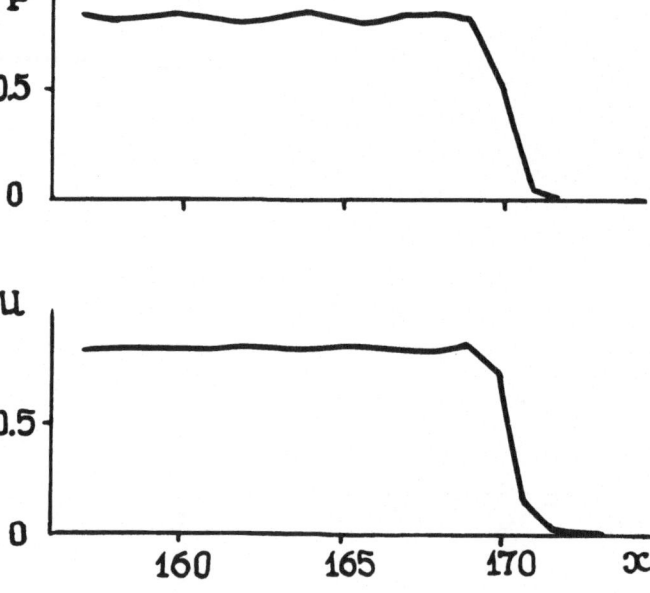

Fig. 8

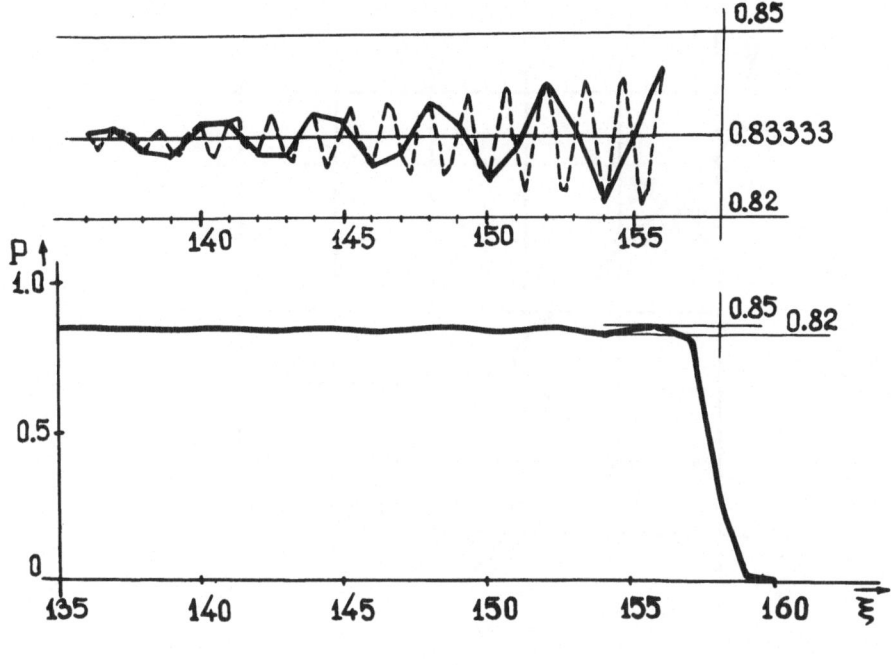

Fig. 9

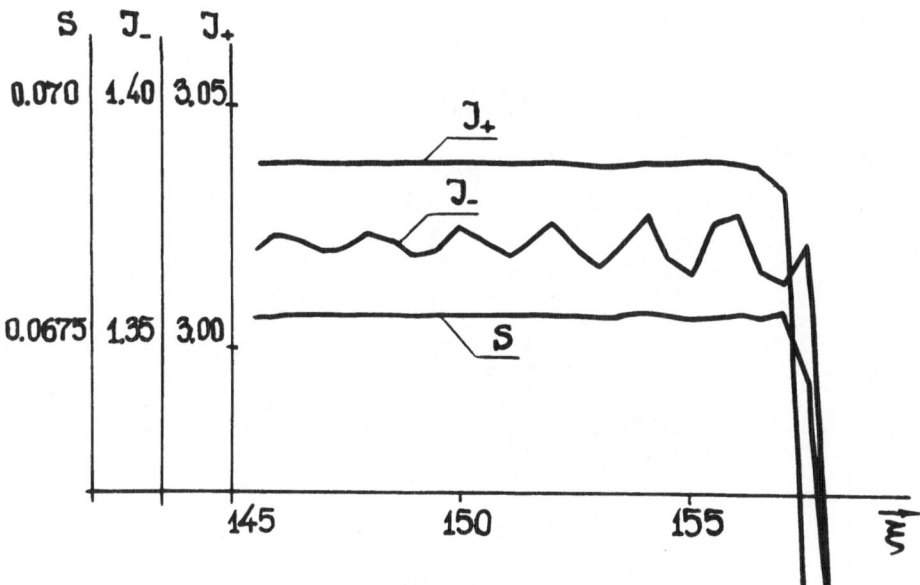

Fig. 10

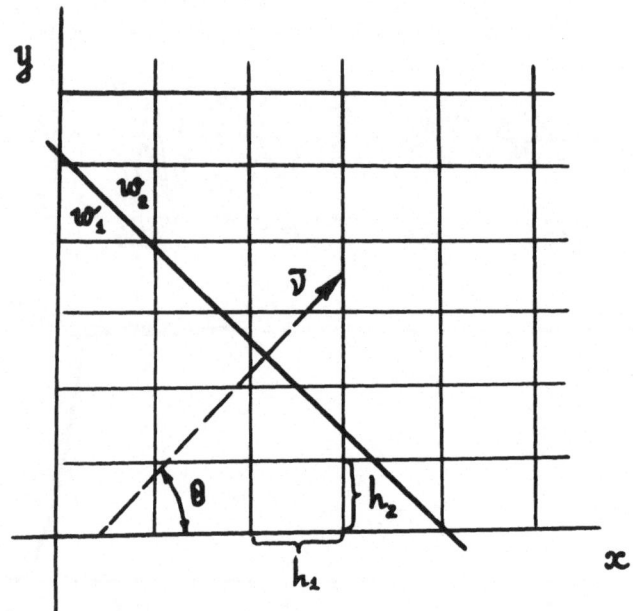

Fig. 11

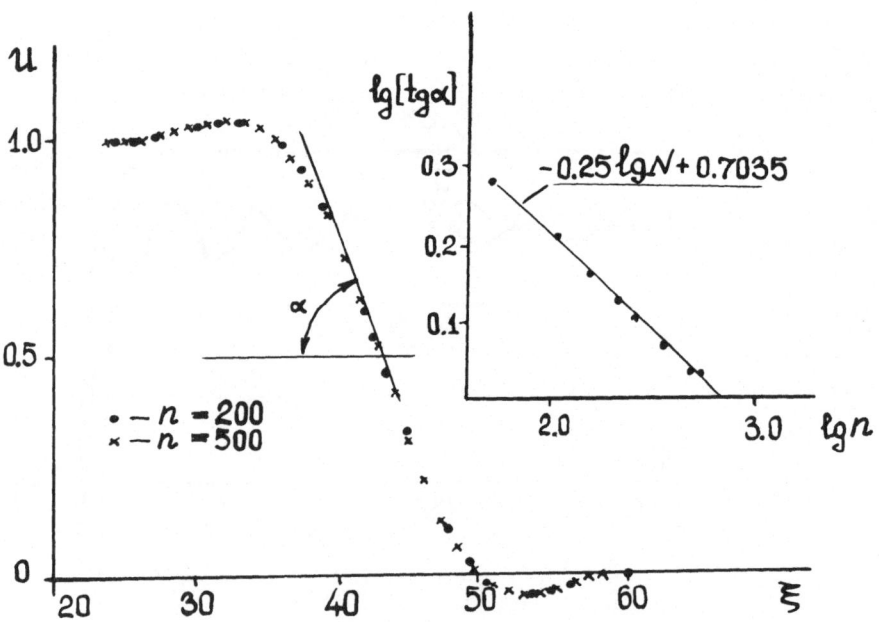

Fig. 12

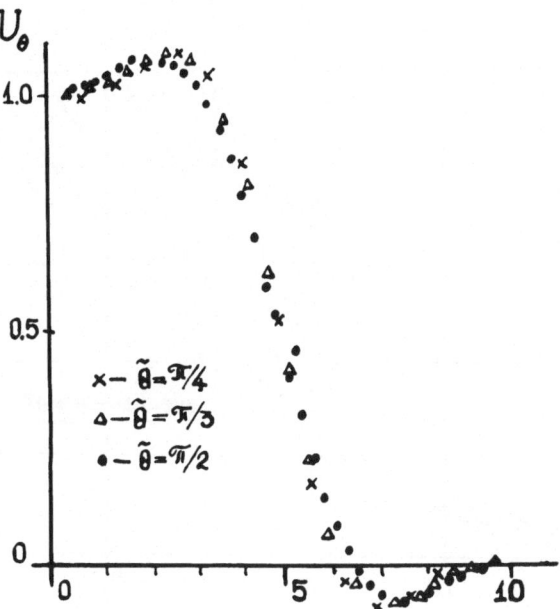

Fig. 13

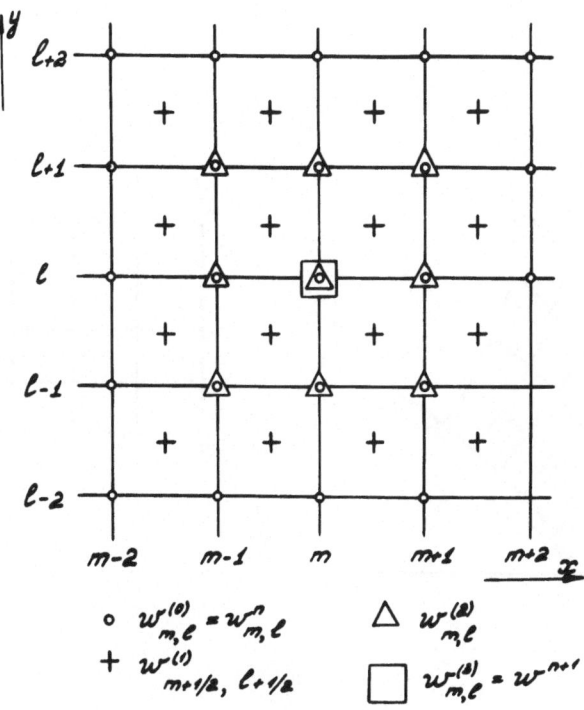

Fig. 14

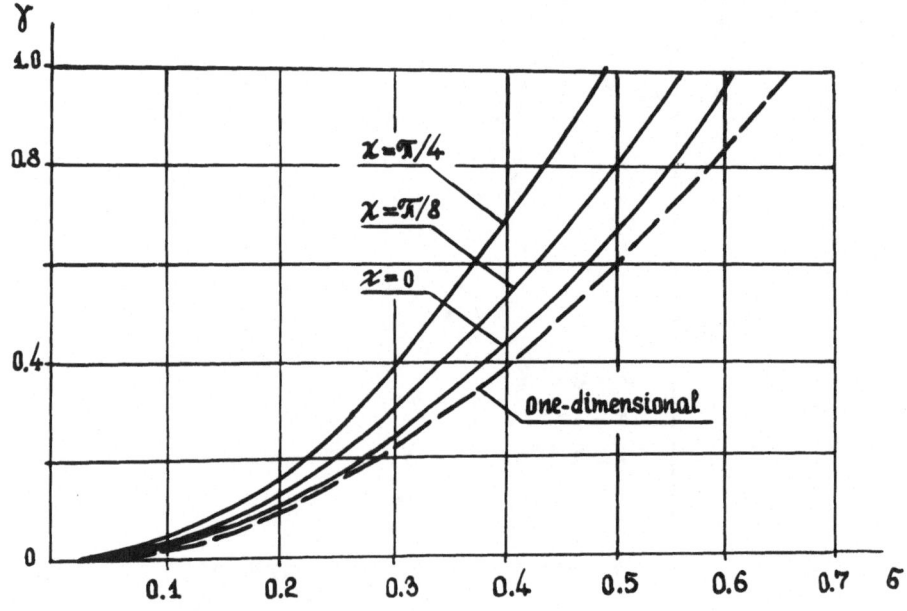

Fig. 15

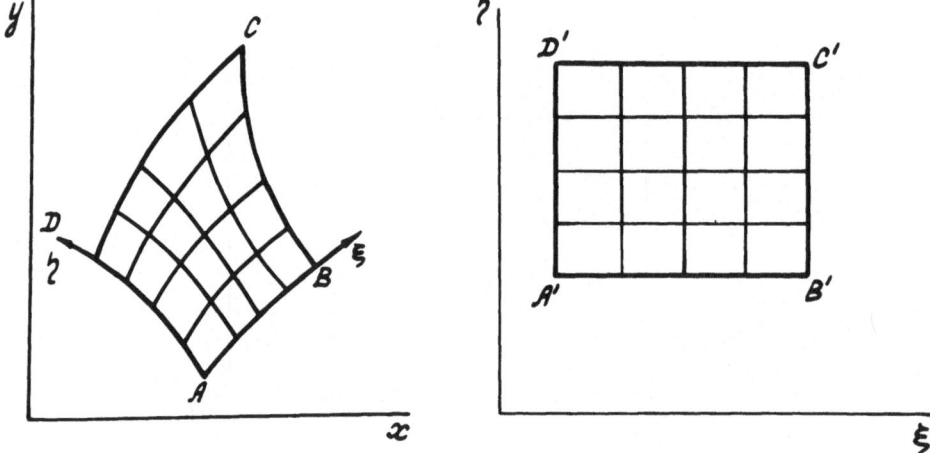

Fig. 16

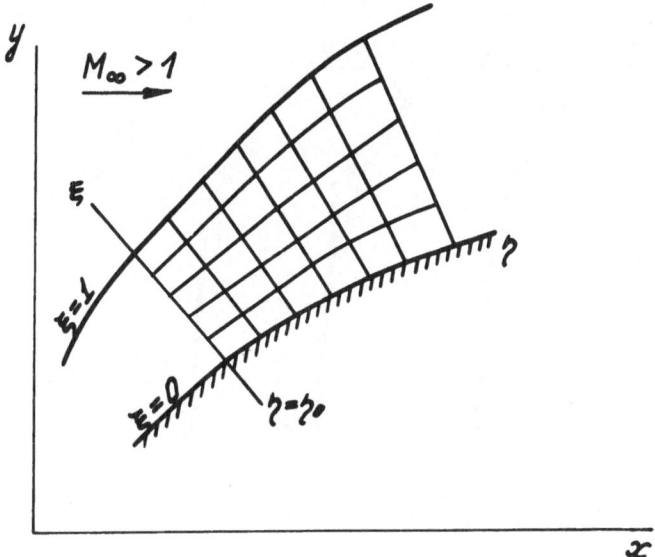

Fig. 17

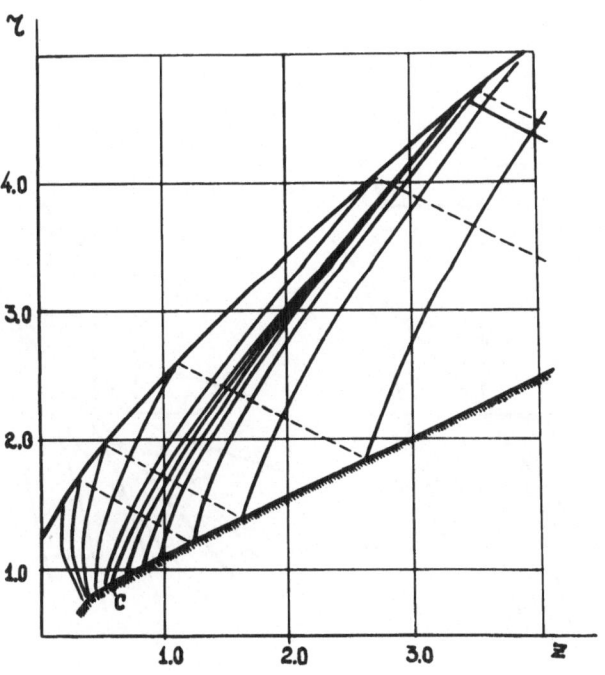

Fig. 18

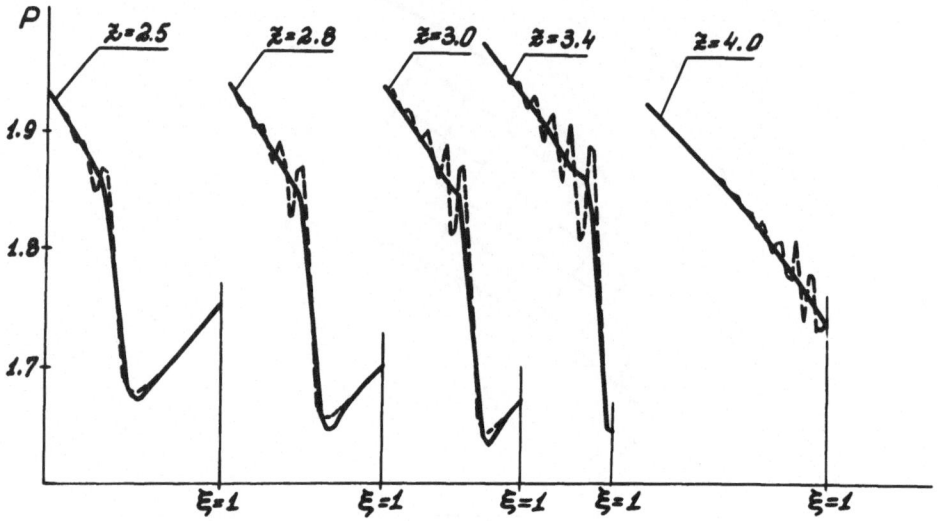

Fig. 19

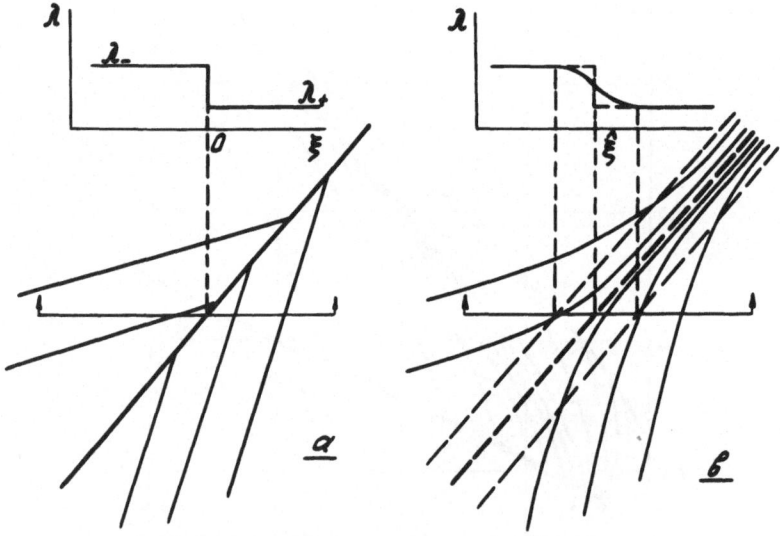

Fig. 20

Studies on the Motion and Decay of a
Vortex Filament

L. Ting
New York

1. Introduction

Mathematical models of inviscid vortex filaments have been frequently employed for the explanation of phenomena in fluid mechanics, aerodynamics and meteorology [1] . The availability of high speed computers and recent experimental evidences of organized vortex-like structures in turbulent shear layers [2, 3] have stimulated the numerical modeling of flow fields by interacting vortices [4-7] .

In many practical flows, in which vorticity effects are important, it is observed that the bulk of vorticity is concentrated in a slender "tube like" region, known as a vortex filament, in which the viscous effects may be important. Away from the filament, the induced flow field is basically inviscid and in agreement with the inviscid theory of a vortex filament with zero cross-sectional area, henceforth to be called a vortex line. The classical inviscid theory of the motion of curved vortex lines has two essential defects. They are: (1) the velocity of the fluid on the vortex line is infinite, and (2) the velocity of the vortex line itself is undefined.

For a two-dimensional potential flow with a vortex of strength Γ located at the point (X,Y), the stream function ψ can be represented [1] by the sum of a stream function representing the background potential flow and that for the vortex alone, i.e.,

$$\psi(x,y) = \psi*(x,y) + \frac{\Gamma}{2\pi} \ell n \; r \qquad (1.1)$$

where r is the distance between (x,y) and (X,Y). The velocity near the vortex behaves as $\Gamma/(2\pi r)$ and becomes infinite as $r \to 0$. The second defect of the inviscid theory is circumvented by the assumption that the vortex point moves with local spatial mean velocity which is also the local velocity of the flow field in absence of the vortex, i.e.,

$$\dot{X} = \psi_y^*(X,Y) \quad \text{and} \quad \dot{Y} = -\psi_x^*(X,Y) \qquad (1.2)$$

where $(\dot{\;})$ represents differentiation with respect to time t.

For a curved vortex line in an inviscid three-dimensional flow field, the velocity $\vec{Q}$ at any point P, which does not lie on the vortex line, is the sum of the induced velocity due to the vortex line, $\vec{Q}_1$ and the velocity $\vec{Q}_2$ of the background flow, i.e.,

$$\vec{Q}(\vec{P},t) = \vec{Q}_1(\vec{P},t) + \vec{Q}_2(\vec{P},t) \qquad (1.3)$$

where $\vec{P}$ is the position vector of the point P. The velocity $\vec{Q}_2$ is finite along the vortex line. If the vortex line is given parametrically by $\vec{X}(s,t)$ where s is the arc length along the line, then the induced velocity at P is given at any time t by the Biot-Savart formula as:

$$\vec{Q}_1(P,t) = -\frac{\Gamma}{4\pi} \int \frac{[\vec{P}-\vec{X}(s',t)]\times d\vec{s}'}{[\vec{P}-\vec{X}(s',t)]^3} . \qquad (1.4)$$

Here Γ is the circulation associated with the vortex line and the direction of s is chosen so that $\Gamma > 0$. We observe that the time t enters (1.4) as a parameter, due to the motion of the vortex line. To determine the behavior of the induced velocity near the vortex line it is convenient to represent the position vector $\vec{P}$ in curvilinear coordinates, r, Θ, s, associated with the point $\vec{X}(s,t)$ on the vortex line, as

$$\vec{P}(x,y,z) = \vec{X}(s,t) + r\hat{r}(\theta,s,t). \qquad (1.5)$$

(see Fig. 1). The point $\vec{X}(s,t)$ on the vortex line is chosen by requiring that r be the shortest distance from the point P to the vortex line. The unit vector $\hat{r}$ is in the direction of $\vec{P} - \vec{X}$. As $r \rightarrow 0$ the integrand in (1.4) becomes singular at $s' = s$. A careful expansion of the integrand near $s' = s$ yields the following behavior of $\vec{Q}_1$ as $r \rightarrow 0$:

$$\vec{Q}_1(\vec{P},t) = \frac{\Gamma}{2\pi r}\hat{\theta} + \frac{\Gamma}{4\pi R(s,t)}\left[\ln\left(\frac{R}{r}\right)\right]\hat{b} + \frac{\Gamma}{4\pi R}(\cos\phi)\hat{\theta} + \vec{Q}_f . \quad (1.6)$$

In (1.6) $\hat{\Theta}$ is the unit circumferential vector, $\hat{b}$ is the unit binormal vector associated with the vortex line at the point $\vec{X}(s,t)$, $R(s,t)$ is the local radius of curvature and Φ is the angle between the normal vector $\hat{n}$ and the vector $\hat{r}$. The vector $\vec{Q}_f(\vec{P},t)$ is part of $\vec{Q}_1(\vec{P},t)$ which has a limit as $r \rightarrow 0$. In (1.6) the first two terms become infinite while the third term has no limit as $r \rightarrow 0$. For the velocity of the vortex line, if the condition for the two-dimensional theory, that $\dot{\vec{X}}$ is the local mean velocity, is imposed, $\dot{\vec{X}}$ is still infinite because the second term in (1.6) remains in the local spatial mean. Therefore, in this inviscid theory, the induced velocity on the curved vortex line is infinite. In addition, the velocity of the vortex line itself is undefined. The usual procedure [1] for overcoming these difficulties is to assume that the core of the vortex filament has some finite size and a prescribed vorticity distribution. In contrast to the two-dimensional problem, the velocity of the filament does depend on the instantaneous vorticity distribution. Since the inviscid theory cannot account for the temporary variation of the vorticity distribution due to viscous effects, it will produce inaccurate results, especially for long time given the correct initial distribution.

For a real fluid the velocity has to be finite everywhere. In a small region where the velocity becomes very large, so does its gradient. Consequently, in such a region viscous terms are no longer negligible and, in fact, they will smooth out the velocity gradient. The defects of the classical inviscid theory for vortex motion can therefore be eliminated if the inviscid solution is identified as the leading term of a matched asymptotic solution of the Navier Stokes equations in the region (the outer region) sufficiently away from the filament. The condition that the velocity should be finite everywhere should enable us to define the velocity of the vortex filament with a decaying vorticity distribution. With this basic premise, the matched asymptotic solutions for a viscous vortex filament submerged in an outer potential flow field were constructed in a series of papers [8-12] from the simple two-dimensional problem to the three-dimensional problem with large circumferential and axial velocity components in the vortex filament. The general procedures and the essential conclusions of the analyses [8-12] are described in the next section. Formulas for the velocity of the vortex filament and the circumferential and axial velocity variations in the vortex filament are presented. The influences of the vortex filament on the background potential flow are of higher order. Comparisons are made between the matched asymptotic solutions and the relevant patched solutions [13, 14, 15] . In addition, several applications of the asymptotic solutions are described.

In section 3, recent investigations for the motion of a vortex filament in a background rotational flow field are outlined. Analyses are presented for the two-dimensional problem. Since the governing equations for the background flow are non-linear and involve the velocity induced by the vortex, the motion of the vortex with a viscous core is now coupled with the temporal variation of the background flow.

In section 4, numerical solutions of Navier-Stokes equations for merging of vortex rings are reported. These results, together with previous results for two-dimensional problems, are employed to establish a practical upper bound for the expansion parameter within which the matched asymptotic solutions are applicable. The numerical results also provide the clues leading to the derivation of several new conservation laws and theoretical results for the viscous merging of vorticities.

2. Vortex Filament Submerged in a Potential Flow

We shall describe in the first subsection 2.1 the physical basis including the relevant spatial and temporal scales and the formulation of the problem in terms of a matched asymptotic analysis. For the relatively simple two-dimensional case, matched asympotic analyses in spatial variables were carried out for solutions with multiple time scales [8] . In 2.2 the analysis is outlined and the essential results, which can be extended to the general three-dimensional case, are stated. In particular, the role of the multiple time scales and the meaning of the solution with only one time scale, the time scale of the background flow, are clarified. Consequently, for the general case of a vortex filament in a three-dimensional potential flow, the matched asymptotic analysis in spatial variables was carried out [11, 12] for solutions involving only the time scale of the background flow. The analysis and the final formulas are presented in 2.3. Several applications are mentioned.

2.1 Formulation

To study the motion and decay of a vortex filament, with circulation Γ submerged in a background potential flow, we shall designate L, U and T as the typical length, velocity and time scales of the background flow respectively. When the background flow is steady we designate T = L/U. We consider the velocity of the background flow and the velocity induced by the vortex filament are of the same order, i.e.

$$U = \Gamma/L \ . \tag{2.1}$$

We shall say a quantity is small (large) when it is much smaller (larger) than the corresponding typical quantity in the background flow.

We shall construct asymptotic solutions for the three-dimensional incompressible, unsteady Navier-Stokes equations with large Reynolds number, i.e.

$$R_e = \Gamma/\nu \gg 1 \tag{2.2}$$

where ν is the kinematic viscosity. They will describe the motion and decay of an initial vorticity distribution concentrated in a slender tube-like region, a vortex filament (Fig. 2). The slenderness implies the length scale of a cross-section of the filament, say, its equivalent radius δ is small, i.e.,

$$\delta \ll L \tag{2.3}$$

We shall define δ as the mean distance from the point of maximum vorticity to a constant vorticity line of e^{-1} times the maximum value.

In the filament, the initial velocity field corresponding to the high vorticity concentration shall have a large circumferential component and may have also a large axial component. From the initial velocity field $\vec{Q}(\vec{P},0)$, we shall assume that a "center line" $\vec{C}(s)$ which lies in the filament can be defined such that only the axial velocity component on $\vec{C}$ can be large, i.e.

$$\vec{Q}(\vec{C},0) \times \hat{\tau} = O(U) \tag{2.4}$$

Here s is the arc length measured along $\vec{C}$ from a given reference point and $\hat{\tau}(s)$ is the unit tangent vector to the curve $\vec{C}$. We shall require that the curve $\vec{C}$ move as a material line so that it will always lie in the filament. We shall designate the coordinates of this line at any time as $\vec{X}(s,t)$, i.e.

$$\vec{X} \cdot \hat{\tau} = 0 \quad \vec{X} \cdot \hat{n} = \vec{Q}(\vec{X},t) \cdot \hat{n}, \quad \vec{X} \cdot \hat{b} = \vec{Q}(\vec{X},t) \cdot \hat{b} \tag{2.5}$$

$$\vec{X}(s,0) = \vec{C}(s)$$

Here $\hat{t}$, $\hat{n}$ and $\hat{b}$ are the unit tangent, normal and binormal vectors to the curve $\vec{X}(s,t)$.

Due to the high vorticity concentration, the effects of viscous terms are important in the vortex filament which will be called the inner region in the matched asymptotic analysis. In the region away from the filament, which will be called the outer region, the reference Reynolds number, R_e, is large and the viscous effects are of the order of $1/R_e$. Furthermore, we note that a potential solution is also a solution of the Navier-Stokes equations, therefore, the flow field in the outer region will remain a potential flow to all orders of $1/R_e$.

We shall assume that at any point P near the vortex filament we can find a unique value s such that the plane N normal to the "center"line, $\vec{X}(s,t)$, of the filament passes through the point P. This will be true if the radius of curvature of the center line of the filament is not too small. The Cartesian coordinate for the point P can then be replaced by the curvilinear coordinates (s,r,Θ) where r and Θ are polar coordinates in the normal plane N with

$$r = |\vec{P} - \vec{X}| \qquad (2.6)$$

Here r is also the minimum distance from P to the center line $\vec{X}(s,t)$. For the solution in the outer region with length scale L, the vortex filament, with $r/L \ll 1$, reduces to the "center" line $\vec{X}(s,t)$ and $\dot{\vec{X}}(s,t)$ represents then the velocity of the filament. Henceforth we shall call $\vec{X}(s,t)$ the equivalent vortex line of the filament.

To analyze the flow in the inner region where $r \ll L$ and to take into account the viscous terms we stretch the spatial variable r by the inverse of the square root of the Reynolds number in the same manner as Prandtl boundary layer theory, i.e., we introduce the stretched inner variable,

$$\bar{r} = r/\varepsilon \qquad (2.7)$$

with

$$\varepsilon = (R_e)^{-1/2} \qquad (2.8)$$

Thus the inner and outer regions and the appropriate inner and outer variables are established and the matched asymptotic analyses can be carried out such that:

(i) At finite distances from the filament the leading term of the outer solution agrees with inviscid theory as a flow due to a moving vortex line $\vec{X}(s,t)$ superimposed on the background potential flow.

(ii) The outer solution is matched with the inner solution and the resulting velocity field has no singularities. The matched solution provides the solution for the entire flow field and in particular gives the velocity of the vortex filament, $\dot{\vec{X}}(s,t)$. This velocity includes the effect of curvature of the equivalent vortex line $\vec{X}(s,t)$ and the leading inner solutions for the circumferential and axial velocity components.

The inner and outer solutions are assumed to have the same time scale, T, i.e., the scale of the background flow. In the inner region, the unsteady term is then of the order of ε^2 relative to the convective terms. Consequently we expect that the velocity of the filament $\dot{\vec{X}}(s,t)$ shall be defined by the analyses without specifying the initial data $\dot{\vec{X}}(s,0)$. Hence, the asymptotic solution with a single time scale represents a special one of the unsteady Navier-Stokes equations because the initial data are required to be consistent with the solution. In general of course, the initial velocity can be arbitrarily assigned. This phenomenon arises when a fixed vortex filament is suddenly released or when a fine vortex filament encounters a sudden gust or crosses over a wave front with discontinuous background velocity. In order to fulfill given initial data, it is necessary to keep the unsteady term, i.e., it has to be of the same order as the convective terms in the inner region. This can be accomplished when we admit a short time variable $\bar{t}$, with

$$\bar{t} = t/\varepsilon^2 \ . \tag{2.9}$$

Since the solution will be oscillatory in $\bar{t}$, the dependence of the solution on the short time variable cannot be restricted to a small initial interval. It is necessary to admit multiple time variables in the solution which shall be called multiple time solutions to distinguish them from the solutions with only one time scale, T. Multiple time scale analyses in conjunction with the matched asymptotic analyses in spatial variables for the two-dimensional problems were carried out in [8] and are outlined in the next subsection.

2.2 Two-Dimensional Problems

The Navier-Stokes equations for an unsteady incompressible flow in terms of the stream function ψ and the vorticity ζ are

$$\zeta_t + \psi_y \zeta_x - \psi_x \zeta_y = \nu \Delta \zeta \tag{2.10}$$

$$\Delta \psi = -\zeta \tag{2.11}$$

The initial vorticity distribution ζ is highly concentrated in a small region called the vortical core accompanied by a large swirl velocity. Inside the core, we can locate a point $C(X_0, Y_0)$ such that the velocity is of the order of U only. This point is called the center of the vortical core and is the location of the equivalent vortex point for the outer solution. We seek the trajectory of this material point, i.e.,

$$\dot{X}(t) = \psi_y(X,Y,t) \quad \dot{Y}(t) = -\psi_x(X,Y,t) \tag{2.12}$$

given

$$X(0) = X_0, \quad Y(0) = Y_0 \tag{2.13}$$

and

$$\dot{X}(0) = U_o, \quad \dot{Y}(0) = V_o. \tag{2.14}$$

In the outer region, i.e. in the region outside of the vortical core, the stream function and the vorticity can be expanded as a power series in ε which is to be defined below. It can be concluded from (2.10 and 2.11) that when the outer flow field is initially irrotational and no vorticity enters through the boundary, the flow in the outer region is a potential flow to any order of ε. In particular the leading terms of the velocity potential $\phi^{(o)}$ for a vortex point of strength Γ located at (X, Y) submerged in a background potential flow is given by (1.1).

For the inner region, the spatial variables x, y will be replaced by the polar coordinates (r, Θ) with respect to the center of the vortical core and $\bar{r}$ is then the stretched variable given by (2.8). We define $\vec{V}$ as the velocity relative to the vortex center, Hence,

$$\vec{V} = 0 \quad \text{at} \quad \bar{r} = 0. \tag{2.15}$$

The symbol $(\sim)$ is introduced to indicate quantities associated with the inner solution with two time variables t and $\bar{t} = t\varepsilon^{-2}$. The stream function $\tilde{\psi}$ for the relative velocity $\vec{V}$ and the vorticity $\tilde{\zeta}$ will now be expanded in power series of ε as follows:

$$\tilde{\psi}(\bar{t}, t, \bar{r}, \theta, \varepsilon) = \tilde{\psi}^{(0)}(\bar{t}, t, \bar{r}, \theta) + \varepsilon \tilde{\psi}^{(1)}(\bar{t}, t, \bar{r}, \theta) + \dots \quad , \tag{2.16a}$$

$$\varepsilon^2 \tilde{\zeta}(\bar{t}, t, \bar{r}, \theta, \varepsilon) = \tilde{\zeta}^{(0)}(\bar{t}, t, \bar{r}, \theta) + \varepsilon \tilde{\zeta}^{(1)}(\bar{t}, t, \bar{r}, \theta) + \dots \quad . \tag{2.16b}$$

By the substitution of (2.16a) and (2.16b) to (2.10) and (2.11) and by equating the coefficients of like powers of ε, the governing equations for $\tilde{\psi}^{(n)}$ and $\tilde{\zeta}^{(n)}$ are obtained. Equation (2.11) leads to

$$\tilde{\zeta}^{(n)} = -\bar{\Delta}\tilde{\psi}^{(n)} \quad \text{for} \quad n = 0, 1, 2, \dots \tag{2.17}$$

where $\bar{\Delta}$ is the Laplacian operator for the polar coordinates $(\bar{r}, \Theta)$. The coefficient of ε^{-2} in (2.10) gives

$$\frac{\partial \tilde{\zeta}^{(0)}}{\partial \bar{t}} + \frac{1}{\bar{r}} \left[\frac{\partial \tilde{\psi}^{(0)}}{\partial \theta} \frac{\partial \tilde{\zeta}^{(0)}}{\partial \bar{r}} - \frac{\partial \tilde{\psi}^{(0)}}{\partial \bar{r}} \frac{\partial \tilde{\zeta}^{(0)}}{\partial \theta} \right] = 0. \tag{2.18}$$

The boundary condition (2.15), becomes

$$\tilde{\psi}^{(0)} = 0, \partial \tilde{\psi}^{(0)} \not{\partial} \bar{r} = 0, \quad \text{at} \quad \bar{r} = 0. \tag{2.19}$$

and the matching condition is

$$\partial \tilde{\psi}^{(0)} / \partial \bar{r} \rightarrow -\Gamma / (2\pi \bar{r}) \quad \text{as} \quad \bar{r} \rightarrow \infty. \tag{2.20}$$

Since we assume that only the circumferential velocity is large, $O(\varepsilon^{-1})$, we have $\tilde{\psi}_\Theta^{(0)} = 0$ and hence $\tilde{\zeta}_\Theta^{(0)} = 0$. Equation (2.18) then reduces to $\tilde{\zeta}_{\bar{t}}^{(0)} = 0$. The leading solutions are axisymmetric and independent of the short time. We write

$$\tilde{\psi}^{(0)}(\bar{t},t,\bar{r},\theta) = \psi^{(0)}(t,\bar{r}) \tag{2.21a}$$

and

$$\tilde{\zeta}^{(0)}(\bar{t},t,\bar{r},\theta) = \zeta^{(0)}(t,\bar{r}). \tag{2.21b}$$

The symbol ($\sim$) is dropped to emphasize that the function depends only on one time variable t. The dependence of $\psi^{(0)}$ and $\zeta^{(0)}$ on t and $\bar{r}$ will be defined below by applying the Θ-periodicity condition to higher order equations in the expansion scheme. It should be noted that (2.21b) requires that the initial distribution of ε^{-2} order vorticity should be axisymmetric. This requirement is consistent with the vortex line structure in which only the circumferential component can be $0(\varepsilon^{-1})$.

The coefficients of ε^{-1} and ε^{0} in (2.10)give, respectively,

$$\partial\tilde{\zeta}^{(1)}/\partial\bar{t} + \tilde{F}_1(\bar{t},t,\bar{r},\theta) = 0, \tag{2.22}$$

$$\partial\tilde{\zeta}^{(2)}/\partial\bar{t} + \tilde{F}_2(\bar{t},t,\bar{r},\theta) = G_2(t,\bar{r}), \tag{2.23}$$

where

$$\tilde{F}_1(\bar{t},t,\bar{r},\theta) = \frac{1}{\bar{r}}\frac{\partial\tilde{\psi}^{(1)}}{\partial\theta}\frac{\partial\zeta^{(0)}}{\partial\bar{r}} - \frac{1}{\bar{r}}\frac{\partial\psi^{(0)}}{\partial\bar{r}}\frac{\partial\tilde{\zeta}^{(1)}}{\partial\theta} ,$$

$$\tilde{F}_2(\bar{t},t,\bar{r},\theta) = \frac{1}{\bar{r}}\frac{\partial\tilde{\psi}^{(2)}}{\partial\theta}\frac{\partial\zeta^{(0)}}{\partial\bar{r}} - \frac{1}{\bar{r}}\frac{\partial\psi^{(0)}}{\partial\bar{r}}\frac{\partial\tilde{\zeta}^{(2)}}{\partial\theta}$$

$$+ \frac{1}{\bar{r}}\frac{\partial\tilde{\psi}^{(1)}}{\partial\theta}\frac{\partial\tilde{\zeta}^{(1)}}{\partial\bar{r}} - \frac{1}{\bar{r}}\frac{\partial\tilde{\psi}^{(1)}}{\partial\bar{r}}\frac{\partial\tilde{\zeta}^{(1)}}{\partial\theta} ,$$

$$G_2(t,\bar{r}) = -\partial\zeta^{(0)}/\partial t + \Gamma\bar{\Delta}\zeta^{(0)}. \tag{2.24}$$

An operator M is introduced to denote the operation of averaging over a long duration of $\bar{t}$; for example,

$$M\tilde{\psi}^{(1)}(\bar{t},t,\bar{r},\theta) = \lim_{T\to\infty} \frac{1}{T}\int_{t\varepsilon^{-2}}^{t\varepsilon^{-2}+T} \tilde{\psi}^{(1)}(\bar{t},t,\bar{r},\theta)d\bar{t} .$$

The condition that all physical quantities, in particular $\tilde{\psi}^{(1)}, \tilde{\zeta}^{(1)}, \tilde{\psi}^{(2)}$, etc. should be finite at any finite time implies that $M\tilde{\psi}^{(1)}, M\tilde{\zeta}^{(1)}, M\tilde{\psi}^{(2)}$, etc. should be finite and $M\tilde{\zeta}_{\bar{t}}^{(1)}$, $M\tilde{\zeta}_{\bar{t}}^{(2)}$, etc. should vanish. By applying the operator M to (2.22) and (2.23), respectively, the following equations are obtained:

$$M\tilde{F}_1 = \frac{1}{\bar{r}}\frac{\partial\zeta^{(0)}}{\partial\bar{r}}\frac{\partial}{\partial\theta}[M\tilde{\psi}^{(1)}] - \frac{1}{\bar{r}}\frac{\partial\psi^{(0)}}{\partial\bar{r}}\frac{\partial}{\partial\theta}[M\tilde{\zeta}^{(1)}] , \tag{2.25}$$

$$M \; \tilde{F}_2 \; = \; \frac{1}{\bar{r}} \; \frac{\partial \tilde{\zeta}^{(0)}}{\partial \bar{r}} \; \frac{\partial}{\partial \theta} \; [M \; \tilde{\psi}^{(2)}] \; - \; \frac{1}{\bar{r}} \; \frac{\partial \tilde{\psi}^{(0)}}{\partial \bar{r}} \; \frac{\partial}{\partial \theta} \; [M \; \tilde{\zeta}^{(2)}]$$

$$+ \; \frac{1}{\bar{r}} \; M \; \{ \frac{\partial \tilde{\psi}^{(1)}}{\partial \theta} \; \frac{\partial \tilde{\zeta}^{(1)}}{\partial \bar{r}} \; - \; \frac{\partial \tilde{\psi}^{(1)}}{\partial \bar{r}} \; \frac{\partial \tilde{\zeta}^{(1)}}{\partial \theta} \} \quad = \; G_2(t,\bar{r}) . \qquad (2.26)$$

It can be shown that $\tilde{F}_2$ is composed of terms which are either derivatives with respect to $\bar{t}$ or Θ; therefore, the average of $M\tilde{F}_2$ with respect to Θ over the period 2π vanishes. Equation (2.26) then yields $G_2 = 0$ which is

$$\bar{r} \bar{\Delta} \tilde{\zeta}^{(0)} (t,\bar{r}) \; - \; \zeta_t^{(0)} (t,\bar{r}) \; = \; 0 \; . \qquad (2.27)$$

This is the governing equation for the diffusion of the leading term of the vorticity distribution with given initial data $\varepsilon^{-2} \zeta^{(0)} (0,\bar{r})$. Equation (2.27) implies conservation of the total vorticity, i.e.

$$2\pi \; \int_0^\infty \; \bar{r} d\bar{r} \zeta^{(0)} \varepsilon^{-2} \; = \; \Gamma \; = \; constant \qquad (2.28a)$$

The solution is [22]

$$\zeta^{(0)} (t,\bar{r}) \; = \; (2\Gamma t)^{-1} \int_0^\infty \; \zeta^{(0)} (0,\rho) e^{-(\rho^2 + \bar{r}^2)/(4\Gamma t)} I_0 (\frac{\rho \bar{r}}{2\Gamma t}) \; \rho d\rho$$

$$(2.28b)$$

where I_0 is the modified Bessel function.

From the vorticity distribution (2.28b) we obtain the circumferential velocity $\psi_r^{(0)} (t, r)$ and (2.22) yields a linear equation for the asymmetric part of the next order solution $\tilde{\psi}^{(1)}$. The boundary conditions at $\bar{r} = 0$ are

$$\tilde{\psi}^{(1)} \; = \; 0 \quad and \quad \tilde{\psi}_{\bar{r}}^{(1)} \; = \; 0. \qquad (2.29)$$

The matching condition is

$$\tilde{\psi}^{(1)} (\bar{t},t,\bar{r}\to\infty,\theta) \; \to \; \bar{r} \{ [\psi_y^* - \tilde{U}^{(0)} (\bar{t},t)] sin\theta$$

$$+ \; [\psi_x^* + \tilde{V}^{(0)} (\bar{t},t)] cos\theta \} \qquad (2.30)$$

Here the background velocity components, ψ_y^* and ψ_x^*, are evaluated at the center of the vortex. $\tilde{X}^{(0)}$, $\tilde{Y}^{(0)}$, which will be functions of $\bar{t}$ and t. $\tilde{U}$ and $\tilde{V}$ are the velocity components of the vortex center, i.e.,

$$\tilde{U} \; \hat{i} + \tilde{V} \; \hat{j} \; = \; (\tilde{X}_t + \varepsilon^{-2}\tilde{X}_{\bar{t}}) \hat{i} + (\tilde{Y}_t + \varepsilon^{-2}\tilde{Y}_{\bar{t}}) \hat{j} \qquad (2.31)$$

The physical condition, that $\tilde{U}$ and $\tilde{V}$ cannot be too large, i.e. $o(\varepsilon^{-1})$, which is also consistent with (2.30), implies

$$\tilde{X}^{(0)}_{\tilde{t}} = \tilde{X}^{(1)}_{\tilde{t}} = \tilde{Y}^{(0)}_{\tilde{t}} = \tilde{Y}^{(1)}_{\tilde{t}} = 0$$

which in turn implies

$$\tilde{X}(\tilde{t},t,\epsilon) = X^{(0)}(t) + \epsilon X^{(1)}(t) + \epsilon^2 \tilde{X}^{(2)}(\tilde{t},t) + \ldots \tag{2.32}$$

and the same for $Y(\tilde{t}, t, \epsilon)$. We can now find the order of magnitude of the differences between the two time solutions and their average with respect to $\bar{t}$. They are

$$\tilde{X} - M\tilde{X} = O(\epsilon^2), \quad \tilde{Y} - M\tilde{Y} = O(\epsilon^2)$$

$$\tilde{U} - M\tilde{U} = O(1), \quad \tilde{V} - M\tilde{V} = O(1) \tag{2.33}$$

In particular we have

$$M\tilde{U} = M(\tilde{X}_t + \epsilon^{-2}\tilde{X}_{\tilde{t}}) = M(\tilde{X}_t)$$

$$= (M\tilde{X})_t = \dot{X}^{(0)}(t) + \epsilon \dot{X}^{(1)}(t) + O(\epsilon^2) \tag{2.34a}$$

and similarly

$$M\tilde{V} = (M\tilde{Y})_t = \dot{Y}^{(0)}(t) + \epsilon \dot{Y}^{(1)}(t) + O(\epsilon^2) \tag{2.34b}$$

$M\tilde{X}, M\tilde{Y}$ shall be called the mean trajectory which is the average of $\tilde{X}, \tilde{Y}$ over a long interval of $\bar{t}$.

Applying the averaging operator M to (2.29) and (2.30) and making use of (2.33, 2.34a,b), we obtain

$$M\tilde{\psi}^{(1)} = 0 \quad (M\tilde{\psi}^{(1)})_{\bar{r}} = 0 \quad \text{at } \bar{r} = 0 \tag{2.35}$$

and

$$M\tilde{\psi}^{(1)} \to \bar{r}\{[\psi^*_y - \dot{X}^{(0)}(t)]\sin\theta + [\psi^*_x + \dot{Y}^{(0)}(t)]\cos\theta\} \tag{2.36}$$

as $\bar{r} \to \infty$. Here ψ^*_x and ψ^*_y are evaluated at the location on the mean trajectory $X^{(0)}(t)$, $Y^{(0)}(t)$. The average stream function $M\tilde{\psi}^{(1)}$ is governed by the second order differential equation (2.25) in spatial variables $\bar{r}, \theta$ and boundary conditions (2.35) and (2.36). By expanding $M\tilde{\psi}^{(1)}$ in a Fourier series in θ, we can conclude that

$$M\tilde{\psi}_\theta^{(1)} \equiv 0 \tag{2.37}$$

and the velocity along the mean trajectory is

$$\dot{X}^{(0)}(t) = (M\tilde{X}^{(0)})_t = \psi_y^*(X^{(0)}, Y^{(0)})$$

$$\dot{Y}^{(0)}(t) = (M\tilde{Y}^{(0)})_t = -\psi_x^*(X^{(0)}, Y^{(0)}) \qquad (2.38)$$

If we assume that the solution has only one time scale T, we can expand the so-called one time inner solutions, the stream function $\psi(t, \bar{r}, \theta, \epsilon)$ and $\zeta(t, \bar{r}, \theta, \epsilon)$, in power series of ϵ and derive the governing equations for $\psi^{(0)}$, $\zeta^{(0)}$, $\psi^{(1)}$, $\zeta^{(1)}$ etc., in the same manner as we did for the two times solution $\tilde{\psi}$ and $\tilde{\zeta}$. Equation (2.21) says that the leading inner two times solution $\tilde{\zeta}^{(0)}$ is independent of t and is the same as the one time solution $\zeta^{(0)}$. They are then defined by (2.28). For the next order one time solution $\psi^{(1)}$ we find that they are governed by the same set of equations (2.25), (2.35) and (2.36) for $M\tilde{\psi}^{(1)}$. Hence (2.37) and (2.38) hold for the one time solution also.

We shall summarize the results of this subsection by the following statements regarding the two times solutions $\tilde{\psi}$, $\tilde{\zeta}$, $\tilde{X}$, $\tilde{Y}$, which satisfy the assigned initial velocity of the center of the vortex, their $\bar{t}$-average over a long duration in $\bar{t}$, $M\tilde{\psi}$ etc. and the one time solutions ψ etc. for which the initial velocity cannot be arbitrarily assigned.

(i) The leading term of the two times inner solution and its $\bar{t}$-average are identical with that of the one time inner solution which depends only on $\bar{r}$ and t and is governed by the simple diffusion equation for an initial axisymmetric vorticity distribution of $O(\epsilon^{-2})$. The total strength of the vorticity, Γ, is conserved.

(ii) The trajectory of the vortex center, $\tilde{X}, \tilde{Y}$, given by the two times solution deviates from its $\bar{t}$-average, the mean trajectory, by at most $O(\epsilon^2)$.

(iii) The mean trajectory differs from that of the one time solution X,Y by at most $O(\epsilon)$. In fact it was shown in [8] that their difference is at most $O(\epsilon^2)$.

(iv) For the outer solution the equivalent point vortex can be located at the center of the vortex core defined by the one time solution, X(t), Y(t). The deviation due to the two times solution will appear at most in ϵ^2-terms.

(v) The leading term of the outer solution is the sum of the background potential flow and the flow induced by the moving point vortex of constant strength Γ.

(vi) The interaction effect of the inner vorticity distribution with the outer solution will be of higher order. For an initial vorticity distribution of finite support or decaying exponentially in $\bar{r}$, it was shown in [8] that the interference effect appeared as a doublet of strength ϵ^4.

(vii) For an assigned initial velocity, the velocity of the vortex center given by the two times solution can deviate from its $\bar{t}$-average, the mean velocity, by an amount of $O(1)$ with a period $O(\epsilon^2 T)$ which changes slowly in t [8, 9].

(viii) The mean velocity and the velocity of the vortex center given by the one time solution have the identical leading terms as defined in (2.38).

(ix) Equation (2.38) says that the velocity of the vortex center in the inner solution, which is also the velocity of the equivalent point vortex in the outer solution, is equal to the local velocity of the background flow and is independent of the inner vorticity distribution. This result confirms the basic assumption of the classical two-dimensional inviscid theory for vortex motion.

Statements (i) to (viii) were shown to be valid also for the axisymmetric case [10, 11] and can be extended to the three-dimensional case. Therefore, when measurements are made with time scale T and length scale L or even ϵL, we will only observe [*] the leading term of the t-average of the two times solution which is equal to the leading term of the one time solution. For the general case of a vortex filament submerged in a three-dimensional flow field, the matched asymptotic analyses were carried out [12] only for the one time solution. The outline of the analyses and the essential results are reported in the next subsection.

2.3 The Three-Dimensional Problems

For an initial flow field resembling a vortex filament submerged in a potential flow field, the center line of the filament, the space curve $\vec{C}(s)$, was defined using the initial large swirl velocity profile in 2.1. The movement of the curve C as a material line is then defined as the center line $\vec{X}(s,t)$ of the moving filament in (2.5). The velocity of the center line $\vec{X}$ can be interpreted as the velocity of the vortex line from the viewpoint of the outer inviscid solution.

We introduce a moving, orthogonal curvilinear coordinate system (s,r,Θ) attached to the center line $\vec{X}(s,t)$. For a point P near $\vec{X}$, the positive vector can be written as

$$\vec{P} = \vec{X}(s,t) + r\hat{r} \tag{2.39}$$

(see Fig. 1 and (2.6)). Here $\hat{r}$ and $\hat{\Theta}$ are the unit radial and circumferential vectors in the plane normal to $\vec{X}(s,t)$. The stretch ratios for the curvilinear system are given by

$$h_1 = 1, \quad h_2 = r, \quad h_3 = \sigma[1 - kr\cos(\theta + \theta_0)] \tag{2.40}$$

where $k(s,t)$ is the curvature of the center line, $\sigma = \vec{X}_s(s,t)$ is the linear strain of the curve, $\phi = \Theta + \Theta_0$ is the angle between $\hat{n}$ and $\hat{r}$. Here $\Theta_0(s,t) = -\int\sigma T ds$ is a function which is used to account for the torsion T of the center line and thereby to give the orthogonal system (r,Θ,s).

---

[*] It should be noted that when the vortex model is applied to meteorological problems, the short time scale $\epsilon^2 T$ and the length scale $\epsilon^2 L$ may be large enough for the effects of the two time solution, i.e. the oscillatory trajectory and velocity fluctuation to be observable.

The velocity $\vec{Q}$ at any point in the flow is

$$\vec{Q} = \dot{\vec{X}} + \vec{V} \qquad (2.41)$$

where the relative velocity $\vec{V}$ is

$$\vec{V} = u\hat{r} + v\hat{\theta} + w\hat{\tau} \qquad (2.42)$$

The Navier-Stokes equations in vector form with density equal to unity are

$$\ddot{\vec{X}} + \left(w - \frac{r}{h_3}\,\hat{r}_t{\cdot}\hat{\tau}\right)\vec{X}_s + \frac{d\vec{V}}{dt} = -\,\nabla P + \frac{v}{h_3}\left(\frac{1}{h_3}\,\vec{X}_s\right)_s + v\Delta\vec{V} \qquad (2.43)$$

where P is the pressure and the continuity equation is

$$r\,(w_s + \vec{X}_s {\cdot}\hat{\tau}) + (ruh_3)_r + (h_3 v)_\theta = 0 \qquad (2.44)$$

From the definitions of the curve $\vec{X}$ and the relative velocity $\vec{V}$, it is required that on $r = 0$

$$u = 0 \quad \text{and} \quad v = 0 \qquad (2.45)$$

We thus have four equations given by (2.43) and (2.44) for the four unknown functions $\vec{V}$, P which depend on r,Θ,s and t, while the three equations in (2.5) supply the extra equations for the motion of the center line, i.e., for $\vec{X}(s,t)$.

To study the initial value problem for these equations, initial values of $\vec{Q}$, $\vec{X}$ and $\dot{\vec{X}}$ must be prescribed. The initial flow field $\vec{Q}$ must be irrotational except in a small neighborhood of $\vec{X}(s,o)$ where there are large swirl and axial flow. In our matched asymptotic analysis, we shall assume that the inner and outer solutions depend on only one time scale T. Therefore the initial value of $\vec{X}$ cannot be prescribed. The physical interpretations of these one time solutions were elaborated in the preceding subsection.

An examination of the expansion of the Biot-Savart integral given in (1.6) with the variable r replaced by $\varepsilon\bar{r}$ indicates that in the inner region the solution in general can be expanded in a power series in both ε and $\ln\varepsilon$. In order to simplify the matching of the inner and outer solutions we shall expand the inner solution in powers of ε, recognizing that the coefficients in these expansions can be power series in $\ln\varepsilon$. By matching like powers of ε we will then determine the dependence on $\ln\varepsilon$.

Hence we assume that in the inner region the relative velocities have expansions in terms of powers of ε in the form

$$u(\bar{r},\theta,s,t,\varepsilon) = u^{(1)}(\bar{r},\theta,s,t) + \varepsilon u^{(2)}(\bar{r},\theta,s,t) + \ldots, \qquad (2.46)$$

$$v(\bar{r},\theta,s,t,\varepsilon) = \varepsilon^{-1}v^{(0)}(\bar{r},\theta,s,t) + v^{(1)}(\bar{r},\theta s,t) + \varepsilon v^{(2)}\ldots \qquad (2.47)$$

$$w(\bar{r},\theta,s,t,\varepsilon) = \varepsilon^{-1}w^{(0)}(\bar{r},\theta,s,t) + w^{(1)}(\bar{r},\theta,s,t) + \varepsilon w^{(2)}\ldots \qquad (2.48)$$

Here and in our subsequent expansions we have not explicitly indicated the dependence of the coefficients on $\ell n\varepsilon$. The leading powers of ε in these expansions were chosen by considering the possibility of matching these velocities with the corresponding outer flow velocities which are taken to be $O(1)$. In addition we allow for large axial flows inside the vortical core. This implies that both the axial and swirl velocities have leading terms proportional to ε^{-1}. We also observe that the momentum equations (2.43) and the fact that the pressure is $O(1)$ in the outer flow require that we choose

$$P(\bar{r},\theta,s,t,\varepsilon) = \varepsilon^{-2}P^{(0)}(\bar{r},\theta,s,t) + \varepsilon^{-1}P^{(1)} + \ldots \tag{2.49}$$

in order to find non trivial velocities $v^{(0)}$ and $w^{(0)}$.

The center line of the filament also depends on the parameter ε. We assume that it has an expansion of the form

$$\vec{X}(s,t,\varepsilon) = \vec{X}^{(0)}(s,t) + \varepsilon\vec{X}^{(1)}(s,t) + \ldots \tag{2.50}$$

The geometric parameters σ, k, and h_s are given in terms of $\vec{X}_s$ and $\vec{X}_{ss}$ and are also expanded in powers of ε.

The boundary condition (2.45) becomes

$$v^{(i)} = 0, \quad u^{(i+1)} = 0, \quad i = 0,1,2,\ldots \text{ at } \bar{r} = 0 \tag{2.51}$$

In addition the solution for the inner flow must be matched to the outer inviscid flow given by (1.6):

$$\vec{Q} = \vec{Q}_1 + \vec{Q}_2 = \frac{\Gamma}{2\pi r}\hat{\theta} + \frac{\Gamma}{4\pi R}\ell n\left(\frac{\bar{R}}{r}\right)\hat{b} + \frac{\Gamma}{4\pi R}\cos\phi\hat{\theta} + \vec{Q}_f + \vec{Q}_2.$$
$$\tag{2.52}$$

Matching (2.52) as $r \to 0$ with the velocities given in (2.46) – (2.48) as $\bar{r} \to \infty$, we get

$$v^{(0)} \sim \frac{\Gamma}{2\pi\bar{r}} + o(\bar{r}^{-n}) \text{ for all } n, \tag{2.53a}$$

$$w^{(0)} \sim o(\bar{r}^{-n}) \text{ for all } n, \tag{2.53b}$$

$$v^{(1)} \sim \frac{\Gamma}{4\pi R^{(0)}}(\cos\phi)\left[\ell n\frac{\bar{R}^{(0)}}{\varepsilon\bar{r}} + 1\right] + (\vec{Q}_0 - \vec{X}^{(0)})\cdot\hat{\theta}^{(0)}, \tag{2.53c}$$

$$u^{(1)} \sim \frac{\Gamma}{4\pi R^{(0)}}(\sin\phi)\ell n\frac{R^{(0)}}{\varepsilon\bar{r}} + (\vec{Q}_0 - \vec{X}^{(0)})\cdot\hat{r}^{(0)}, \tag{2.53d}$$

$$w^{(1)} \sim \vec{Q}_0\cdot\hat{r}^{(0)}. \tag{2.53e}$$

where $R^{(0)} = 1/k^{(0)}$ is the leading term of the radius of curvature of the reference line. Here $\vec{Q}_0$ $= \vec{Q}_2(\vec{X}) + \lim_{r \to 0}\vec{Q}_f$.

Formulas for the evaluation of the last term, the finite part of the Biot-Savat integral, are derived in Appendix C of [12] .

In deriving (2.53a) and (2.53b) we have assumed that the initial vorticity distribution and axial velocity are either of compact support or decay exponentially to zero as $\bar{r} \to \infty$.

Substituting the expansions (2.46) to (2.50) the momentum equation (2.43) and the continuity equation (2.44) and setting coefficients of corresponding powers of ε equal to zero, we get sets of governing equations for $v^{(i)}$, $w^{(i)}$ etc. for $i = 0,1,... $.

From the first set of governing equations, we obtain

$$v_\theta^{(0)} = 0, \quad w_\theta^{(0)} = 0, \quad P_\theta^{(0)} = 0$$

and

$$P_r^{(0)} = [v^{(0)}]^2/\bar{r}. \tag{2.54}$$

Averaging the second set of governing equations with respect to Θ over its period 2π, we obtain

$$(\bar{r} \, u^{(1)})_{\bar{r}} + \frac{\bar{r}}{\sigma^{(0)}} \, w_s^{(0)} = 0. \tag{2.55a}$$

$$\frac{w^{(0)}}{\bar{r}} <u^{(1)}> + \frac{w^{(0)} w_s^{(0)}}{\sigma^{(0)}} = -\frac{P_s^{(0)}}{\sigma^{(0)}} \tag{2.55b}$$

$$\frac{1}{\bar{r}} (\bar{r} v^{(0)})_{\bar{r}} <u^{(1)}> + \frac{w^{(0)} v_s^{(0)}}{\sigma^{(0)}} = 0. \tag{2.55c}$$

and

$$-\frac{2v^{(0)}}{\bar{r}} = <v^{(1)}> = -<P^{(1)}>_{\bar{r}} \tag{2.55d}$$

Here we have used the notation

$$<f> = \int_0^{2\pi} f d\theta/(2\pi) \ .$$

In the matching condition, equation (2.53a), we note that the leading term in the circumferential velocity is independent of the axial location. Motivated by this we look for solutions in the inner region $v^{(0)}$ independent of s, i.e. we shall assume in this subsection that

$$v_s^{(0)} = 0. \tag{2.56a}$$

It follows from (2.55) that

$$\langle u^{(1)} \rangle = 0$$

and then

$$w_s^{(0)} = 0. \tag{2.56b}$$

Conversely, if we assume that (2.56b) is true, equations (2.55a) and (2.51) yield $\langle u^{(1)} \rangle = 0$ and (2.55b) yields $P_s^{(0)} = 0$. From this last equation of (2.54), we arrive at (2.56a).

We have shown the following consistency relationship:

$$v_s^{(0)} = 0 \text{ if and only if } w_s^{(0)} = 0 \tag{2.57}$$

This is also true in the inviscid case since viscosity does not enter at this stage of our analysis. This relationship should be observed when an outer potential flow is matched or patched with an inner rotational flow [13, 14, 15]. If one assigns a swirl velocity profile independent of s in the inviscid theory, one cannot assign an axial velocity profile which depends on s. If there is no large axial velocity, we conclude from (2.57) that $v^{(0)}$ and hence $\zeta^{(0)}$ should be independent of s. This implies that we cannot patch in the vortical core an axial vorticity distribution which is stretched according to the local radius of curvature which depends on s.

In the remaining part of this paper we shall assume that (2.56a) and hence (2.56b) are valid. Thus we note that the inner flow to leading order, ε^{-1} is independent of Θ and s and we must consider initial velocity fields which possess this property.

We now use (2.56a and b) to rewrite the second set of governing equations to define the Θ-dependence of $u^{(1)}$, $w^{(1)}$, $v^{(1)}$, and $p^{(1)}$. From matching conditions (2.53) on $u^{(1)}$ and $v^{(1)}$, we obtain the formula for $\vec{X}$. By averaging the third set of the governing equations with respect to Θ over its period 2π, we obtained finally the governing equations for the leading terms of the inner structure, i.e., $v^{(0)}(\bar{r},t)$ and $w^{(0)}(\bar{r},t)$ and complete the set of equations for the leading terms of the inner structure and the velocity of the vortex line $\vec{X}^{(0)}$. They are:

$$w_t^{(0)} - \frac{\Gamma}{\bar{r}} (\bar{r} w^{(0)})_{\bar{r}} = \frac{1}{2} \bar{r}^3 \left[\frac{w^{(0)}}{\bar{r}^2}\right]_{\bar{r}} \frac{\dot{s}^{(0)}}{s^{(0)}}, \tag{2.58}$$

$$\zeta_t^{(0)} - \frac{\Gamma}{\bar{r}} (\bar{r} \zeta^{(0)})_{\bar{r}} = \frac{1}{2} \frac{(\bar{r}^2 \zeta^{(0)})_{\bar{r}}}{\bar{r}} \frac{\dot{s}^{(0)}}{s^{(0)}}, \tag{2.59}$$

$$\vec{X}^{(0)} \cdot \hat{\tau}^{(0)} = 0, \tag{2.60}$$

$$\vec{X}^{(0)} \cdot \hat{n}^{(0)} = \vec{Q}_0 \cdot \hat{n}^{(0)}, \tag{2.61}$$

and
$$\vec{\dot{X}}(0) \cdot \hat{b}(0) = \vec{Q}_0 \cdot \hat{b}(0) + \frac{k^{(0)} \Gamma}{4\pi} \ln \frac{R^{(0)}}{\epsilon S_0} + C_1(s,t), \tag{2.62}$$

where
$$C_1(s,t) = \frac{k^{(0)} \Gamma}{2\pi} \{\frac{1}{2} \lim_{\bar{r} \to \infty} (\frac{4\pi^2}{r^2} \int_0^{\bar{r}} \xi (v^{(0)})^2 \, d\xi - \ln \bar{r}/S_0)$$
$$- \frac{1}{4} - \frac{4\pi^2}{r^2} \int_0^\infty \xi (w^{(0)})^2 \, d\xi\} . \tag{2.63}$$

In addition we note that

$$v^{(0)} = \frac{1}{\bar{r}} \int_0^{\bar{r}} \xi \zeta^{(0)} (\xi,t) d\xi \tag{2.64}$$

and that

$$s^{(0)}(t) = \int_0^{S_0} (\vec{X}_\xi^{(0)} (\xi,t) \cdot \vec{X}_\xi^{(0)} (\xi,t))^{1/2} \, d\xi . \tag{2.65}$$

We observe that the $\ln \epsilon$ term only appears explicitly in the velocity of the vortex line in the binormal direction. This implies that $S(t,\epsilon)$ has an expansion in powers of ϵ with leading order term $S^{(0)}(t)$ independent of $\ln \epsilon$. Thus $S^{(0)}(t)$ and $\dot{S}^{(0)}(t)$ are order one terms. We denote the initial length $S^{(0)}(0)$ as S_0.

Equations (2.58, 2.59) define the inner structure which is coupled to the movement of the vortex filament through its total length $S^{(0)}(t)$. Equations (2.60 - 2.62) define the velocity of the vortex line, $\vec{X}(s,t)$. The binormal component of the velocity depends on the inner structure through the term $C_1(s,t)$.

The term on the right side of (2.58), and (2.59), represents the inviscid effect on the structure due to the stretching of the filament. When $\dot{S}^{(0)} = 0$, equations (2.58) and (2.59) become the simple two-dimensional axisymmetric viscous diffusion equation.

We can conclude from (2.58) and (2.59) respectively two conservation laws:

$$\frac{d}{dt} \int_0^\infty 2\pi \bar{r} \zeta^{(0)} d\bar{r} = \frac{d\Gamma}{dt} = 0 \tag{2.66}$$

and

$$\frac{d}{dt} \{[s^{(0)}]^2 \int_0^\infty 2\pi \bar{r} w^{(0)} d\bar{r}\} = \frac{d}{dt} \{[s^{(0)}]^2 m(t)\} = 0 \tag{2.67}$$

The first one states that the circulation is conserved while the second implies that the axial mass flux m times the square of the total length is conserved.

We shall introduce new independent variables τ_1 and η by

$$\tau_2 = \frac{1}{s^{(0)}(t)} \int_0^t s^{(0)}(\xi) d\xi + \tau_{20}, \qquad \eta = \frac{\bar{r}}{\sqrt{4\Gamma\tau_2(t)}} \qquad (2.68a)$$

and

$$\tau_1 = s^{(0)} \tau_2 . \qquad (2.68b)$$

so that the unknown $s^{(0)}(t)$ is absorbed and both equations (2.58) and (2.59) become separable. τ_{20} is a positive constant whose choice is at our disposal. A method for picking τ_{20} is discussed below. The solutions of (2.58) and (2.59) in terms of the new variables are

$$w^{(0)} = \frac{1}{s^{(0)}} e^{-\eta^2} \sum_{n=0}^{\infty} C_n L_n(\eta^2) \tau_1^{-(n+1)} \qquad , \qquad (2.69)$$

$$\zeta^{(0)} = s^{(0)} e^{-\eta^2} \sum_{n=0}^{\infty} D_n L_n(\eta^2) \tau_1^{-(n+1)} \qquad , \qquad (2.70)$$

where L_n, $n = 0,1,...$, are the Laguerre polynomials and C_n, D_n are constants which can be determined in terms of the initial values of $w^{(0)}$, $\zeta^{(0)}$ and $s^{(0)}$ as

$$C_n = 2s_0 \tau_{10}^{n+1} \int_0^{\infty} w^{(0)}(\eta\sqrt{4\Gamma\tau_{20}},0) L_n(\eta^2) \eta \, d\eta \qquad (2.71a)$$

$$D_n = \frac{2\tau_{10}^{n+1}}{s_0} \int_0^{\infty} \zeta^{(0)}(\eta\sqrt{4\Gamma\tau_{20}},0) L_n(\eta^2) \eta \, d\eta \qquad (2.71b)$$

where we have taken the initial time as 0. We shall pick the constant τ_{20} such that $D_1 = 0$ so that the vorticity distribution approaches the similarity solution, the first term, $n = 0$, in (2.70) within the shortest time. τ_{20} is then called the optimum time shift which was discussed in detail in [11] and also in [23].

Equations (2.60 - 2.62) and the equation

$$\frac{d\tau_1}{dt} = \tau_{20} \dot{s}^{(0)} + s^{(0)} \qquad , \qquad (2.72)$$

which is equivalent to (2.68), can be solved as a simultaneous system for the unknowns $\vec{X}^{(0)}$ and $\tau_1(t)$ with the values of $w^{(0)}(\bar{r},t)$ and $\zeta^{(0)}$ being simultaneously determined from (2.69) and (2.70). We observe that in these equations the only independent variables are s and t. They can be solved numerically as an initial value problem by the procedure used in the axially symmetric problem [11, 16].

It is interesting to note that the terms in (2.69) and (2.70) corresponding to n = 0 give the long time behavior of the solution. They yield the similarity solutions

$$w^{(0)} = \frac{C_0}{(S^{(0)})\tau_1} e^{-\eta^2} = \frac{(S_0/S^{(0)})^2}{2\Gamma\tau_2(t)} e^{-\eta^2} \int_0^\infty w^{(0)}(\bar{r},0)\bar{r}\ d\bar{r} \quad (2.73a)$$

$$\zeta^{(0)} = \frac{S^{(0)}D_0}{\tau_1} e^{-\eta^2} = \frac{1}{2\Gamma\tau_2(t)} e^{-\eta^2} \int_o^\infty \zeta^{(0)}(\bar{r},0)\bar{r}\ d\bar{r} \ . \quad (2.73b)$$

These equations can be rewritten by use of the circulation and the initial axial mass flux (2.66), (2.67), as

$$w^{(0)} = \frac{m(0)}{4\pi\Gamma\tau_2} [\frac{S_0}{S^{(0)}(t)}]^2 e^{-\eta^2} \quad (2.74a)$$

$$\zeta^{(0)} = \frac{1}{4\pi\tau_2(t)} e^{-\eta^2} \ . \quad (2.74b)$$

They represent the inner solutions for a filament created at $\tau_2 = 0$, with zero core radius, i.e., the initial data for $\zeta^{(0)}$ and $w^{(0)}$ at $\tau_2 = 0$ are delta functions in $\bar{r}$.The age of the filament at t = 0 is, therefore, $\tau_2(0) = \tau_{20}$. At each instant, the function $\exp(-\eta^2)$ in (2.74) represents the dependence of $w^{(0)}$ and $\zeta^{(0)}$ on $\bar{r}$. We can now define an effective size of the filament by the condition that the local value is equal to e^{-1} times the maximum value, i.e. at $\eta = 1$. The effective size of the core is

$$\delta(t) = (4\nu\tau_2(t))^{1/2} \quad (2.75)$$

The velocity of the vortex filament in the binormal direction (2.62), which depends on the inner structure, can be rewritten as

$$\dot{\vec{X}}^{(0)} \cdot \hat{b}^{(0)} = \vec{Q}_0 \cdot \hat{b}^{(0)} + \frac{k^{(0)}\Gamma}{4\pi} [\ell n \frac{R^{(0)}}{\delta} - .558 - \frac{m^2(0)}{\Gamma^3\tau_2} (\frac{S_0}{S^{(0)}(t)})^4] \quad (2.76)$$

We observe from (2.75) that if the inner structure is similar, then the dependence of the velocity of the filament on the inner structure is completely determined by two constants namely, the initial axial mass flux m(0) and the initial age τ_{20} of the core. In a real problem, the initial data for the inner velocity distribution of a viscous core are not available. Thus the study of the motion of a vortex filament with unknown initial core structure reduces to the study of a one parameter family of similarity distributions. Examples using the above theory for the interaction of two circular vortex rings with non overlapping viscous cores were presented in [16]. The passage of a circular vortex ring over a rigid sphere was analyzed in [17] when the viscous core does not merge with the boundary layer along the sphere. These examples of simple model problems demonstrate the influence of viscous core structure on the motion of vortex rings.

The flow model consisting of axially symmetric vortex rings with viscous cores submerged in a uniform stream was employed [18, 19] to simulate the fluctuation of the pressure field in the vicinity of a circular jet. The time interval between the shedding of successive vortices is taken to be a random variable with a probability distribution chosen to match that from experiments. It is found that up to 5 diameters downstream of the jet exit, statistics of the computed pressure field are in good agreement with experimental results. Statistical comparisons are provided for the overall sound pressure level, the peak amplitude, and the Strouhal number based on the peak frequency of the pressure signals. The five diameter distance is about the length of the potential core of the jet. No comparison was made beyond the potential core since vortex ring-type structures are rarely seen there.

3. Vortex Filament Submerged in a Rotational Flow

Since vorticity is no longer confined to the neighborhood of the filament, we shall formulate the problem and then outline the analysis. We begin with the simple two-dimensional case.

3.1 Formulation of the Problem

Let us consider an initial vorticity distribution $\zeta(x,y,o)$ which can be split into two parts

$$\zeta(x,y,o) = f_1(x,y) + f_2(\bar{x},\bar{y}). \tag{3.1}$$

f_1 is the initial vorticity of the background rotational flow. It is distributed with the reference length scale L and its magnitude is the order of U/L where U is the reference velocity of the background flow. The part f_2 represents a concentrated distribution near a point $C(X(0), Y(0))$ henceforth referred to as the vortex point. It is of compact support or decays exponentially in $\bar{r}$ where $\bar{r}$ is the distance from C on a small length scale εL. We write f_2 as a function of the stretched variables, $\bar{x},\bar{y}$, with

$$\bar{x} = (x - X)/\varepsilon, \quad \bar{y} = (y - Y)/\varepsilon \tag{3.2}$$

The total strength of f_2 is assumed to be of the order UL, i.e.,

$$\int_{-\infty}^{\infty} \int f_2 \, dx \, dy = \Gamma = O(UL) \tag{3.3}$$

or

$$\int_{-\infty}^{\infty} \int f_2 \, d\bar{x} \, d\bar{y} \, \varepsilon^2 = O(UL). $$

Therefore f_2 is of the order $\varepsilon^{-2} U/L$. We write

$$f_2 = \varepsilon^{-2} \bar{f}_2 \quad \text{with } \bar{f}_2 = O(1). \tag{3.4}$$

Thus we defined an initial vorticity distribution which resembles a highly concentrated vortical core submerged in a rotational flow. To take into account the viscous effects inside the core, we identify the small parameter ε as

$$\varepsilon = \frac{1}{\sqrt{R_e}} = \sqrt{\frac{\nu}{\Gamma}} \tag{3.5}$$

We shall seek solutions of the unsteady Navier-Stokes equations with large Reynolds number subjected to the initial condition of (3.1) and appropriate boundary conditions. For example, we may specify an upstream shear flow $U(y)$, i.e.,

$$\vec{Q} = U(y) \hat{i}$$

as $x \to -\infty$. If there is a rigid body present in the flow (see Fig. 3), we assume that the shortest distance from the vortex point to the body is at least of the order L. Consequently the diffusion in the vortical core and that in the boundary layer along the body surface will interact only indirectly, i.e., in their higher order terms.

The vorticity distribution $\zeta(x,y,t)$ and the stream function $\psi(x,y,t)$ are governed by these two equations

$$\zeta_t + u\zeta_x + v\zeta_y = \nu\Delta\zeta \tag{3.6}$$

and

$$\Delta\psi = -\zeta \tag{3.7}$$

We seek solutions $\zeta(x,y,t)$ and $\psi(x,y,t)$ for $t>0$ with only one time scale L/U. Because of the viscous terms, the velocity (u,v) in the flow field shall remain finite and single valued. This physical condition shall enable us to define the velocity of the vortex point $(\dot{X}(t), \dot{Y}(t))$.

3.2 Outline of the Analysis

The special form of the initial data suggest that we write the solution as a composite of two length scales,

$$\zeta(x,y,t,\varepsilon) = \zeta_1(x,y,t,\varepsilon) + \varepsilon^{-2}\bar{\zeta}_2(\bar{x},\bar{y},t,\varepsilon) \tag{3.8}$$

such that at $t = 0$,

$$\zeta_1 = f_1(x,y) \tag{3.9a}$$

and

$$\zeta_2 = \varepsilon^2 f_2(\bar{x},\bar{y}) = \bar{f}_2(\bar{x},\bar{y}) \tag{3.9b}$$

Likewise we write

$$\psi(x,y,t,\varepsilon) = \psi_1(x,y,t,\varepsilon) + \bar{\psi}_2(\bar{x},\bar{y},t,\varepsilon) \tag{3.10}$$

ψ_2 is the stream function induced by the vorticity distribution ζ_2,

$$\bar{\psi}_2 = \frac{1}{4\pi} \int\int_{-\infty}^{\infty} \zeta_2(\bar{\xi}, \eta, t, \epsilon) \, \ell n(\xi^2 + \eta^2) d\xi \, d\eta \qquad (3.11)$$

It is a particular integral of the equation,

$$\bar{\Delta\psi}_2 = \bar{\zeta}_2$$

or

$$\Delta\bar{\psi}_2 = \epsilon^{-2}\bar{\zeta}_2 \cdot \qquad (3.12)$$

Consequently ψ_1 is a solution of the equation

$$\Delta\psi_1 = \zeta_1 \qquad (3.13)$$

subjected to the boundary conditions for the velocity field of $\psi_1 + \bar{\psi}_2$.

We shall decompose the velocity components

$$u = u_1(x,y,t,\epsilon) + \epsilon^{-1}\bar{u}_2(\bar{x},\bar{y},t,\epsilon), \quad v = v_1(x,y,t,\epsilon) + \epsilon^{-1}\bar{v}_2(\bar{x},\bar{y},t,\epsilon)$$

$$(3.14)$$

and relate them to ψ_1 and $\bar{\psi}_2$

$$u_1 = \psi_{1,y}, \quad v_1 = -\psi_{1,x}, \qquad (3.15a)$$

$$\bar{u}_2 = \bar{\psi}_{2,\bar{y}}, \quad \bar{v}_2 = -\bar{\psi}_{2,\bar{y}} \cdot \qquad (3.15b)$$

In the multiple scale formulations, we have the identity on ζ,

$$\zeta(X,Y,t,\epsilon) = \zeta_1(X,Y,t,\epsilon) + \epsilon^{-2}\bar{\zeta}_2(0,0,t,\epsilon) \qquad (3.16)$$

and note that similar identities hold for the quantities ψ, u, and v.

The vorticity equation (3.6) can be split into two equations for ζ_1 and $\bar{\zeta}_2$,

$$\zeta_{1,t} + (u_1 + \epsilon^{-1}\bar{u}_2)\zeta_{1,x} + (v_1 + \epsilon^{-1}\bar{v}_2)\zeta_{1,y} = \Gamma\epsilon^2\Delta\zeta_1 \qquad (3.17)$$

and

$$\epsilon^2\bar{\zeta}_{2,t} + [\epsilon(u_1 - \dot{X}) + \bar{u}_2]\bar{\zeta}_{2,\bar{x}} + [\epsilon(v_1 - \dot{Y}) + \bar{v}_2]\bar{\zeta}_{2,\bar{y}} = \Gamma\epsilon \, \bar{\Delta\zeta}_2$$

$$(3.18)$$

They are coupled because of the nonlinear convective terms. In the preceding section, the background flow was a potential flow, i.e., $\zeta_1 = 0$ and $\Delta\psi_1 = 0$, and the leading term of the background flow is not altered by the presence of the vortex. This is no longer the case. Although (3.18) is the same as that for the inner solution in 2., equation (3.17) represents an additional equation which governs the variation of the background vorticity distribution. We shall uncouple (3.18) and (3.17) by applying the method of matched asymptotic expansions to

(3.18). Analytical solutions will be constructed for the leading terms of $\bar{\zeta}_2(\bar{x},\bar{y},t,\varepsilon)$ and $X(t,\varepsilon)$, $Y(t,\varepsilon)$. They will depend on the values of $u_1^{(0)}$ and $v_1^{(0)}$ at $(X^{(0)}, Y^{(0)})$. Here the superscript (0) denotes "the leading term of". We shall then solve the leading equations of (3.17), (3.13) and (3.15) for $\zeta_1^{(0)}$, $\psi_1^{(0)}$, $u_1^{(0)}$ and $v_1^{(0)}$.

We note that in section 2 the two lowest order matching conditions, (2.20) and (2.30), between the inner and outer solutions involve only the values of the background velocity at $(X^{(0)}, Y^{(0)})$ and not its gradients. Therefore, the conclusions in section 2 regarding only the first two orders of the inner solution should remain valid whether the background flow is rotational or irrotational. In particular we have

$$\bar{\zeta}_{2,\theta}^{(0)} = 0 \tag{3.19}$$

$$\bar{\zeta}_{2,t}^{(0)} = \Gamma\Delta\bar{\zeta}_2^{(0)} \tag{3.20}$$

and

$$\bar{\zeta}_{2,\theta}^{(1)} = 0 \tag{3.21}$$

Equation (3.19) is consistent with the condition that the flow near the vortex point has large swirl velocity while the radial velocity is $O(1)$. Therefore, the initial data $\bar{f}_2$ has to be a function of $\bar{r} = \varepsilon r$ only and the inner structure $\bar{\zeta}_2^{(0)}$ is related to its initial data by (2.28b). Since $\bar{u}_2$ and $\bar{v}_2$ are defined as the velocity induced by $\bar{\zeta}_2$, (3.11),(3.15b), equations (3.19), and (3.21) yield:

$$\bar{u}_2^{(0)} = \bar{v}_2^{(0)} = 0 \tag{3.22}$$

and

$$\bar{u}_2^{(1)} = \bar{v}_2^{(1)} = 0 \tag{3.23}$$

at $\bar{x} = 0$, $\bar{y} = 0$. The matching condition (2.30) then yields

$$\dot{X}^{(0)} = u_1^{(0)} (X,Y,t) \tag{3.24a}$$

and

$$\dot{Y}^{(0)} = v_1^{(0)} (X,Y,t) \tag{3.24b}$$

In contrast to the case when the background flow is potential, we now have to construct the unsteady background flow. The latter has length scale L, i.e. $O(1)$. We shall use the symbol ($\sim$) to denote the leading terms of ζ_1, ψ_1, u_1 and v_1, e.g.

$$\tilde{\zeta} = \zeta_1^{(0)} (x,y,t) \tag{3.25}$$

The governing equation for $\tilde{\zeta}$ is obtained from (3.17). It is a convective type equation,

$$\tilde{\zeta}_t + (\tilde{u} + \varepsilon^{-1}\bar{u}_2)\tilde{\zeta}_x + (\tilde{v} + \varepsilon^{-1}\bar{v}_2)\tilde{\zeta}_y = 0 \tag{3.26}$$

This equation in general can be solved, by a finite difference method, simultaneously with the integration of the ordinary differential equations (3.24a,b) for the trajectory $\hat{X}(t)$, $\hat{Y}(t)$ of the vortex point.

We shall outline a numerical scheme so that the spatial and temporal step sizes are independent of ε. We note that

$$\varepsilon^{-1}(u_2\hat{i} + v_2\hat{j}) \sim \frac{\hat{\Gamma\theta}}{2\pi\bar{r}\varepsilon} + 0(\frac{1}{r}) = \frac{\hat{\Gamma\theta}}{2\pi r} + 0(1) \tag{3.27}$$

as $\bar{r}/L \gg 1$ while $r/L \ll 1$. Consequently the coefficients of $\tilde{\zeta}_x$ and $\tilde{\zeta}_y$ in (3.26) are of the order of U, i.e. $0(1)$, outside of the ε-neighborhood of the vortex point (X,Y).

In addition at $\bar{x} = 0$, $\bar{y} = 0$ i.e. at $x = X$, $y = Y$, we use (3.16), (3.22), and (3.23), to reduce (3.26) to

$$\tilde{\zeta}_t + \tilde{u}\tilde{\zeta}_x + \tilde{v}\tilde{\zeta}_y = 0 \tag{3.28}$$

For the leading term $\tilde{\zeta}$, equation (3.28) holds along $x = \tilde{X}$, $y = \tilde{Y}$. On account of (3.24), equation (3.28) then takes the form

$$\tilde{\zeta}_t + \dot{X}\tilde{\zeta}_x + \dot{Y}\tilde{\zeta}_y = 0$$

This equation implies that $\tilde{\zeta}$ is constant along the trajectory, i.e.

$$\tilde{\zeta}(\tilde{X},\tilde{Y},t) = \tilde{\zeta}(X_0,Y_0,0) = f_1(X_0,Y_0) \tag{3.29}$$

In contrast to $\tilde{\zeta}_2$, $\tilde{\zeta}$ is a smooth function with length scale L which is $0(1)$. We can pick a grid size

$$\Delta x = \Delta y \gg \varepsilon. \tag{3.30}$$

We shall use $\tilde{\zeta}_{ij}^k$ to denote $\tilde{\zeta}(x_i, y_j, t_k)$. At an instant t_k and at a grid point outside of the ε-neighborhood of the vortex point $\tilde{X}^k$, $\tilde{Y}^k$ we know from (3.27) that the coefficients of $\tilde{\zeta}_x$ and $\tilde{\zeta}_y$ in (3.26) are $0(1)$. Thus we can select a time step,

$$\Delta t \gg \varepsilon. \tag{3.31}$$

The finite difference equation of (3.26) for the assigned step sizes can be applied to all the interior grid points except perhaps one point (i,j) which at the instant t_k lies in an ε-neighborhood of the vortex point $\tilde{X}^k$, $\tilde{Y}^k$ (see Fig. 4). To supply the missing equation at (i,j) we shall make use of the fact that $\tilde{\zeta}$ is "smooth" and approximate $\tilde{\zeta}$ at $(\tilde{X}^k,\tilde{Y}^k)$ by a finite Taylor series with respect to (i,j), e.g.,

$$\tilde{\zeta}(\tilde{X}^k,\tilde{Y}^k,t_k) = \tilde{\zeta}_{ij}^k + (\tilde{X}^k - x_1)(\tilde{\zeta}_x)_{ij}^k + (\tilde{Y}^k - y_j)(\tilde{\zeta}_y)_{ij}^k \tag{3.32}$$

The left side of the equation is known from (3.29), consequently, equation (3.32) provides missing the finite difference equation at (i,j).

The detailed analysis and numerical examples and the extension to three-dimensional problems will be reported elsewhere.

4. Numerical Studies of the Merging of Vortex Rings

Our analysis in sections 2 and 3 were carried out for an asymptotic expansion with a small parameter ε, which is the ratio of the effective radius δ of the viscous vortical core to the reference length scale L of the background flow, i.e.,

$$\varepsilon = \delta/L \ll 1 . \tag{4.1}$$

In a physical problem, the reference length L can be, for example, the shortest of the following scale lengths: L_1, local radius of curvature of the center line of the vortex filament, L_2, the distance between two adjacent vortices or filaments and L_3, the distance between a filament and the boundary of the flow.

As is well known the asymptotic solutions remain quite accurate even when ε is not too small. To determine an upper bound on ε, we shall compare asymptotic solutions with fully numerical solutions of the unsteady Navier-Stokes equations for several canonical problems, namely: (i) self merging of a vortex filament when $\delta/L_1 = 0(1)$ (ii) merging of two adjacent vortices or filaments when $\delta/L_2 = 0(1)$ and (iii) merging of a filament with a boundary layer when $\delta/L_3 = 0(1)$.

Numerical solutions of the two-dimensional Navier-Stokes equations were obtained for problem (ii), i.e. the merging of two vortices [20]. The numerical results show that the asymptotic solutions are quite accurate even for $\varepsilon = 1/2$. They also suggested several theoretical investigations which yield the following results:

$$\frac{d\Gamma}{dt} = \frac{d}{dt} \int\limits_{-\infty}^{\infty}\!\!\int \zeta \; dxdy = 0 \tag{4.2}$$

$$\frac{d}{dt} \int\limits_{-\infty}^{\infty}\!\!\int x\zeta \; dxdy = 0 \qquad \frac{d}{dt} \int\limits_{-\infty}^{\infty}\!\!\int y\zeta \; dxdy = 0 \tag{4.3}$$

$$\frac{dI}{dt} = \frac{d}{dt} \int\!\!\int (x^2 + y^2)\zeta \; dxdy = \nu \int\!\!\int \zeta \; dxdy \tag{4.4}$$

when the initial vorticity distribution decays exponentially in $(x^2 + y^2)$. The first two statements say that the total vorticity Γ and its first moments are conserved for both viscous and inviscid flows. If the total vorticity Γ is different from zero, we can pick the origin such that the first moments vanish. Equation (4.4) yields the linear dependence of the second polar moment I on t,

$$I(t) = \Gamma\nu t + I(0) \tag{4.5}$$

If either $\Gamma = 0$ or $\nu = 0$, I is then conserved. The proofs for (4.2) - (4.4) and their relevance to the optimum similarity solution of the linear diffusion equation will be reported elsewhere [24]. In this section, we shall report numerical studies of problems (i) and (ii) for axisymmetric flows in sections 4.1 and 4.2, respectively. The numerical results establish the practical upper bound on ε for the accuracy of the asymptotic solutions and also motivate new theoretical and numerical investigations. We would like to bring out the salient role played by the asymptotic solutions. They make it possible for us to identify appropriate canonical problems for numerical study and pinpoint from the numerical and graphical outputs, the few key constants or curves from which we can draw qualitative or physical conclusions.

4.1 Self Merging of a Vortex Ring

The solutions of the initial value problem for the self merging of an isolated circular vortex filament depend on the Reynolds number Γ_0/ν and the infinitely many generalized Fourier coefficients necessary for the specification of the initial vorticity profile. Certainly, we cannot make a complete parametric study. We observed in the asymptotic analysis in section 2 and in [8-12] that the diffused vorticity distribution approaches an optimum similarity solution before the effective core size is doubled. We recall that the effective core size δ is defined as the distance over which the vorticity decays from the maximum by a factor of e. We also learn that when $\delta/L \ll 1$, the vorticity diffusion equation in the asymptotic theory is two-dimensional in nature and the invariant condition, in the total strength of the vorticity over a cross-section of the filament should hold. In cylindrical coordinates r, ϕ, z with z as the axis of the ring, we observe the following condition in the moments of the vorticity distribution in a meridianal plane,

$$I_n(t) = \int_{-\infty}^{\infty} dz \int_0^{\infty} dr \ r^n \zeta \sim R_0^n \int_{-\infty}^{\infty} \int d\bar{x} d\bar{y} (\varepsilon^2 \zeta) = R_0^n \Gamma_0 \qquad (4.6)$$

where Γ_0 is the initial strength of the ring when $\delta/R_0 \to 0$, $\bar{x}$ and $\bar{y}$ are the stretched variables with respect to the center of the core (R_0, Z) i.e. $\bar{x} = (z-Z)/\varepsilon$, $\bar{y} = (r-R_0)/\varepsilon$ and $\varepsilon = \delta_0/R_0 \ll 1$, (see Fig. 4). Here the reference length is R_0. The vortex ring will move parallel to the z axis with constant radius R_0 and its inner structure will soon approach that of an optimum similarity solution long before the core size grows to a value δ_0 comparable to R_0. After that instant of time, say $t = t_0$, we shall begin our numerical program to study the self-merging process. On first consideration one might assume that the solution depends on two parameters Γ/ν and δ_0/R_0. If we identify t_0 as the life time of the ring, that is at $t = 0$ the optimum similarity solution has zero core radius, the numerical solution of the Navier-Stokes equations beginning at $t = t_0$ should be equivalent to that beginning at $t = 0$ with zero core radius. It is then clear that the numerical solution for $t > t_0$ should depend on only one parameter, the Reynolds number Γ/ν, and the solution can in general be written in the dimensionless form,

$$\frac{\zeta}{\Gamma_o/R_o^2} = f\left(\frac{\Gamma}{\nu} , \frac{t}{t_o} , \frac{z}{R_o} , \frac{r}{R_o}\right) \tag{4.7}$$

with $\delta_o = (4 \nu t_o)^{1/2}$ from (2.77).

We start the numerical solution of the Navier-Stokes equations for an initial similarity profile with small δ_o/R_o at $t = t_o$ and test the accuracy of the asymptotic solutions by checking on the moments of the vorticity from the numerical computations. Also we notice that the maximum vorticity in a similarity solution decays as t^{-1} i.e.

$$\zeta_{max}(t)/\zeta_{max}(t_o) = t_o/t \tag{4.8}$$

When the numerical solutions begin to deviate from (4.6) and (4.8), we arrive at the useful limit of the asymptotic solution and the onset of self-merging.

Numerical solutions were obtained for $\Gamma/\nu = 4\pi$ and $\delta_o/R_o = 1/4$. Fig. 5 shows that the maximum vorticity begins to deviate from the asymptotic solution, equation (4.8), when $t/t_o \sim 4$ with $\delta(t)/R_o \sim 1/2$. Fig. 6 shows that about the same time, the total vorticity I_o begins to deviate from the constant initial value Γ_o. Figs. 7, 8, 9 show the constant vorticity lines at $t/t_o = 2.92$, 3,92 and 16.25 respectively. We see that the constant vorticity lines begin to deviate slightly at $t/t_o = 3.92$ and are quite different from circles at $t/t_o = 16.25$.

It is clear from Fig. 5 that the second moment $I_2(t)$ remains constant long after the onset of merging. This suggests that the second moment should be conserved for an axisymmetric viscous flow field. We shall now prove this assertion.

For an axisymmetric flow, the vorticity $\zeta(t,r,z)$ in the circumferential direction obeys the equation,

$$\zeta_t + u\zeta_r + w\zeta_z - u\zeta/r = \nu[\zeta_{rr} + \zeta_r/r + \zeta_{zz} - \zeta/r^2] . \tag{4.9}$$

Here u and w are the radial and axial velocity components. They are related to the stream ψ as follows

$$u = -\psi_z/r \quad \text{and} \quad w = \psi_r/r . \tag{4.10}$$

The stream function is then related to ζ by the equation

$$\psi_{rr} + \psi_{zz} - \psi_r/r = -r\zeta \tag{4.11}$$

We consider the flow field, ζ and ψ being induced by an initial vorticity distribution, i.e., we impose the initial condition at $t = 0$,

$$\zeta(0,r,z) = f(r,z) \tag{4.12}$$

where f is of bounded support or decays exponentially in $(r^2 + z^2)^{1/2}$. Consequently, the vorticity should decay exponentially in distance from the diffused vorticity distribution for $t > 0$, i.e.

$$\zeta = o(d^{-n}) \quad \text{for all } n, \tag{4.13}$$

where

$$d^2 = r^2 + (z - Z(t))^2$$

and $z = Z(t)$ is the center plane of the vorticity distribution with

$$Z(t) = \frac{\int_{-\infty}^{\infty} dz \int_0^{\infty} dr\ zr^2 \zeta(t,r,z)}{\int_{-\infty}^{\infty} dz \int_0^{\infty} dr\ r^2 \zeta(t,r,z)}. \tag{4.14}$$

$Z(t)$ accounts for the movement of the vorticity distribution and can be large for large t.

From (4.11) we can express ψ in terms of ζ [1] ,

$$\psi = \frac{r}{4\pi} \int_{-\infty}^{\infty} dz' \int_0^{\infty} dr' \int_0^{2\pi} \frac{\zeta(t,r',z')}{\rho}\ r' \cos\theta\ d\theta \tag{4.15}$$

where $\rho^2 = (z - z')^2 + r^2 + r'^2 - 2rr' \cos\theta$. To prove our final result, we need the behavior of u and v at large distances d. We note

$$\rho^{-1} = d^{-1}\{1 + [(z'-z)(\tfrac{z-Z}{d}) + r'(\tfrac{r}{d})\cos\theta]d^{-1} + O(d^{-2})\} \tag{4.16}$$

and from (4.15), (4.10) we obtain

$$\psi = (1/4)(r/d)^2 d^{-1} \int_{-\infty}^{\infty} dz' \int_0^{\infty} dr'\ (r')^2 \zeta(t,r',z') + O(d^{-2})$$

$$u = O(d^{-2}) \quad \text{and} \quad v = O(d^{-2}) \tag{4.17}$$

We multiply (4.9) by r^2 and make use of the continuity equation to obtain

$$(r^2\zeta)_t + (r^2 u\zeta)_r + (r^2 w\zeta)_z - 2ru\zeta = \nu[\ r^2 \zeta_r)_r + (r^2\zeta)_{zz} - (r\zeta)_r]$$

Integrating both sides of this equation over the upper zr plane, and using the conditions (4.13) and (4.17) for large z and r, we obtain

$$\frac{d}{dt} \int_{-\infty}^{\infty} dz \int_0^{\infty} dr\ r^2 \zeta(t,r,z) = 2 \int_{-\infty}^{\infty} dz \int_0^{\infty} dr\ ru\zeta$$

$$= \frac{1}{2\pi} \int_{-\infty}^{\infty} dz \int_{-\infty}^{\infty} dz' \int_0^{\infty} dr \int_0^{\infty} dr' \int_0^{2\pi} \frac{\zeta(t,r,z)\zeta(t,r',z')(z-z')rr'}{\rho^3} \cos\theta\ d\theta$$

The last term should remain unchanged when we interchange the integration variables r with r' and z with z' but it changes sign after these interchanges. Consequently the last term is zero and we have proven that the second moment is conserved, i.e.,

$$\frac{d}{dt} I_2(t) = \frac{d}{dt} \int_{-\infty}^{\infty} dz \int_{0}^{\infty} dr\, r^2 \zeta(t,r,z) = 0$$

$$I_2(t) = \text{constant} = I_2(0) = \int_{-\infty}^{\infty} dz \int_{0}^{\infty} dr\, r^2 f(r,z) \qquad (4.18)$$

From (4.16) we have at a large distance d from the instantaneous center (0,Z),

$$\psi(t,r,z) = \frac{r^2}{4\pi d^3} (\pi I_2) + O(d^{-2}) \qquad (4.19)$$

The leading term represents the stream function of a doublet with constant strength πI_2 located at (0,Z).

When the initial data f(r,z) represents a vortex ring of strength Γ_o, radius R_o and zero core radius, we have $\pi I_2 = \Gamma_o(\pi R_o^2)$. Under inviscid theory, the vortex ring is equivalent to a doublet distribution of constant strength Γ_o over the circular disc bounded by the ring. Therefore for the far field we see a doublet of strength $\Gamma_o(\pi R_o^2)$ located at the center of the ring. The viscous terms have to be included to define the velocity of the center (0,Z) of the vortex ring in both the asympotic and numerical analyses.

As initiated by the numerical data, we have derived the theoretical results (4.18) and (4.19). They say: (i) For an arbitrary initial axisymmetry vorticity distribution, its second moment I_2, which is the integral of $r^2 \zeta$ over the upper zr plane, is independent of t. (ii) At a far distance from the diffused vorticity distribution, the leading term is that of a doublet of constant strength πI_2 located at the center (0,Z(t)) of the distribution. Viscous terms are needed to define Z(t).

4.2 Interaction of Two Vortex Rings

Fig. 10 shows the initial geometry of two vortex rings. The minimum distance L_2 between them is assumed to be much larger than their core radii δ_i, i = 1, 2, so that their vortical cores are clearly distinct. Since we are studying the mutual interaction and merging of these two rings instead of the selfmerging of a single ring, we choose the initial geometry such that

$$\delta_i/R_i \ll 1 \qquad (4.20)$$

prior to and during the merging of the vortical cores of the rings.

When $\delta_1 + \delta_2$ remains much less than L_2, the motion and decay of the rings can be described by asymptotic solutions. The velocity of each ring is composed of two parts -- one is the self induced velocity which is the dominant term and the other is the local velocity induced by the other ring.

In the special case of two initially identical rings, the self induced velocities are the same. The relatively small local velocity at one ring induced by the other gradually changes the radii of the rings which in turn causes a difference in their self induced velocities. Fig. 11 shows such an example with $\Gamma_i/v = 10^6$, $Z_2 - Z_1 = R_i = 100\,\delta_i$, at $t = t_o$. From the trajectories of the two rings shown in Fig. 11, we see that by the time one ring passes through the other, for example, at $\tilde{t} = \Gamma_1 (t-t_o)/R_1^2 (t_o) \sim 3$ and at $\tilde{t} \sim 9$, the difference in the radii of the rings is already much larger than the core radius. During the interval from $\tilde{t} = 0$ to $\tilde{t} = 12$, the core radii grows by 50% but the sum $\delta_1 + \delta_2$ remains much smaller than their minimum separation distance L_2. Consequently the asymptotic solutions are valid for that interval and the merging of the vortical cores will take place at a much later time.

Figures 12-15 show the constant vorticity lines for two vortex rings at $t/t_o = 1.05$, 1.55, 2.45, and 5.95. At $t = t_o$ the initial conditions are $10\,\delta_i = R_i = R_o$, $\Gamma_2/v = 4\pi$, $\Gamma_1 = 2\,\Gamma_2$ and $Z_2 - Z_1 = 3\,\delta_i$. Because of the difference in the initial strengths, the self induced velocity of the ring $i = 1$ is twice that of the ring, $i = 2$. The merging of two vortical cores takes place soon afterwards without any noticeable change in the radius of the rings. Since the effective size of the vorticity distribution from $t = t_o$ to $5.95\,t_o$ remains small relative to the ring radius, the merging is two-dimensional in nature, i.e., the moments $I_n(t)$ remain nearly constant in agreement with (4.6). In Fig. 12, we see that the constant velocity lines are circles and the asymptotic solutions shoul be valid for $t/t_o < 1.05$. In Fig. 13 we can see the onset of merging at $t/t_o = 1.55$. In Fig. 14 we see only one local maximum. This can be interpreted as the weaker vortex ring loses its identity at $t = 2.45\,t_o$. In Fig. 15 at $t = 5.95\,t_o$, the image of the weaker ring diseappers completely. Originally we were expecting the stronger ring to pass through the weaker one and then separate from it with or without an exchange of vorticity. From the numerical results for larger t/t_o we conclude that these two rings merge into a single ring and remain that way.

In this subsection we have studied two limiting cases. In the first case, Fig. 11, the two rings are passing over each other without merging of their vortical cores. In the second case, Fig. 12-15, the two vortex rings merge into a single one. In general the problem depends on six parameters Γ_1/v, Γ_1/Γ_2, δ_1/δ_2, $(Z_2 - Z_1)/R_1$, R_1/R_2 and δ_1/R_1. The last parameter can again be omitted if the initial vorticity distribution for each ring is similar. We should be able to keep four of the remaining five parameters constant and adjust the fifth one, say R_1/R_2. We expect that as R_1/R_2 decreases from unity (as in our second example), there will be a finite range of R_1/R_2 where these two rings will merge partially and then separate into two distinct rings possibly with some exchange of strength since their total strength is conserved. For even smaller values of R_1/R_2, we should then expect the ring $i = 1$ to pass through the ring $i = 2$ without merging. Numerical studies to verify these cases are of interest in particular for cases of partial merging.

5. Summary

A review of asymptotic analyses for the study of the motion and decay of a vortex filament submerged in a background potential flow is presented. Emphasis is placed on physical intuitions motivating the analyses and on the physical interpretation of the asymptotic solutions. The extension of the analyses to a vortex filament submerged in a rotational flow is outlined.

The asymptotic analyses identify various canonical problems which require numerical solutions of the Navier-Stokes equations. These numerical studies in turn initiate new theoretical analyses.

Acknowledgments

The author wishes to acknowledge the collaboration of Professor Andrew Callegari, Professor Max Gunzburger, Mrs. Fanny Kung, Dr. Eddie Liu and Dr. Chee Tung, in particular the recent efforts of Professor Callegari and Dr. Liu in sections 3 and 4. This research is partially supported by the Office of Naval Research.

6. References

[1] L a m b, H., Hydrodynamics, Dover Publ. (1949).

[2] W i l l m a r t h, W.W.: Structure of turbulence in boundary layers. Advances in Applied Mechanics, 15 (1975), pp. 159-254.

[3] C a n t w e l l, B., C o l e s, D., D i n n o t a k i s, P.: Structure and entrainment in the plane of symmetry of a turbulent spot. J. Fluid Mech., 87, (1978), pp. 641-672.

[4] C h e r i n, A.J., B e r n a r d, P.S.: Discretization of a vortex sheet with an example of roll-up. J. Comput. Phys. 13 (1973), pp. 423-429.

[5] L e o n a r d, A.: Numerical simulation of interacting, three-dimensional vortex filaments. Lecture Notes in Phys. 35, Springer-Verlag, Berlin, (1975), pp. 245-250.

[6] M e n g, J.C.S., T h o m s o n, J.A.L.: Numerical studies of some non-linear hydrodynamics problems by discrete vortex element methods. JFM 84 (1978), pp. 433-454.

[7] C o u e t, B., B u r r e m a n, O., L e o n a r d, A.: Simulation of three-dimensional turbulent flows with a vortex in cell method. Inst. for Plasma Research SU-IPR Report 715, Stanford Univ., (1979).

[8] T i n g, L., T u n g, C.: Motion and decay of a vortex in a non-uniform stream. Physics of Fluids 8, (1965), pp. 1039-1051.

[9] G u n z b u r g e r, M.: Long time behaviour of a decaying vortex. Z. Angew. Math. Mech. 53 (1973), pp. 751-760.

[10] T u n g, C., T i n g, L.: Motion and decay of a vortex ring. Phys. Fluids 10 (1967), pp. 901-910.

[11] T i n g, L.: Studies of the motion and decay of vortices. Aircraft Wake Turbulence and its Detection. Edited by J.H. Olsen, A. Goldburg and M. Rogers, Plenum Publ. (1971), pp. 11-39.

[12] C a l l e g a r i, A.J., T i n g, L.: Motion of a curved vortes filament with decaying vortical core and axial velocity. SIAM J. Appl. Math. $\underline{35}$, (1978), pp. 148-175.

[13] S a f f m a n, P.G.: The velocity of viscous vortex rings. Studies Appl. Math. $\underline{49}$ (1970), pp. 371-380.

[14] M oo r e, D.W., S a f f m a n,P.G.: The motion of a vortex filament with axial flow. Philos. Trans. Roy. Soc. London, Ser. A , $\underline{1226}$, (1972), 403-429.

[15] W i d n a l l, S.E.: The structure and dynamics of vortex filaments. Annual Review of Fluid Mechanics, Annual Review, Palo Alto, CAL., (1975), pp. 141-165.

[16] G u n z b u r g e r, M.: Motion of decaying vortex rings with non-similar vorticity distributions. J. Engrg. Math. $\underline{6}$ (1972), pp. 53-61.

[17] W a n g, H.C.: The motion of a vortex ring in the presence of a rigid sphere. Annual Report of the Institute of Physics, Academia Sinica, Taiwan (1970), pp. 85-93.

[18] L i u, C.H., M a e s t r e l l o, L., G u n z b u r g e r, M.D.: Simulation by vortex rings of unsteady pressure field near a jet. Progress in Astronautics and Aeronautics-Aeroacoustics: Jet Noise, Combustion and Core Engine Noise. Edited by I.R. Schwartz, AIAA, New York, $\underline{43}$, pp. 47-64.

[19] F u n g, Y.T., L i u, C.H., G u n z b u r g e r, M.D.: Simulation of the pressure field near a jet by randomly distributed vortex rings. AIAA J. $\underline{17}$ (1979), pp. 553-557.

[20] L o, K.C.R., T i n g, L.: Studies of merging of two vortices. Phys. Fluids $\underline{19}$ (1976), pp. 912-913.

[21] van D y k e, M.: Perturbation methods in fluid mechanics. Academic Press, Stanford, CAL. $\underline{248}$ (1975), pp. 7798.

[22] C a r s l a w, H.S., J a e g e r, J.C.: Conduction of heat in solid. Oxford Univ. Press, London (1959).

[23] K l e i n s t e i n, G., T i n g, L.: Optimum one term solutions for heat conduction problems. ZAMM, $\underline{51}$, (1971), pp. 1-16.

[24] T i n g, L.: Integral invariants and decay laws of vorticity distributions, to be published.

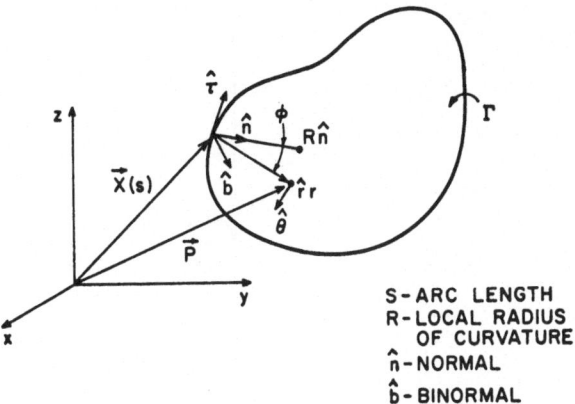

Figure 1. Coordinate system for a vortex line.

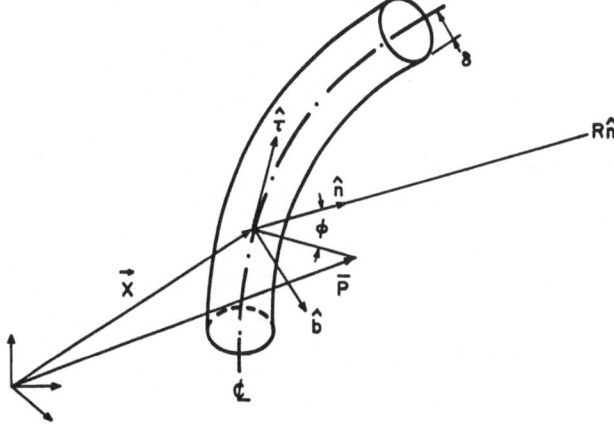

Figure 2. A segment of a vortex filament and its center line for the inner region, the equivalent vortex line for the outer region.

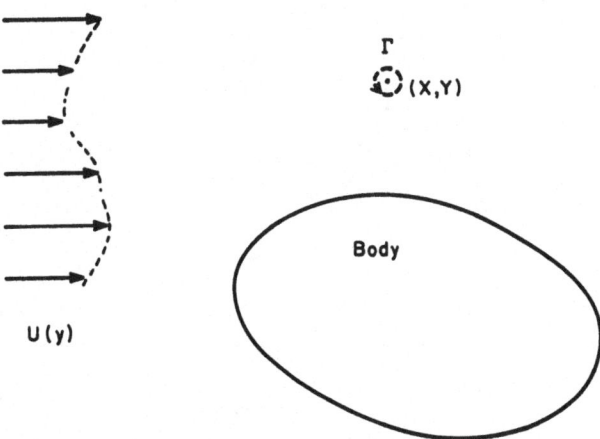

Figure 3. A vortex-like structure in a rotational flow field.

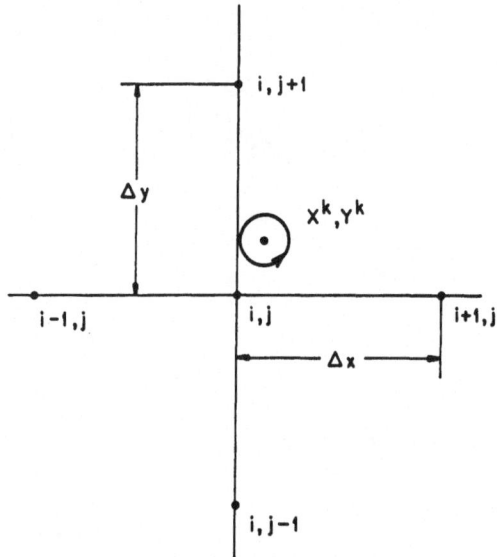

Figure 4. Finite difference scheme at a point near the vortical core of a vortex.

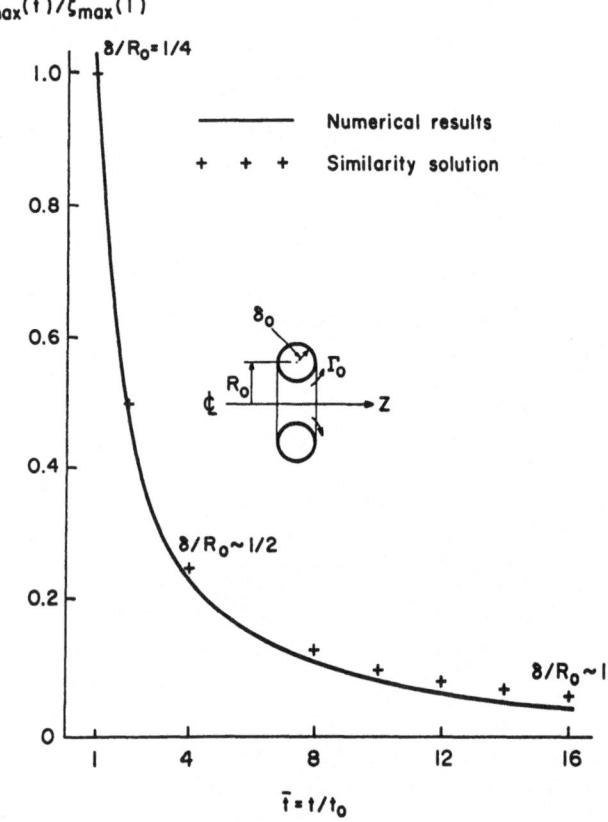

Figure 5. The decay of the maximum vorticity for the self merging of a vortex ring.

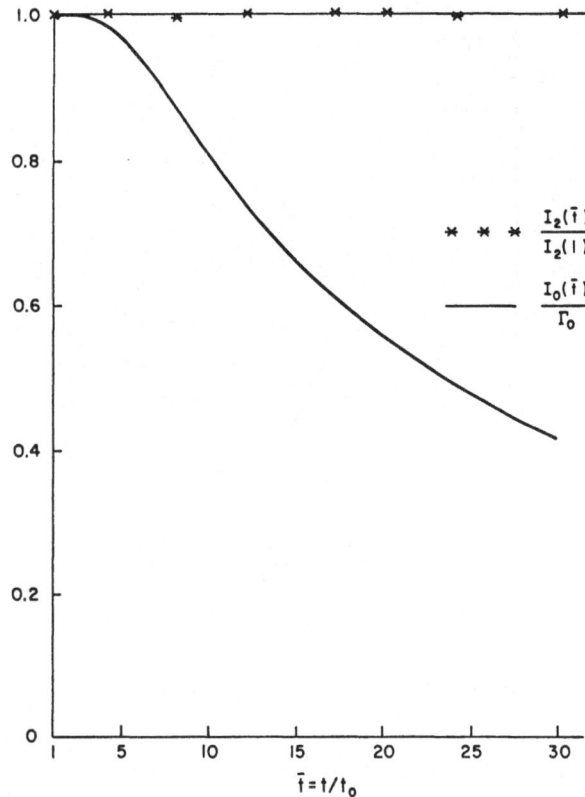

Figure 6. The variation of the total vorticity and its second moment.

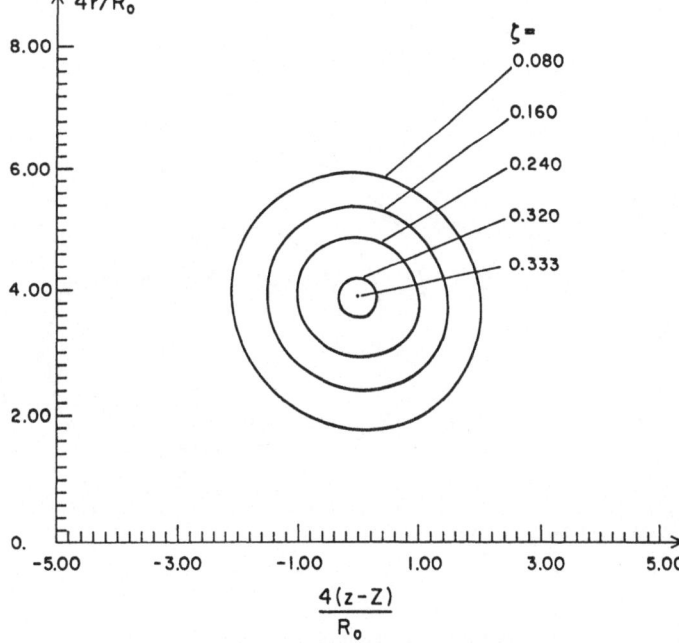

Figure 7. Self merging of a vortex ring at $t/t_0 = 2.92$.

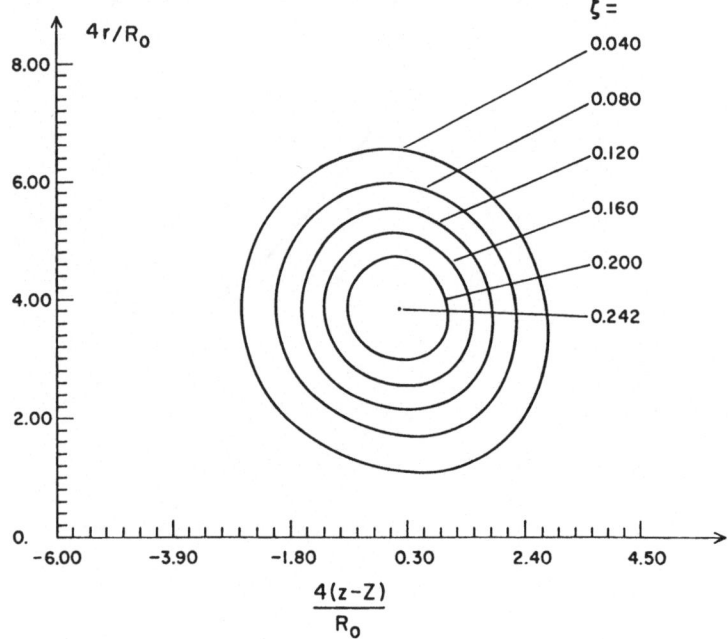

Figure 8. Self merging of a vortex ring at $t/t_0 = 3.92$.

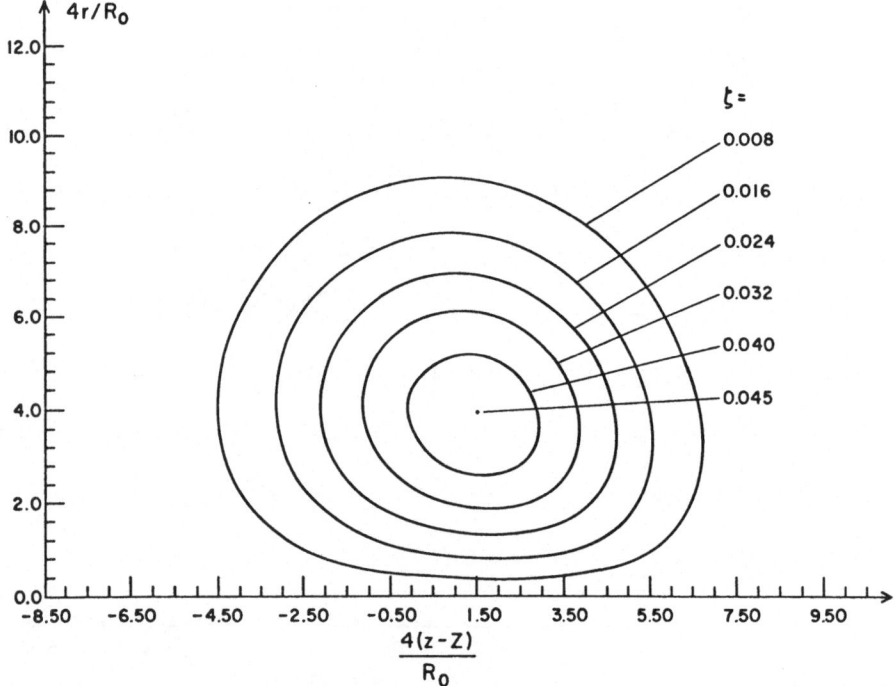

Figure 9. Self merging of a vortex ring at $t/t_0 = 16.25$.

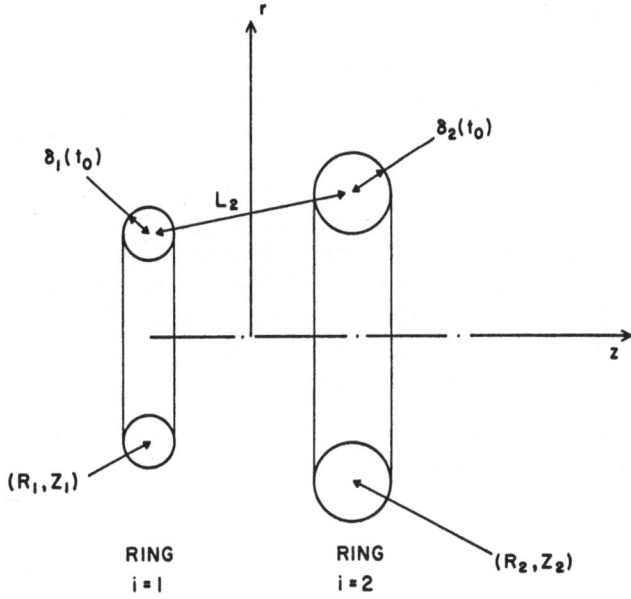

Figure 10. Initial geometry of two coaxial vortex rings.

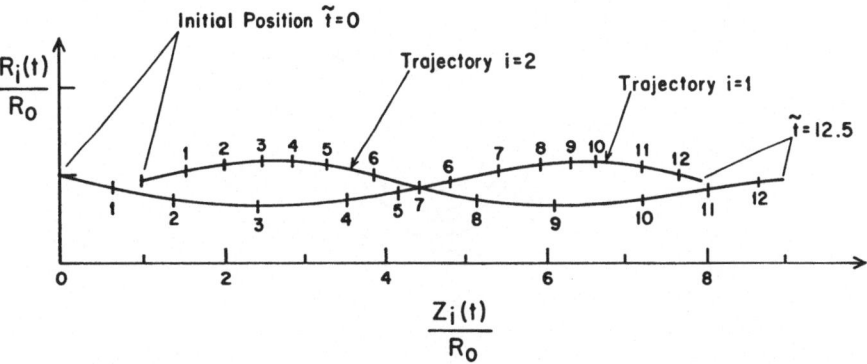

Figure 11. Trajectory of two vortex rings with their minimum distance much larger than their vortical core sizes.

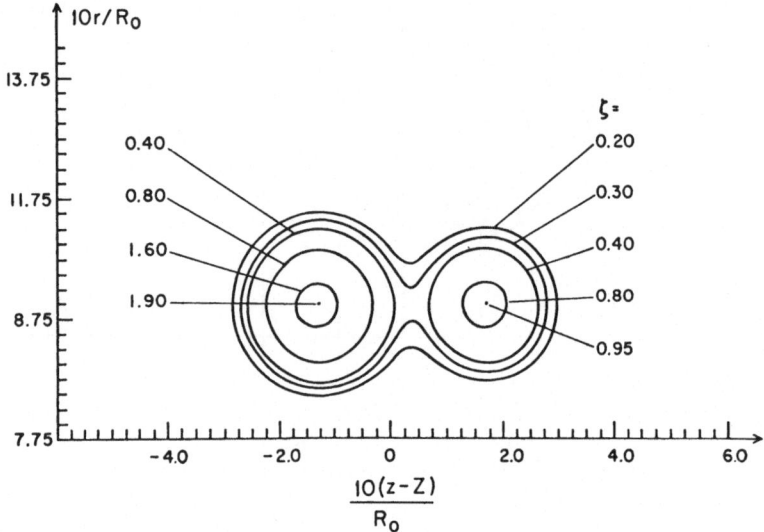

Figure 12. Merging of two vortex rings at $t/t_0 = 1.05$.

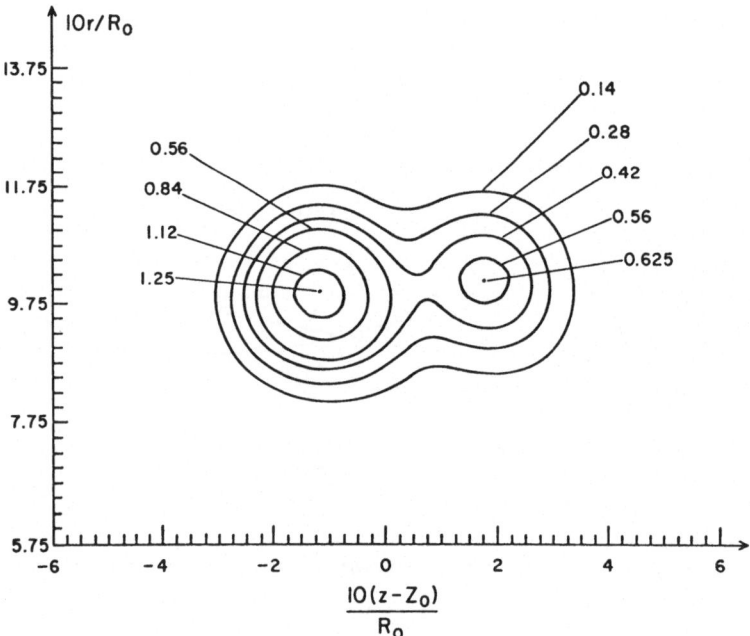

Figure 13. Merging of two vortex rings at $t/t_0 = 1.55$.

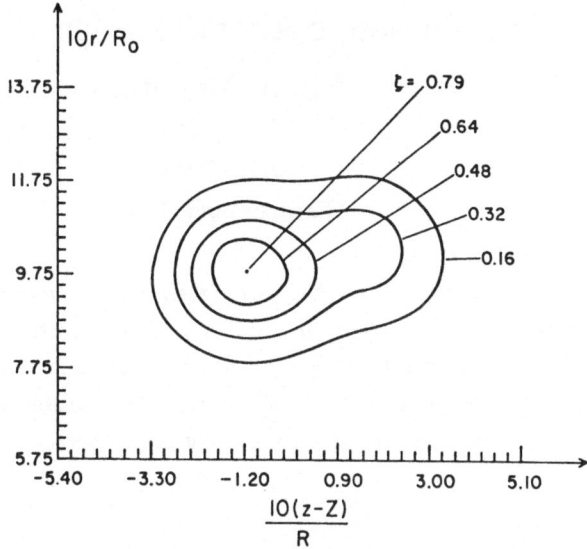

Figure 14. Merging of two vortex rings at $t/t_o = 2.45$.

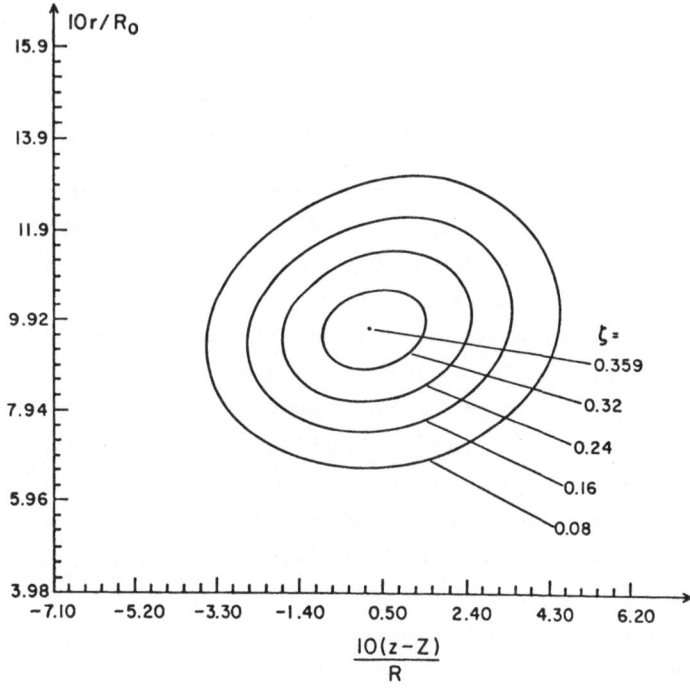

Figure 15. Merging of two vortex rings at $t/t_o = 5.95$.

The Contribution to Aircraft Design of Research in Fluid Dynamics

J. E. Green

Farnborough Hants

The paper is concerned with the essential part played by fluid dynamics research in the progress of aircraft design. Recent research in the UK on three particular topics - vortex flows, boundary layers, viscous effects on wings - is reviewed to illustrate the theme. The paper concludes with a brief discussion of future directions in computational and experimental fluid dynamics and of the facilities needed to ensure continued progress.

1. Introduction

Ladies and Gentlemen, although the title of my paper is "The Contribution to Aircraft Design of Research in Fluid Dynamics", I should like to begin with the birds. Fig. 1 shows falco tinnunxulus, der Turmfalke in this country, the kestrel in England. It is an impressive flying machine, aerodynamically efficient and well adapted to its operational role. It has a good high speed dash, high manoeuvrability, capacity for sustained hover and it can carry an appreciable payload. What is more it looks good. Conscious research in fluid dynamics made no contribution to its development, but then the development phase was rather long - about 60 million years!

The lesson we should draw from the birds is that, although man has long dreamt of emulating their flight, it was not until he applied his intellect to the underlying problems in fluid dynamics that he made progress. Fig. 2, taken from the remarkable three-part paper entitled "On Aerial Navigation" which was published by Sir George Cayley [1] in 1809-1810, illustrates an early stage in the evolution of our understanding. The upper sketch shows Cayley's explanation of gliding bird flight, with the wing inclined at a small angle of attack to the airstream and both lift and drag forces indicated. The lower sketch shows schematically his concept of how these principles might be applied to a payload-carrying machine. Cayley's paper includes a description of a large glider, embodying these principles, which he built and flew successfully. It was in the form of a classical aeroplane, with wings to generate lift and a fuselage which carried horizontal and vertical tail surfaces to stabilise and trim the machine. In describing his experiments with this glider, Cayley says, "It was very beautiful to see this noble white b i r d sail majestically from the top of the hill to any given point on the plain below it, according to the set of its rudder, merely by its own weight...". Many here today will have experienced a similar response to the sight of a flying machine. In his next sentence Cayley says, "Every man acquainted with experiments upon a large scale well knows how leisurely fact follows theory...", a comment with which those of us who operate large wind tunnels will agree, I am sure.

Since the time of Cayley the progress that has been made in aerodynamic design has depended on an improved understanding of the airflow in both mathematical and physical terms. Theory and experiment have of necessity advanced hand-in-hand, and deeper understanding of the s t r u c t u r e of the flow has been at the heart of the most important steps forward. An outstanding example here is Prandtl's formulation [2] in 1904 of the concept of the boundary layer, which provided us with a model of the structure of the flow about aircraft which is still the key to aerodynamic design in 1980.

Today, I shall review some more recent research in fluid dynamics which has contributed, or may in the future contribute, to the evolution of the flying machine. My purpose is to stress the value of, and continuing need for, research to make clear the structure of the flow. Time does not permit me to be comprehensive so I shall confine myself to three topics which illustrate the nature of the problems still to be solved. These three topics are:

- vortex flows
- boundary layers
- viscous effects on wings.

To limit my scope further, I shall illustrate my talk primarily with examples drawn from recent work in Britain. I shall conclude by looking briefly to the future - the problems to be solved and the facilities needed to do the job.

2. Vortex Flows

The first topic I have chosen is vortex flows. Although the relationship between bound vorticity and lift has long been understood, the concept of using the suction generated by strong streamwise vortices as a means of producing lift is relatively recent and marked a new departure in aerodynamic design philosophy.

Fig. 3, due to Maltby, illustrates schematically the complex form of the vortex sheet generated by flow separation from the sharp leading edges of a slender wing. The vortices are of sufficient strength to produce high levels of suction on the upper surface of the wing and hence to generate a substantial amount of non-linear lift. Despite its complexity, the flow pattern was found by those who first investigated it to be stable and to persist over a wide range of angle of attack. This benign aerodynamic behaviour gave to a slender winged aircraft very stable and predictable handling characteristics in low speed flight. Thus it was demonstrated that a slender-winged aircraft, designed primarily for cruise at supersonic speeds, could, by means of the high vortex lift achieved at low speed, attain an acceptable airfield performance without the need for variable sweep of the wings or for complex high lift devices [3] . As a result (Fig. 4) Concorde was born - an aircraft of revolutionary design, whose beginnings lay in the under-standing of the structure of the flow about a simple, idealised but novel form of lifting surface.

Research in the UK on the aerodynamics of slencer wings had been inspired chiefly by Dietrich Küchemann and it was he who, in 1971, advanced the concept illustrated in Fig. 5 of a slender wing combined with a swept wing. His aim was to overcome the deficiency in lift on the centre-line of the classical swept wing, by creating over the centre section of the wing a strong vortex lift produced by the highly swept leading edges of the centre section. In making this proposal, Küchemann foreshadowed the development of the strake - although the latter actually emerged [4, 5] from experimental work, in other countries, aimed at a slightly different objective. Fig. 6 illustrates one of the consequences of the strake which had not been foreseen by Küchemann when he made his proposal. It is found [6] that, at a moderate angle of attack, the addition of a strake can induce separation of the flow over a wing on which, in the absence of a strake, the flow is fully attached. The lower part of the figure illustrates the surface oil flow on the wing upper surface, without the strake, at an angle of attack of 14 degrees. It is clear that the flow is fully attached. The upper photograph shows that the addition of a strake to the wing, at the same angle of attack, generates an inboard region of flow dominated by the strake vortex, outboard of which flow separation occurs at the leading edge. This type of flow has been studied theoretically by Fiddes and Smith [6] and they have provided, in Fig. 7, an explanation of the behaviour. Fig. 7 shows the variation across the span of a straked wing of the upwash and sidewash induced at the wing leading edge by the strake vortex. As can be seen, the wing outboard of the vortex has an increased effective angle of attack (greater than its geometric value) due to the upwash. There is also a large sidewash velocity near the kink in the planform, under the vortex, so that the direction of flow here is more along the leading edge, i. e. the wing is at an effectively higher sweep. This combination of increased effective sweep and increased effective angle of attack appears to be sufficient to take the wing from a condition at which attached flow can be sustained over its outboard part to one in which the flow is caused to separate at the leading edge.

However, strakes are considered primarily as a means of extending the range of angle of attack over which a wing will operate satisfactorily [7] . That is, their application is to extending the lifting range of a wing on which, in the absence of a strake, the flow would be stalled. Fig. 8 illustrates the effect of strakes on a wing of low aspect ratio, fitted with trailing-edge high-lift devices. The surface oil flow pattern at lower left is for the wing without strake, at 22 degrees angle of attack, and illustrates the extensive region of separation over the main wing surface. The surface oil flow on the right-hand side shows that, with a strake fitted, separation is still present on the outboard part of the wing, but inboard, beneath the vortex from the strake, is a region of well-behaved and essentially attached flow. In the upper part of the figure we see how the curve of lift against incidence is extended substantially by the addition of the strake to the wing. On the left-hand side we see that, whether the flaps are set at 0 degrees or at 40 degrees, the addition of a strake extends the incidence range over which lift continues to increase by approximately 12-15 degrees and produces a significant increase in the maximum lift coefficient. The right-hand curves show however that similar benefits can be obtained by fitting the wing with a leading-edge slat instead of a strake.The flow mechanisms by which these two results are produced are quite different but the end result is broadly similar, and we must obviously understand the consequences of these differences between the two types of flow if we are fully to assess the relative merits of two such configurations.

Fig. 9 shows a Harrier aircraft which has been used in a flight research programme to demonstrate how the addition of a strake to an existing aircraft can extend the incidence range over which the aircraft handling characteristics are acceptable. The flight trials were preceded by a wind-tunnel programme to develop a suitable form of strake, and some results from this programme are shown in Fig. 10. The upper curves illustrate how, at Mach numbers of 0.6 and 0.8, the lift curve is extended significantly and the kink associated with stalling of the wing is removed by the addition of strakes. The middle curves show how the addition of strakes eliminates the onset of large rolling moments associated with asymmetric stall on the unstraked wing, while the lower curves illustrate the appreciable reduction strakes provide in the unsteady wing root bending moment - buffet excitation in effect. Flight trials have confirmed the benefits predicted from these wind tunnel tests, and we can report strong pilot approval for this particular example of applied fluid dynamics.

A factor which simplifies theoretical analysis of the flowfield about aircraft with slender wings or with strakes is the existence of a sharp leading edge which fixes the separation line from which the vortex sheet develops. A more difficult theoretical problem, but one with appreciable practical significance, is the separation of vortices from smooth bodies. This problem, illustrated in Fig. 11 for the case of vortices separating from a conical nose at incidence, is one on which we have been working recently at RAE. An analysis by Smith [8] has shown that, for inviscid flow, the vortex sheet separating from a smooth surface must take the form given by the equation in Fig. 11. Fiddes [9] has now combined: this local analysis by Smith; a vortex sheet model of the inviscid flow; a laminar boundary layer calculation; and a triple-deck model [10] of the viscous-inviscid interation in the vicinity of separation. He has thus produced a rational treatment of the separation of a laminar boundary layer from a cone to form a symmetrical pair of vortices, taking account of the interaction between the external flow field and the development of the boundary layer. Fig. 12 gives the results of Fiddes-analysis for a cone set at an angle of incidence which is three times the cone semi-angle. The predicted variation of separation position with Reynolds number is indicated as a solid line and the trend is shown to be in good agreement with experiment [11]. Also shown are predictions of separation position obtained by neglecting the vortex sheet [12] (the upper, chain dotted line), and by incorporating the vortex sheet but neglecting the finite Reynolds number effects introduced by the triple deck model [9] (smooth separation). This analysis markes only the beginning, for the same approach is about to be applied to flows with asymmetric separations and could be extended to shapes other than conical ones. Flows with turbulent boundary layers could be similarly treated, by incorporating an appropriate multi-deck model for the turbulent separation, and in principle secondary vortex separations could be included. The approach, therefore, offers the potential for predicting the substantial side forces and yawing moments which can develop, as a result of flow asymmetry, on slender-nosed aircraft at high angles of attack. We can envisage a broadly similar approach being applied to more general problems of three-dimensional separation, and it is likely to become an essential element of future numerical treatments of complex separated flows, other than those which rely on solution of the Navier-Stokes equation over the entire flow field.

3. Boundary Layers

My second topic is boundary layers. Prandtl [2] proposed the concept of the boundary layer in 1904, and since that time boundary layers have provided research workers with a seemingly limitless supply of questions which must be answered. The past 15 years has been a particularly active period in turbulent boundary layer research. During this time it has become generally recognised that the turbulence structure is determined by what has happened upstream rather than by the local mean velocity profile, and a number of boundary layer prediction methods have been developed (see e. g. Ref. 13) in which one or more turbulence transport equations are used to evaluate the Reynolds stress terms in the equations of motion. The period has also seen several noteworthy, even beautiful, experimental investigations of eddy structure within the boundary layer, and work of this nature will undoubtedly continue [14] . Work on the important process of transition has progressed slowly and there are some who might say that, even today, our understanding does not go far beyond an appreciation that the phenomenon is one of extraordinary complexity and subtlety [15].

In the early part of this period, the emphasis in RAE work was in obtaining data in supersonic flows at high Reynolds number, in response to the need for accurate friction drag predictions for Concorde [16]. This work included data on excrescence drag [17]. Later we became concerned with prediction methods for turbulent boundary layers in two-dimensional and three-dimensional compressible flows, for application particularly in wing design [18-20] . In the later of these methods we followed Bradshaw [21] , first by use of a turbulence transport equation, second by incorporating allowances for the so-called secondary influences on the turbulence structure - for example, longitudinal curvature, flow convergence or divergence, streamwise density gradients (dilatation) - which appear to be of more than secondary importance in many flows of practical importance. With emphasis in wing design methods being on speed of computation, we have concentrated on an integral boundary-layer method which has been developed to apply to boundary layers and wakes in both two-dimensional and three-dimensional flows. The three-dimensional method [19] works in a general, flexible coordinate system, but relies on a simple modelling of crossflow velocity profiles. This method has performed well in tests [22] against experiment and against other prediction methods, including finite difference methods, and is in general use in our wing design programmes. A three-dimensional finite difference method has also been developed, working in the same general coordinate system as the integral method, and the assessment of alternative turbulence models will be one of its future uses.

Experimental work on turbulent boundary layers over this period has covered a wide range of problems, with emphasis in RAE being on validating or improving the boundary-layer prediction methods used in transonic wing design. There have been studies of the boundary layers and wakes on large scale aerofoils and swept wings [23-25] and one particular feature of such flows, the interaction between a normal shock wave and a turbulent boundary layer, has been studied at large scale and in some detail[26] . Fig. 13 shows the experimental set up used in this study. The experiments were done in the RAE 3ft x 3ft supersonic tunnel, using laser anemometry, as well as conventional pitot rakes, to survey the flow. An example is given in Fig. 14 of visualisation of the shock/boundary-layer interaction by schlieren photography and, beneath it, a mapping of the same flow field as given by laser anemometry. The Mach number ahead of the shock is 1.4. Among the details brought out by this mapping are the small tongue of sonic flow downstream of the foot of the shock, and the strength of the vortex sheet or shear layer, visible in the schlieren photograph, which emerges from the bifurcation in the shock. Comparisons between the flow field measured in this experiment and computations by current Navier-Stokes codes are expected to provide a useful test of the turbulence models used in these codes.

Another of our recent experimental investigations [27] has been into equilibrium turbulent boundary layers. This investigation has been carried out in the 1.2 m x 0.3 m boundary-layer tunnel at RAE Bedford. Figs. 15 and 16 illustrate the seven different equilibrium flows which were studied, with pressure gradients ranging from favourable (Flow 1) to strongly adverse (Flow 7). Fig. 15 shows the linear growth of momentum thickness characteristic of equilibrium flows, the growth rate in the severest adverse pressure gradient being some 30 times that in the most favourable gradient: in Fig. 16 the shapes of the equilibrium velocity profiles are also shown - the profiles are non-dimensionalised with respect to streamwise distance, and again the pronounced difference in rate of growth can be seen. The relationship between profile shape and pressure gradient found in these experiments is in good agreement with the results of previous work and with the 'equilibrium locus' incorporated in our integral prediction method [20]. Shearstress measurements brought to light other features of these flows which have a significant bearing on the turbulence modelling for current and future boundary-layer or Navier-Stokes codes: the main conclusions we have drawn from this study are as follows:

(1) the law of the wall for the velocity profile holds good, even in the stongest pressure gradients;

(2) the simple mixing length formulation for shear stress d o e s   n o t   hold good in all pressure gradients;

(3) the simple dissipation length formulation for turbulent energy dissipation d o e s hold good in all pressure gradients;

(4) the rates of lateral diffusion of turbulent kinetic energy and turbulent shear stress in equilibrium flows are proportional to the gradient of the diffused quantity rather than to its magnitude.

One recent development to our integral boundary-layer method, illustrated in Fig. 17, has been the extension [28] of the concept of equilibrium turbulent boundary layers to embrace separated flows, analogous to the similarity solutions with reversed flow that have been found for laminar boundary layers. Fig. 17 shows a hypothetical equilibrium locus of the pressure gradient parameter m and shape parameter J. The separate curves are for different values of

Reynolds number. The upper part of the graph applies to separated flows and has been derived by extrapolation of data for attached boundary layers. The top right hand corner of the graph (m = o, J = 1.0) corresponds to the free shear layer over a semi-infinite cavity. We believe the concept is a valid one, useful in the development of engineering prediction methods. So far we have not succeeded in setting up wholly satisfactory equilibrium separated flow in our boundary layer tunnel, but hope in due course to do so. An integral boundary-layer method incorporating the locus of Fig. 17 has been programmed in an inverse mode, using the streamwise distribution of displacement thickness rather than pressure as boundary condition, and tested successfully against a range of two-dimensional flows with regions of separation [28]. In Fig. 18a + b the predictions of this method are compared with results from the experiment on shock and boundary layer interaction, which was shown in Fig. 14, for an initial Mach number of 1.4. Fig. 18a compares the predicted and measured Mach number distributions, when displacement thickness is the input to the calculation. Fig. 18b compares predicted and measured distributions of momentum thickness, shape parameter and skin friction. The general agreement with experiment is encouraging and invites incorporation of these ideas into aerodynamic design methods.

Finally, to illustrate the problems we face in modelling three-dimensional boundary layers, Fig. 19 shows some measurements [29], made in our boundary-layer tunnel, of the flow approaching separation ahead of a forward facing step which was swept at 45° to the stream. The curves above the axis show how the velocity vector at various constant heights in the boundary layer is progressively deflected away from the step (ß positive) as the step is approached, the limiting streamline at the surface (the uppermost curve) being turned parallel to the step (ß = 45°), to become the separation line, at a distance from the step equal approximately to one step height. There is greater interest however in the direction of the velocity-gradient and shear-stress vectors. Both of these are deflected increasingly towards the step (ß negative) but by markedly different amounts. The shear stress lags appreciably behind the velocity gradient - i. e. the shear, which is the lowest curve - such that their directions differ by approximately 20° by the time separation is reached. Eddy viscosity models cannot represent this behaviour and we see that it is necessary to uncouple the shear stress and the shear, not only in magnitude (which we know already that we have to do for two-dimensional boundary layers) but in direction also.

4. Viscous Effects on Wings

My third topic I have called 'viscous effects on wings'.

An important advance in the past decade has been the development of wing design methods in which proper account is taken of viscous effects - i. e. of the influence of boundary layer and wake growth on the pressure distribution over the wing. The earliest work at RAE on this topic [30] involved coupling together, by an iterative procedure, a method for two-dimensional, subcritical potential flow [31] with an integral method for predicting the aerofoil boundary layer [18] . Since then there have been, inter alia, the following developments:

the subcritical potential flow method has been replaced by the transonic small perturbation method [32] , and later by the method of Garabedian and Korn [33] for solving the full potential equation;

the treatment of the viscous flow has been extended to include the wake, and the original entrainment method for the boundary layer has been replaced by the lag entrainment method [20] ;

curvature of the boundary layers and wake has been included in modelling the effect of the viscous layers on the inviscid flow [34] ;

a method has been developed for the treatment of multiple aerofoils [35] ;

three-dimensional counterparts for the inviscid [36] and boundary layer [19] methods have been combined to give prediction methods for wings and wing-body combinations at transonic speeds, including allowance for aeroelastic effects.

Fig. 20 shows some calculations by Collyer [34] which illustrate the significance of viscous effects on a modern aerofoil at transonic conditions. They show the changes in pressure distribution from inviscid flow (the outermost, chain dotted line) as the modelling of viscous effects is progressively refined. The first viscous approximation (the innermost, dotted line) models only the displacement thickness of the boundary layer on the aerofoil surface: the next (the broken line) also includes the displacement effect of the wake: finally in the 'fully viscous' calculation (the solid line) account has been taken of the effects of the curvature of both aerofoil surface and wake.

Table 1 below gives the corresponding changes calculated for lift and trailing edge pressure, the bottom line giving the percentage changes in C_L with each successive refinement of the model. It is easy to see why, for modern aerofoils, viscous effects need to be represented correctly in our design methods.

	Inviscid	δ^*on aerofoil surface only	δ^* on aerofoil and wake	Full curvature effects included
C_p at t.e.	0.364	0.261	0.211	0.219
C_L	0.922	0.663	0.721	0.699
Change in C_L from previous solution	–	-28%	+9%	-3%

Fig. 21 illustrates the treatment of three-dimensional transonic problems. This compares the results of calculations by Firmin [37] with measurements on a wing suitable for a high-speed long-range transport aircraft. The wing was mounted in a low position on an axisymmetrical fuselage. The calculations were done using the transonic small perturbation method with a full three-dimensional method for the wing boundary layers and wake. Agreement with experiment is encouraging, particularly over the inner wing and in the vicinity of the planform kink.

Fig. 22 shows a comparison between experiment and calculations by Butter and Williams [35] (using a method with the acronym MAVIS) for a three element aerofoil in low-speed flow. The agreement is generally good, except for the local suction peak on the upper fore part of the flap. For the pressure distribution over the flap the differences are shown between an inviscid calculation (the upper, broken line), a calculation allowing only for the displacement effects of boundary layer and wake (the intermediate, dotted line), and a calculation including also the effects of wake curvature (the solid line). Again, the need to represent viscous effects correctly in our design methods is apparent.

Although the advances I have described represent a great step forward in our aerodynamic design capability, I could give many examples from recent experience to illustrate complex viscous effects which are beyond the scope of our present prediction methods. As illustrated in Fig. 23, the influence of Reynolds number on separated flow over swept wings at transonic speeds is particularly difficult to forecast. It involves highly three-dimensional flows, with the possiblity of interaction between trailing edge separation and shock induced separation, and the probability that the relative significance of these two types of separation will change as Reynolds number varies. This figure, which is a sketch of surface oil flow patterns at two Reynolds numbers on a research wing tested by Weeks [38] in the RAE 8ft x 8ft tunnel, illustrates this point. In the upper diagram, corresponding to the lower Reynolds number, the flow over the outer part of the wing is well separated. In the lower diagram, however, note the appearance at the higher Reynolds number of a narrow region of attached flow near the wing tip and, beyond it, a small region of shock-induced separation.

Fig. 24 gives us another indication of the complex interaction between viscous and compressiblity effects, this time the results from recent tests by Woodward [39] on a high-lift wing in the RAE 5m pressurized low-speed tunnel. Maximum lift coefficient is plotted against Reynolds number at constant Mach number, and against Mach number at constant Reynolds number (the horizontal scale of the graph is a composite one which includes both Mach number and Reynolds number). Increasing Reynolds number leads to a significant increase in maximum lift whilst increasing Mach number has the opposite effect. The adverse effect of increasing Mach number is apparent even from a free stream Mach number as low as 0.1. The two dotted lines in this figure are of interest in showing the change in maximum lift, at two different levels of tunnel pressure, as wind speed is increased (thereby increasing Reynolds number and Mach number simultaneously). Variation of wind speed has sometimes in the past been used as a means of assessing the effect of Reynolds number on high lift wings. It is clear from this graph that such results are likely to include adverse Mach number effects and in consequence will be misleading.

5. Future Directions

I should like now to look briefly to the future. And in considering the directions our work might take, I should like to highlight two topics of particular importance: computational fluid dynamics, and new experimental facilities and techniques.

The first of these, computational fluid dynamics, has made dramatic progress over the past decade, and we may expect developments over the next 10 years to be at least as impressive. Greatly reduced costs of computer hardware, major advances in computer architecture and development of improved algorithms all contribute to this progress. Chapman [40] has recently reviewed the field, describing most of the noteworthy achievements of recent years and providing an admirable assessment of future potential, to which little can be added. Table 2, which is taken from Chapman's paper, summarises past and future developments in terms of the type of equations to be solved, the complexity of configuration for which a solution is sought and the class of computer needed for the job. However, it omits one particular class of method which is now a key element in aerodynamic design procedures in several countries. This is the type of method, for which I showed some results in earlier figures, which is based on a combination of a non-linear inviscid method with a thin-shear-layer treatment of the viscous flow. This class of method is likely in the foreseeable future to remain more economical of computer time than one which solves the Reynolds-averaged Navier-Stokes equations and therefore, for a given power of computer, will be capable of application to more complicated geometries than can be treated by the Navier-Stokes codes.

Considering the likely developments of these two types of method - i. e. the viscous-inviscid interaction method and the Navier-Stokes code - we can see that, in addition to the development of more efficient algorithms, improved turbulence modelling is needed for both types of method. This requires conceptual advances in our underständing of turbulence, and further experiments to verify or negate the concepts we are already using. In addition, the viscous-inviscid interaction methods, which rely on a thin-shear-layer representation of the viscous flow, will require further advances in modelling the overall flow structure, particularly in dealing with regions of separated flow which have a strongly three-dimensional character. The likelihood that the global flow features of three-dimensional separation will prove exceedingly difficult to model explicitly in any very general way is one of the arguments for placing reliance on Reynolds-averaged Navier-Stokes codes. I am sure, however, that both types of method are needed, and I have some personal sympathy with methods involving thin-shear-layer models because they are likely to call for greater effort on the part of the research worker to understand the global structure of the flow. With solutions of the Navier-Stokes equations, there is perhaps a temptation to allow computation to become a substitute for thought, to accept computer output in place of physical insight. It can easily happen, but will be avoided by those who make understanding the fluid dynamics their first priority. Before leaving this subject it is worth adding the comment that, as our theoretical methods become more complex, so it becomes increasingly difficult to validate them against experiment. That is to say, it is often much easier for the theoretician to introduce new unknowns into his flow model than it is for the experimenter to measure these unknowns accurately. Nevertheless experiments, carefully performed, must always be the final test of the truthfulness of our models - particularly when turbulence is involved.

This thought brings me to my second important topic for the future, experimental facilities and techniques. Experience over the past 10-15 years has shown the need [41] for a major investment in certain new types of aerodynamic test facility, one of which is typified by the RAE 5m wind tunnel [42] , seen from the air in Fig. 25. The 5m, which came into operation in 1978, is a pressurized low speed tunnel intended primarily for research and development work on the low speed aerodynamics of modern aircraft and their complex high-lift systems. The pressure range of the tunnel, from 1-3 atmospheres, enables the performance envelope shown in Fig. 26 to be achieved. Thus tests can be performed over a range of Reynolds number at a constant Mach number or over a range of Mach number at constant Reynolds number, to allow discrimination between scale and compressibility effects as shown in Fig. 24 previously. A model of the A300B Airbus mounted in the tunnel is shown in Fig. 27. At the highest tunnel Reynolds numbers, which are approximately quarter full scale for this model, we have found good agreement with flight test results, particularly in the way in which the stall pattern develops over the wing. However, testing over the full range of tunnel pressure showed the same kind of interaction between scale and compressibility effects which has been found on the research model referred to in Fig. 24. These results have emphasised the importance of doing low speed tests of high-lift configurations at the c o r r e c t M a c h n u m b e r , and have shown that misleading results can be obtained if Reynolds numbers are not sufficiently high.They have confirmed that we were right in our decision to build the tunnel and indicate that, besides using the tunnel for development testing in support of specific aircraft projects, we must also use it to study the fluid dynamics of scale and compressibility effects at a more basic level.

Although certain kinds of fluid dynamics research can be done only in a major wind tunnel such as the 5m, we must be thankful that a great deal of basic work is possible in smaller and considerably cheaper facilities. Fig. 28 shows the 1.2m x 0.3m boundary-layer tunnel which we built at Bedford at the same time as the 5m tunnel was under construction at Farnborough. It has proved very useful for studies of boundary-layer behaviour and structure - in previous sections I have cited two examples of the experimental work in the tunnel - and is the type of facility which is well suited to a university research laboratory.

The great advances that have been made [43] in techniques for flow measurement, notably in laser anemometry, and the similarly impressive developments there have been in the technology of signal processing and data analywis, open the way for powerful investigations of flow structure in simple facilities of this kind. In providing data needed to develop or refine the turbulence and shear layer modelling employed in flow-field computation methods, such fundamental experimental studies have a crucial role to play in the advancement of aerodynamic design. There is one proviso however - we must not allow ourselves to become mesmerised by the wonders of modern instrumentation to such a degree that we neglect the fluid dynamics. Data gathering and analysis, like computation, is no proper substitute for thought.

My final figure, 29, really is a look into the future -an artist's impression of ETW, the European Transonic Wind Tunnel [44] . This is a project on which the Federal Republic of Germany, France, the Netherlands and the United Kingdom are working together. A preliminary design study of the project, under a four-nation Technical Group based in Amsterdam, is now virtually complete and the ground is being prepared for agreements to cover the final design an construction of the facility. ETW will be a pressurised, fan-driven tunnel, using nitrogen at cryogenic temperature as its working fluid and enabling testing at full-scale Reynolds numbers over a stubstantial part of the subsonic and transonic flight envelopes of both civil and military aircraft. To facilitate model interchange, the tunnel circuit will be above ground level - models will be lowered vertically from the test section to a rigging area below. The circuit will be housed in the thermally insulated concrete case seen in the left of the picture: the tall chimney stack is the nitrogen exhaust. This tunnel will complement the major low-speed tunnels which have recently come into operation, 5m, F1 Fauga [45] and DNW [46] , and will be the largest and probably most important single investment in aerodynamic testing to have been made in Europe. Although its primary role is expected to be in support of the design and development of new projects, research at high Reynolds numbers will also be an important aspect of its work. Such research, linked closely with the development of flow-field computation methods, will provide the foundation for the substantial advances in transonic aerodynamic design that undoubtedly lie ahead.

Important though ETW is to Europe, I wish to end by talking about people rather than about grandiose test facilities. I have already noted the important work that there is to be done in laboratory facilities of a more modest scale, and I have stressed the need for understanding, for creative thought, and for theory and experiment to advance hand-in-hand. In the last analysis, our large wind tunnels and our super computers will be of limited value unless the individuals who use them are of the right calibre, capable of, and interested in, understanding the structure of the flows which they seek to manipulate. It is for this reason that the opening of the new laboratories of the Institute here in Aachen is an important occasion. The Laboratories provide not only facilities for some of the basic research essential to our continued progress but, more important, they are a training ground for the aerodynamicists and fluid dynamicists of tomorrow. Since the time of Cayley, fluid dynamics research has attracted workers of high calibre, some of them scientists of the greatest distinction, and their discoveries have made possible the amazing achievements of aviation in the present century. The new laboratories of the AIA have their part to play in the continuation of this tradition.

Table 2

stage	computed results	when results first obtained			computer required for practical 3-D calculations
		2-D airfoil body rev	simple 3-D b rev at α wing	practical 3-D wing body	
I linearised inviscid	pressure dist vortex drag sup wave drag	1930	1940s	1968	medium power 1970s computers
II nonlinear inviscid	above, plus: transonic flow supersonic flow hypersonic flow	1971	1973	1976	current super-computers 50 x above
III Navier-Stokes Re-averaged model all scales of turbulence	above, plus: separated flow total drag performance buffeting, buzz	1975	1978	early 1980s	40 x current super-computers (NASF)
IV large eddy simulation model subgrid-scale turbulence	above, plus: aerodynamic noise transition surface pressure fluctuations	early 1980s	mid 1980s	1990s	at least 100 x NASF

6. References

[1] G i b b s - S m i t h, C.H.: Sir George Cayley's Aeronautics 1796-1855. HMSO London (1962).

[2] P r a n d t l, L.: Über Flüssigkeitsbewegung bei sehr kleiner Reibung. Proceedings III International Math. Congress, Heidelberg (1904), pp. 484-491.

[3] K ü c h e m a n n, D.: The aerodynamic design of aircraft. Pergamon (1978).

[4] B u c k n e r, J.K., H i l l, P.W.: Aerodynamic design evolution of the YF-16. AIAA Paper 74-935 (1974).

[5] P a t i e r n o, J.: YF-17 design concepts. AIAA Paper 74-936 (1974).

[6] F i d d e s, S.P., S m i t h, J.H.B.: Strake-induced separation from the leading edges of wings of moderate sweep. AGARD-CP-247, Paper 7 (1979).

[7] M o s s, G.F.: Some UK research studies of the use of wing body strakes on combat aircraft configurations at high angles of attack. AGARD-CP-247, Paper 4 (1979).

[8] S m i t h, J.H.B.: Behaviour of a vortex sheet separating from a smooth surface. RAE Technical Report 77058 (1977).

[9] F i d d e s, S.P.: A theory of the separated flow past a slender elliptic cone at incidence. Paper 30, AGARD FDP Symposium on Computation of Viscous-Inviscid Interactions, Colorado Springs, 29 September to 1 October 1980.

[10] S m i t h, F.T.: Three-dimensional viscous and inviscid separation of a vortex sheet from a smooth non-slender body. RAE Technical Report 7805 (1978).

[11] R a i n b i r d, W.J., C r a b b e, R.S., J e r e w i c z, L.S.: A water tunnel investigation of the flow separation about circular cones at incidence. NAE (Canada) Aeronautical Report LR-385 (1963).

[12] C o o k e, J.C.: The laminar boundary layer on an inclined cone. RAE Technical Report 65178 (1965).

[13] K l i n e, S.J. et al.: Computation of turbulent boundary layers. AFOSR-IFP-Stanford Conference Proceedings (1968).

[14] K l i n e, S.J. et al.: Turbulent boundary layers - experiments, theory and modelling. AGARD FDP-CP-271 (1980).

[15] K l i n e, S.J. et al.: Laminar-turbulent transition. AGARD FDP-CP-224 (1977).

[16] W i n t e r, K.G., G a u d e t, L.: Turbulent boundary layer studies at high Reynolds numbers at Mach numbers between 0.2 and 2.8. ARC R&M 3712 (1970).

[17] G a u d e t, L., W i n t e r, K.G.: Measurements of the drag of some characteristic aircraft excrescences immersed in turbulent boundary layers. AGARD-CP-124, Paper 4 (1973).

[18] G r e e n, J.E.: Application of Head's entrainment method to the prediction of turbulent boundary layers and wakes in compressible flows. ARC R&M 3788 (1972).

[19] S m i t h, P.D.: An integral prediction method for three-dimensional compressible turbulent boundary layers. ARC R&M 3739 (1972).

[20] G r e e n, J.E., W e e k s, D.J., B r o o m a n, J.W.F.: Prediction of turbulent boundary layers and wakes in compressible flow by a lag-entrainment method. ARC R&M 3791 (1973).

[21] B r a d s h a w, P., F e r r i s s, D.H., A t t w e l l, N.P.: Calculation of turbulent boundary layer development using the turbulent energy equation. J. Fluid Mech., _28_, (1967), pp. 593-616.

[22] E a s t, L.F.: Computation of three-dimensional turbulent boundary layers (Euromech 60, Trondheim 1975). FFA TN AE-1211.

[23] C o o k, T.A.: Measurements of the boundary layer and wake of two aerfoil sections at high Reynolds numbers and high subsonic Mach numbers. ARC R&M 3722 (1971).

[24] C o o k, P.H., M c D o n a l d, M.A., F i r m i n, M.C.P.: Aerofoil RAE 2822 - pressure distributions, and boundary layer and wake measurements. AGARD-AR-138, Paper A6 (1979).

[25] C o o k, P.H., M c D o n a l d, M.A., F i r m i n, M.C.P.: Wind tunnel measurements of the mean flow in the turbulent boundary layer and wake in the region of the trailing edge of a swept wing at subsonic speeds. RAE Technical Report 79062 (1979).

[26] E a s t, L.F.: The application of a laser anemometer to the investigation of shock-wave boundary-layer interactions. AGARD-CP-193, Paper 5 (1976).

[27] E a s t, L.F., S a w y e r, W.G.: An investigation of the structure of equilibrium turbulent boundary layers. AGARD-CP-271 (1979).

[28] E a s t, L.F., S m i t h, P.D., M e r r y m a n, P.J.: Prediction of the development of separated turbulent boundary layers by the lag-entrainment method. RAE Technical Report 77046 (1977).

[29] E a s t, L.F., S a w y e r, W.G.: Measurements of the turbulence ahead of 45° swept step using a double split-film probe. RAE Technical Report 79136 (1979).

[30] F i r m i n, M.C.P.: Calculation of the pressure distribution, lift and drag on aerofoils at subcritical conditions. Part I. Interim method. RAE Technical Report 72235 (1973).

[31] B l o c k l e y, R.H.: A second-order method for estimating the sub-critical pressure distribution on a two-dimensional aerofoil in compressible inviscid flow. ESDU TD Memo 72025 (1974).

[32] A l b o n e, C.M., C a t h e r a l l, D., H a l l, M.G., J o y c e, Gaynor: An improved numerical method for solving the transonic small perturbation equations for flow past a lifting aerofoil. RAE Technical Report 74056 (1974).

[33] G a r a b e d i a n, P.R., K o r n, D.G.: Analysis of transonic aerofoils. Comm. of Pure and Applied Maths., _24_, (1971), pp. 841-851.

[34] C o l l y e r, M.R.: An extension of the method of Garabedian and Korn for the calculation of transonic flow past an aerofoil to include the effects of a boundary layer and wake. RAE Technical Report 77194 (1977).

[35] B u t t e r, D.J., W i l l i a m s, B.R.: The development and application of a method for calculating the viscous flow about high lift aerofoils. Paper 25, AGARD FDP Symposium on Computation of Viscous-Inviscid Interactions, Colorado Springs, 29 September to 1 October 1980.

[36] A l b o n e, C.M., H a l l, M.G., J o y c e, Gaynor: Numerical solutions for transonic flows past wingbody combinations. IUTAM Symposium Transonicum II, Göttingen (1975).

[37] F i r m i n, M.C.P.: Calculations of transonic flow over wing-body combinations with an allowance for viscous effects. Paper 8, AGARD FDP Symposium on Computation of Viscous-Inviscid Interactions, Colorado Springs, 29 September to 1 October 1980.

[38] W e e k s, D.J.: RAE, Unpublished.

[39] W o o d w a r d, D.S.: RAE, Unpublished.

[40] C h a p m a n, Dean R.: Computational aerodynamics development and outlook. AIAA Journal, 17, (1979), pp. 1293-1313.

[41] C h a p m a n, Dean R.: The need for large wind tunnels in Europe. Report of the Large Wind Tunnels Working Group, AGARD-AR-60, December 1972.

[42] S p e n c e, A., W o o d w a r d, D.S., C a i g e r, M.T., S a d l e r, A.J., J e f f e r y, R.W.: The RAE 5 meter pressurised low speed wind tunnel. 11th Congress of ICAS, Lisbon, 11-16 September 1978.

[43] S p e n c e, A., W o o d w a r d, D.S., C a i g e r, M.T., S a d l e r, A.J., J e f f e r y, R.W.: Applications of non-intrusive instrumentation in fluid flow research. AGARD-CP-193.

[44] H a r t z u i k e r, J.P., N o r t h, R.J.: The European Transonic Wind Tunnel (ETW) for high Reynolds number testing. 11th Congress of ICAS, Lisbon, 11-16 September 1978.

[45] P i e r r e, M.: Soufflerie subsonique pressurisée F1 du Centre du Fauga-Mauzac de l'ONERA. 11th Congress of ICAS, Lisbon, 11-16 September 1978.

[46] S e i d e l, M., J a a r s m a, F.: The German-Dutch low speed wind tunnel DNW. Aeronautical Journal, April 1978, pp. 167-173.

Figure 1. Kestrel (Photo by J.B. and S. Bottomley provided by Ardea London).

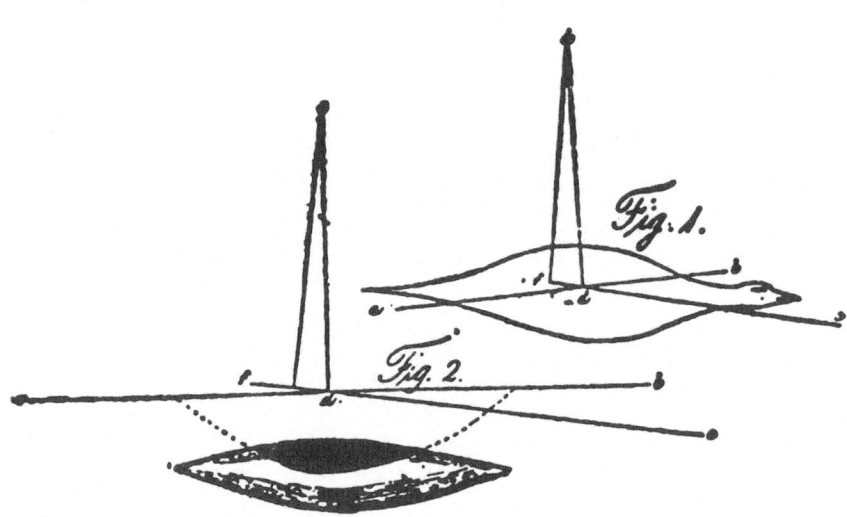

Figure 2. Diagrams illustrating fixed wing flight (Sir George Cayley 1809).

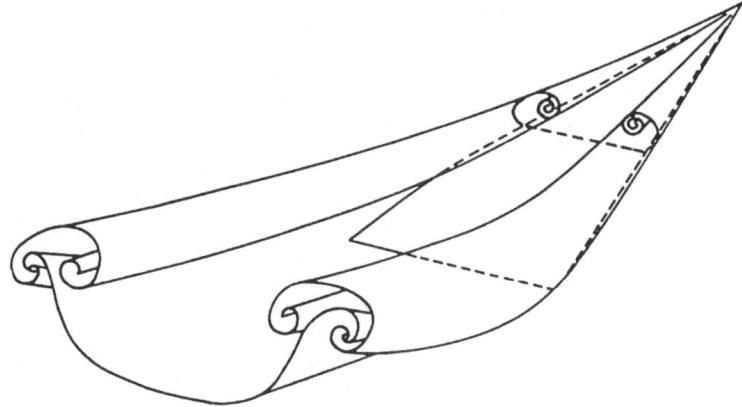

Figure 3. Model of the flow past a lifting slender wing (Maltby).

Figure 4. Concorde

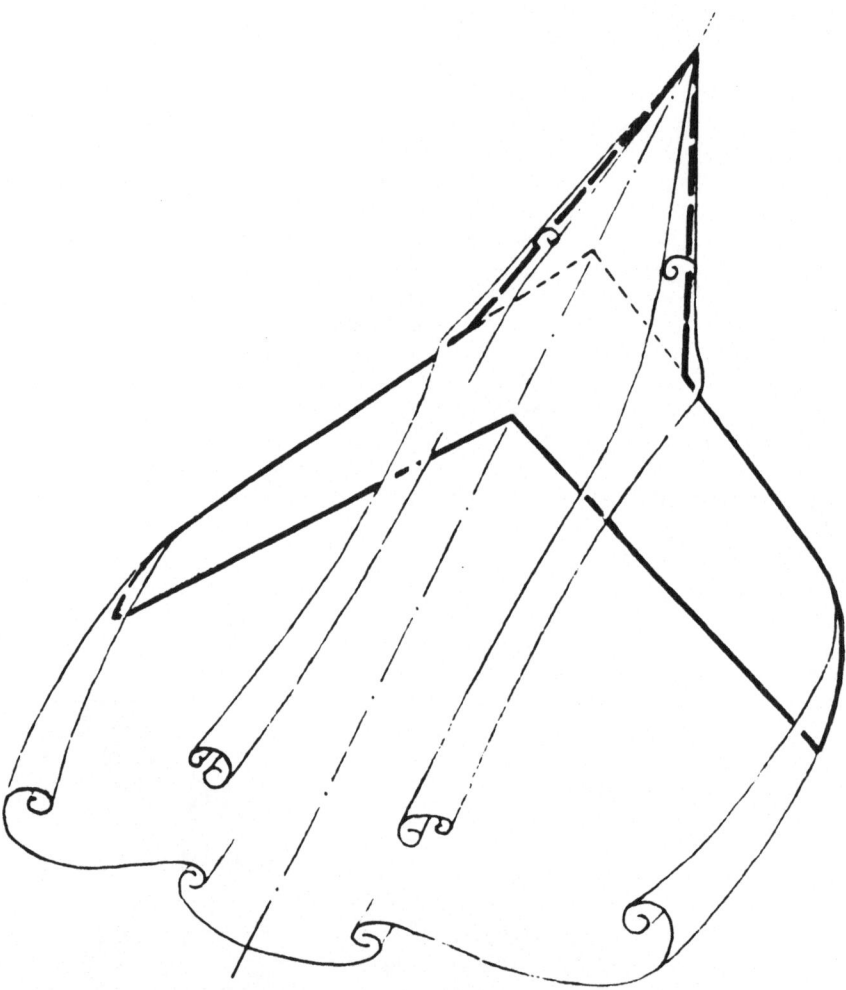

Figure 5. Sketch of vortex sheets shed in combined flow (Küchemann 1971).

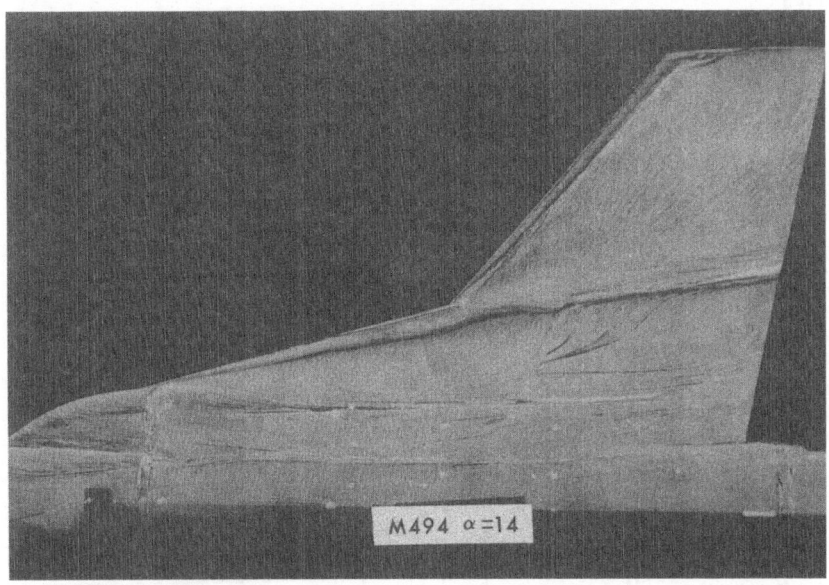

Figure 6. Surface oil-flow on swept wing, low speed, 14° incidence, with and without strake.

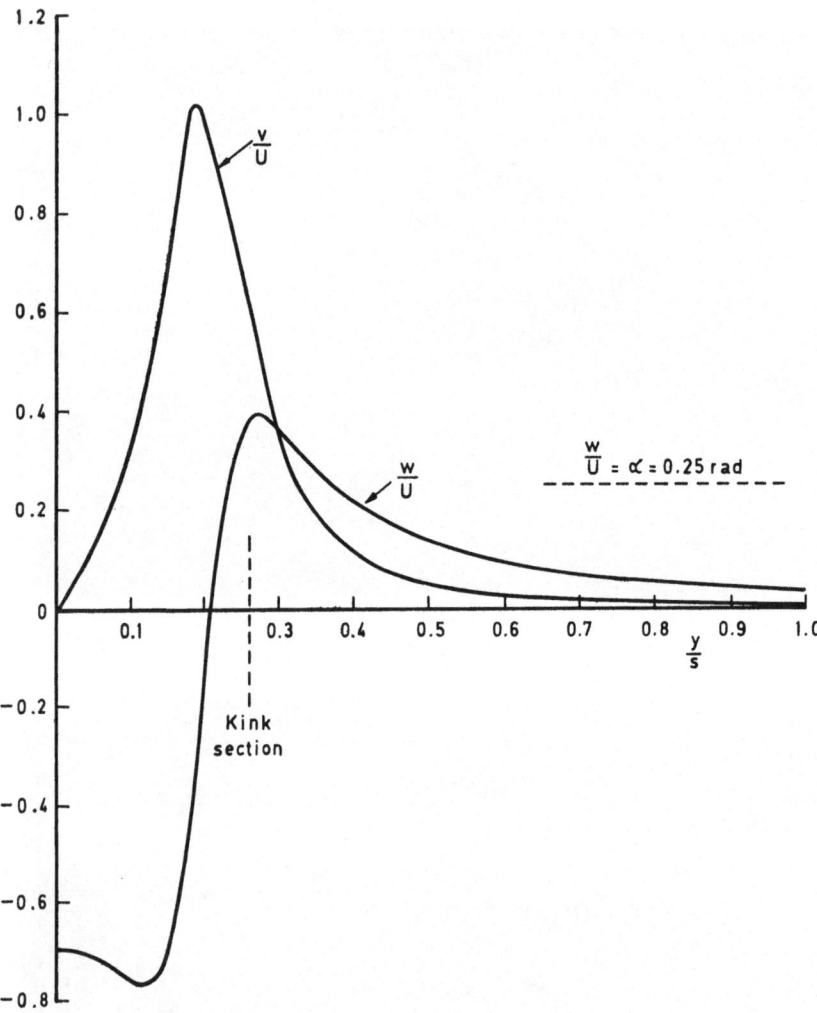

Figure 7. Spanwise variation of strake-induced upwash and sidewash at leading edge of a swept wing (Smith and Fiddes 1977).

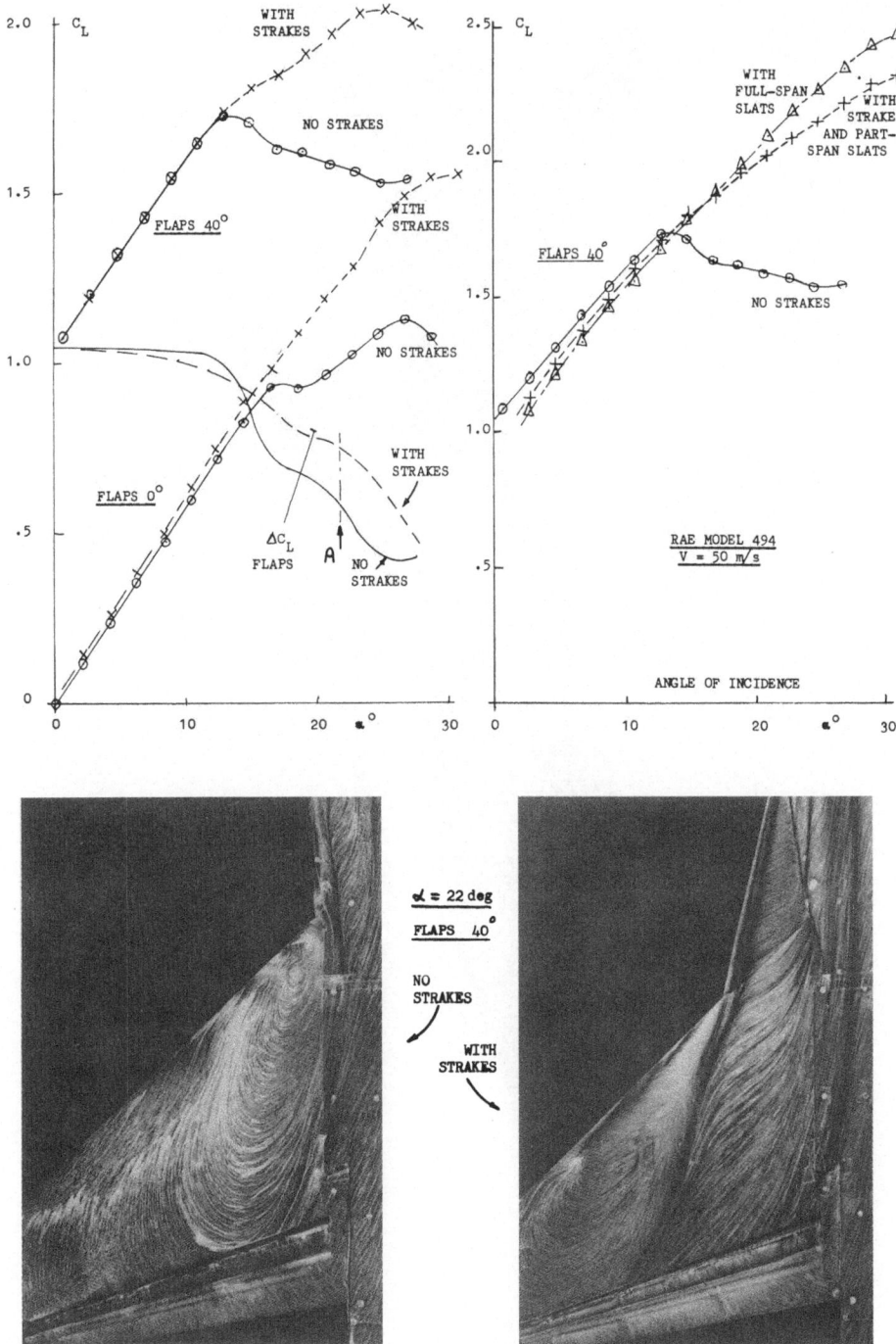

Figure 8. Effect of strakes on low aspect ratio wing with double-slotted flaps.

Figure 9. Harrier aircraft fitted with strakes.

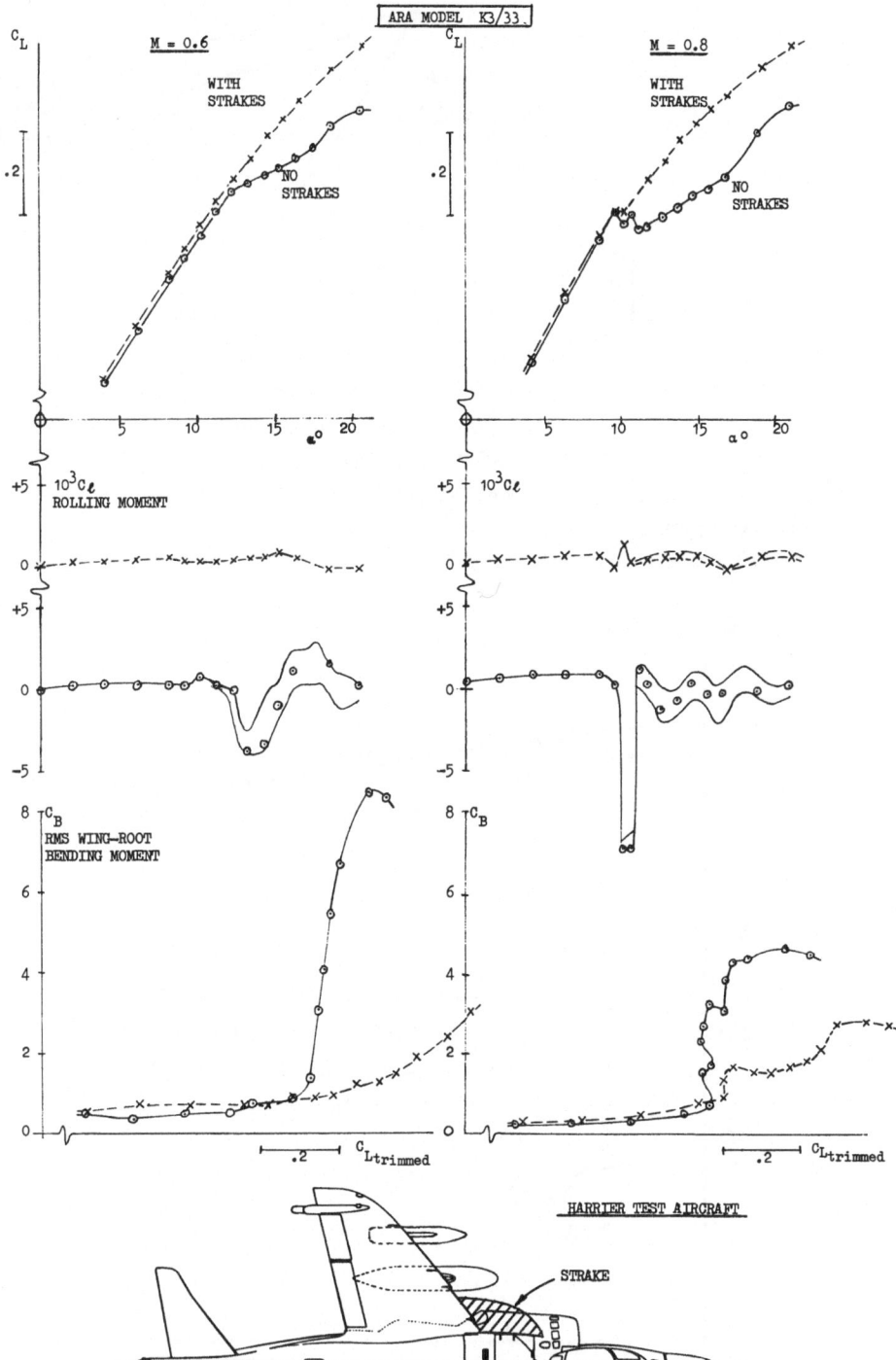

Figure 10. Effect of strakes on wing buffeting and unsteady rolling.

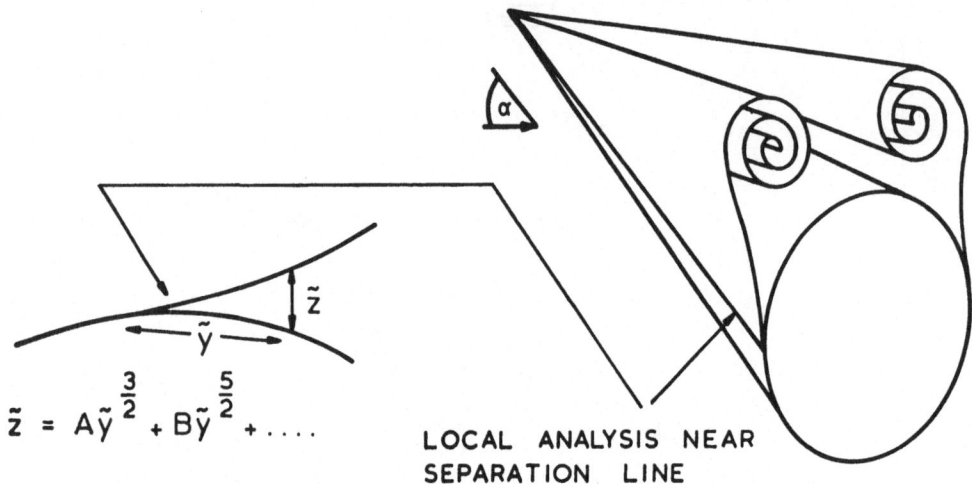

$$\tilde{z} = A\tilde{y}^{\frac{3}{2}} + B\tilde{y}^{\frac{5}{2}} + \dots$$

LOCAL ANALYSIS NEAR
SEPARATION LINE

Figure 11. Results of local analysis of vortex separation from a smooth surface (Smith 1977).

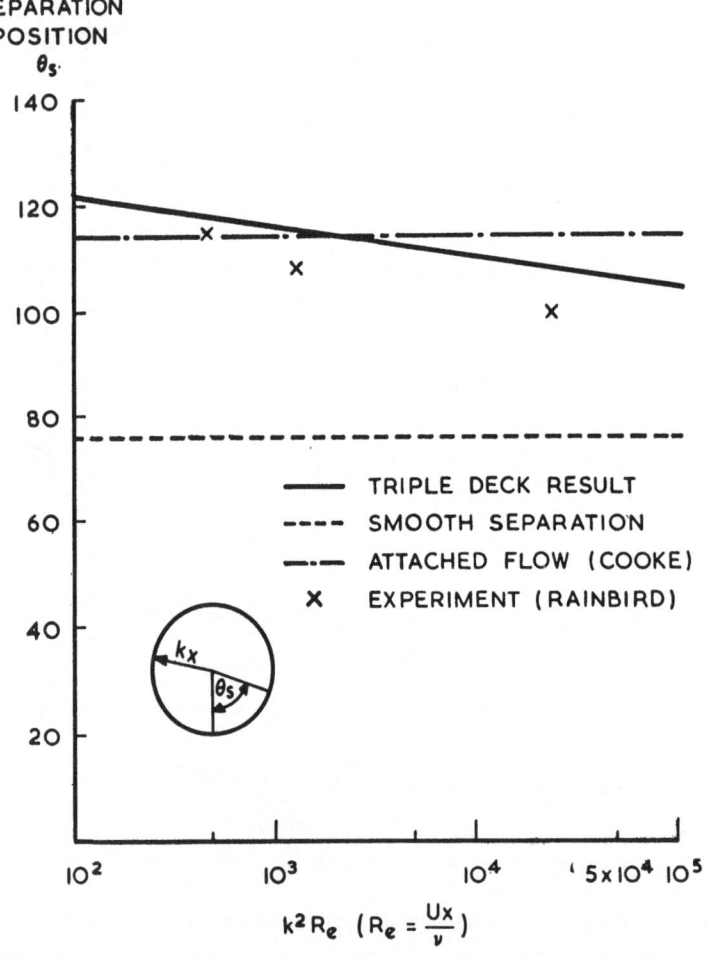

Figure 12. Predicted variation of separation position with Reynolds number for a circular cone at $\alpha = 3k$ (Fiddes 1980).

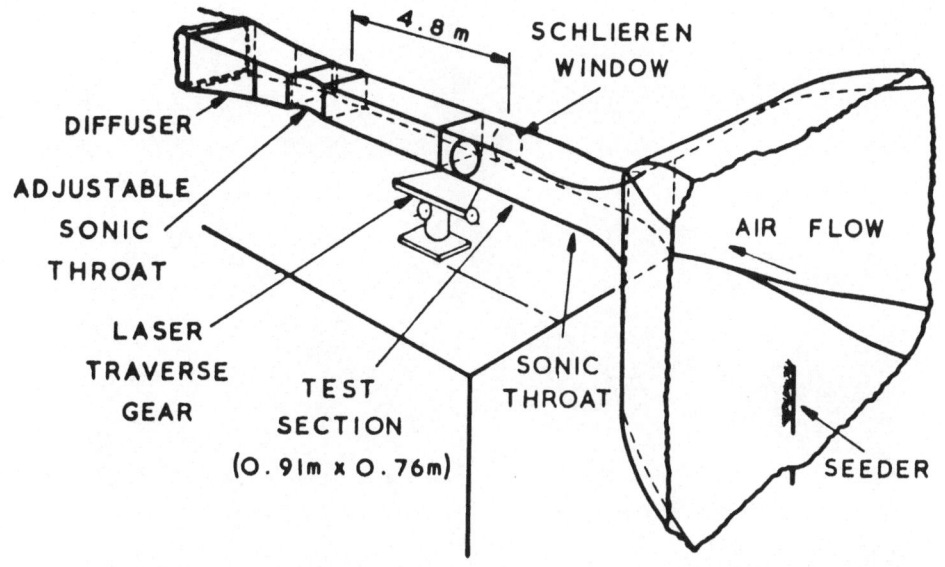

Figure 13. Experimental set up for investigation of shock and boundary layer interaction in RAE 3ft x 3ft tunnel (East 1976).

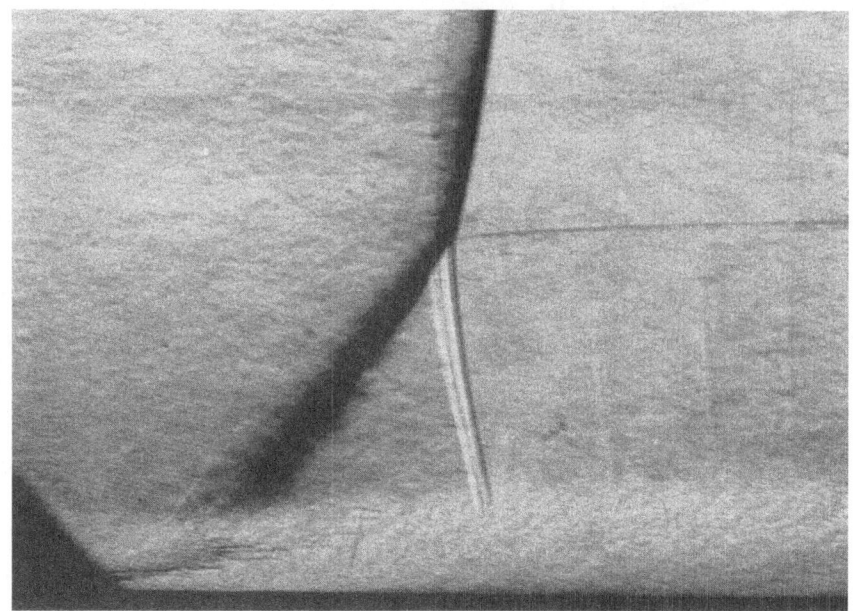

(a) Schlieren photograph

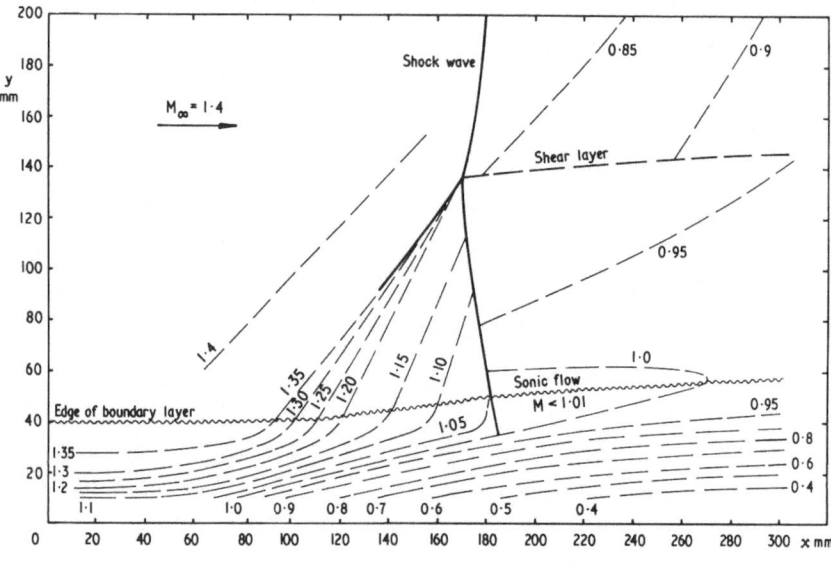

(b) Mapping of flow field by laser anemometers

Figure 14. Study of interaction between a normal shock wave and a turbulent boundary layer at M = 1.4 (East 1976).

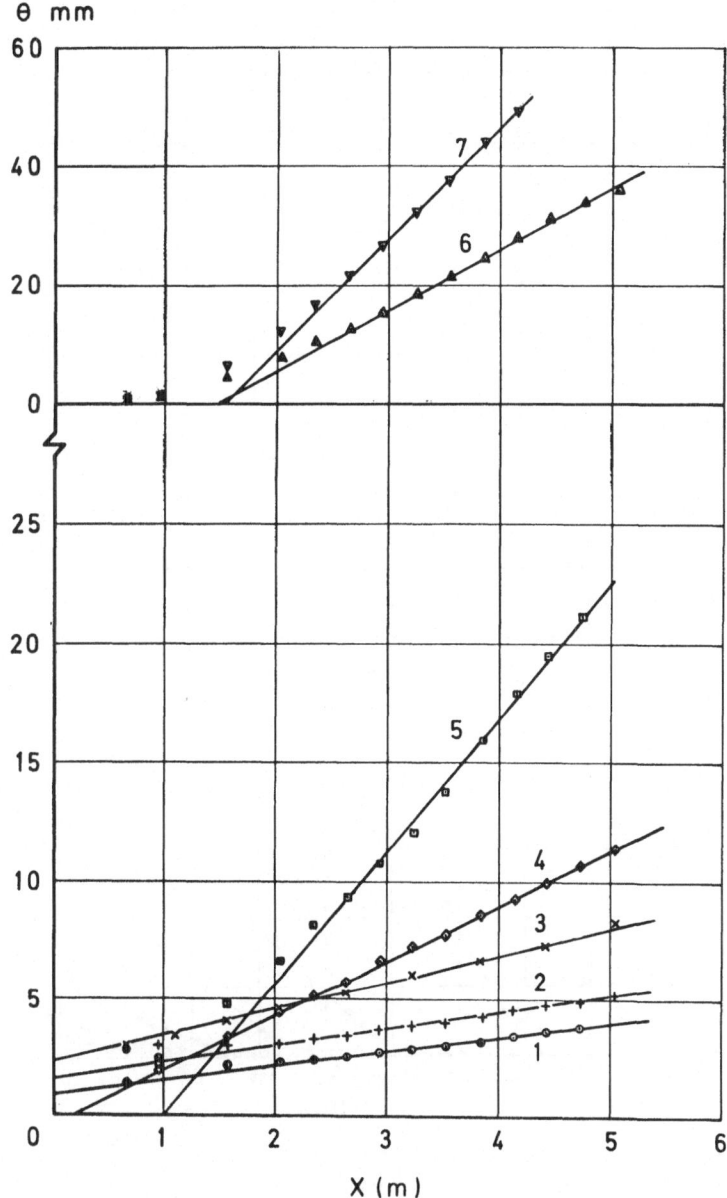

Figure 15. Growth of momentum thickness of a turbulent boundary layer in seven equilibrium flows (East, et al. 1979).

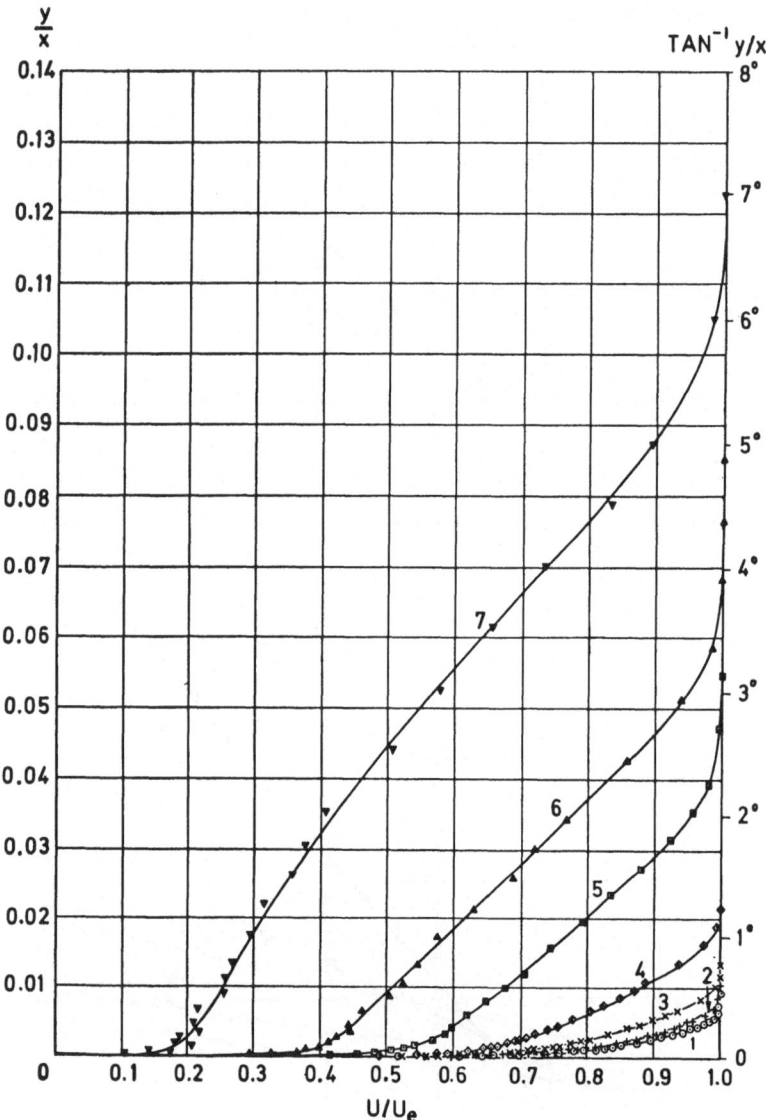

Figure 16. Velocity profiles in equilibrium boundary layers (East, et al. 1979).

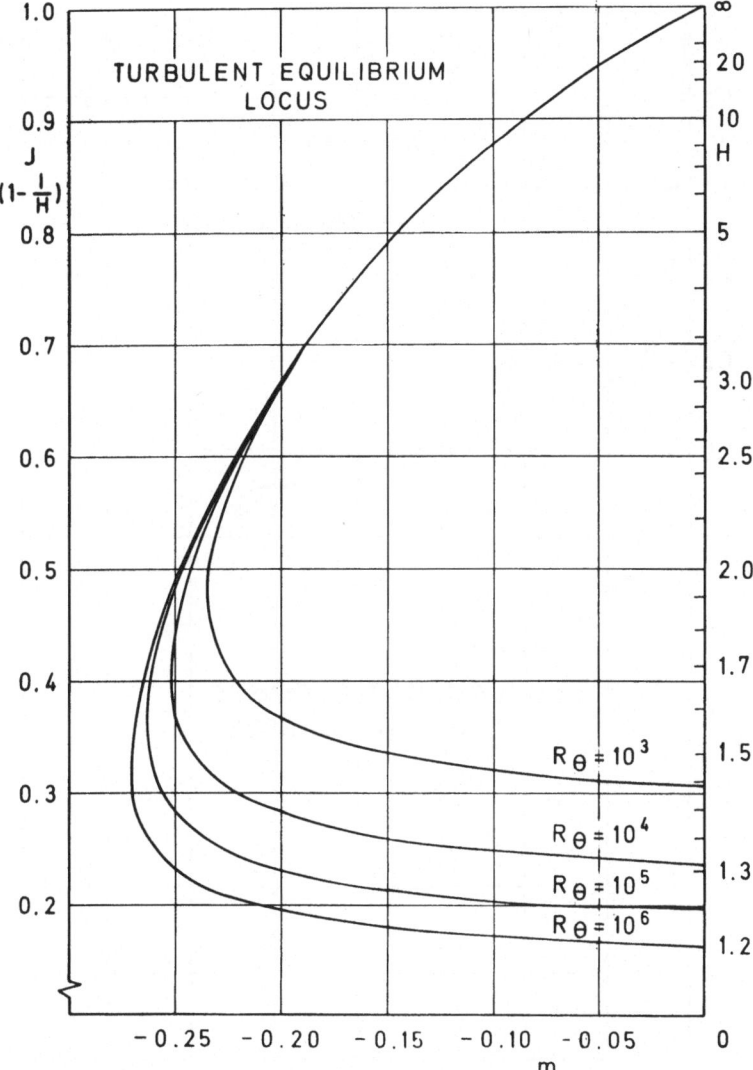

Figure 17. The shape parameter for equilibrium turbulent boundary layers in attached and separated flows (East, et al. 1977).

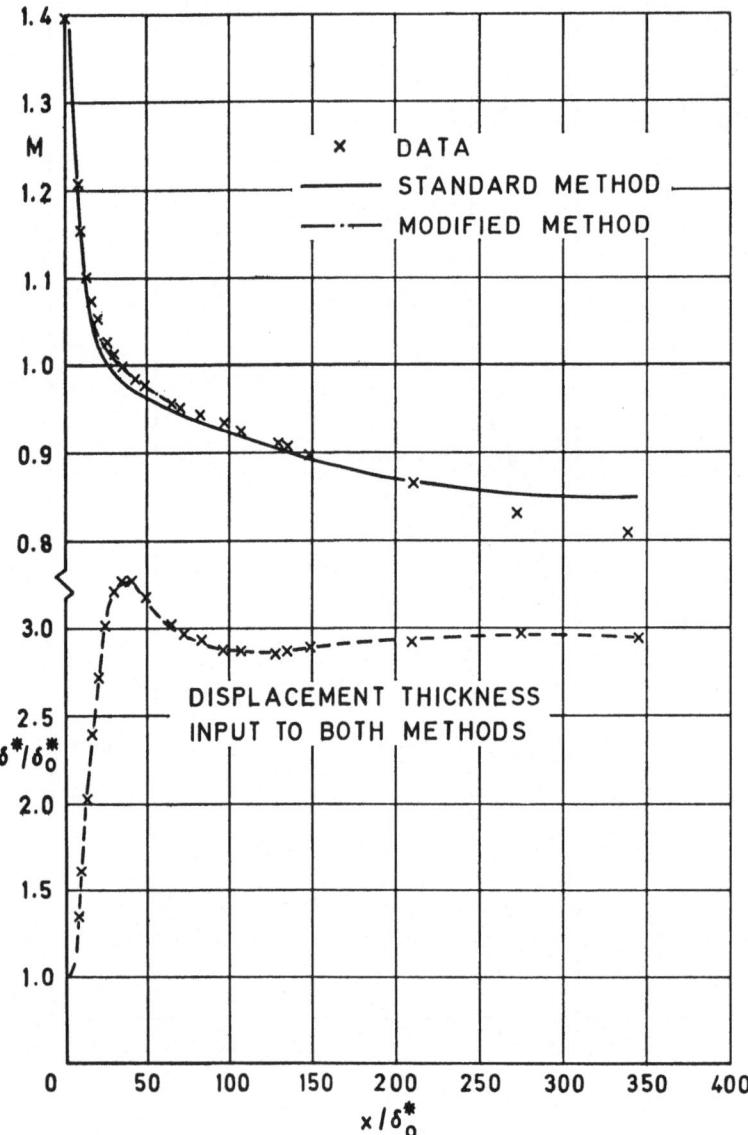

Figure 18a. Prediction of shock wave and boundary layer interaction at $M_\infty = 1.4$ (East, et al. 1977).

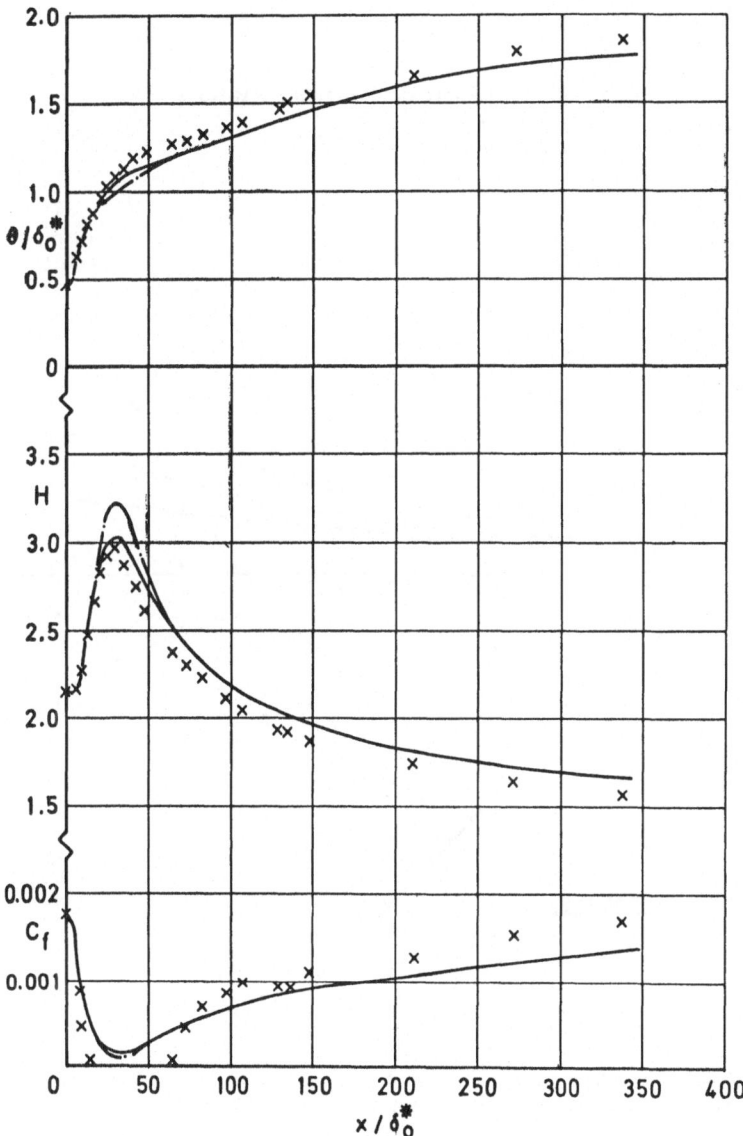

Figure 18b. Prediction of shock wave and boundary layer interaction at $M_\infty = 1.4$ (East, et al. 1977).

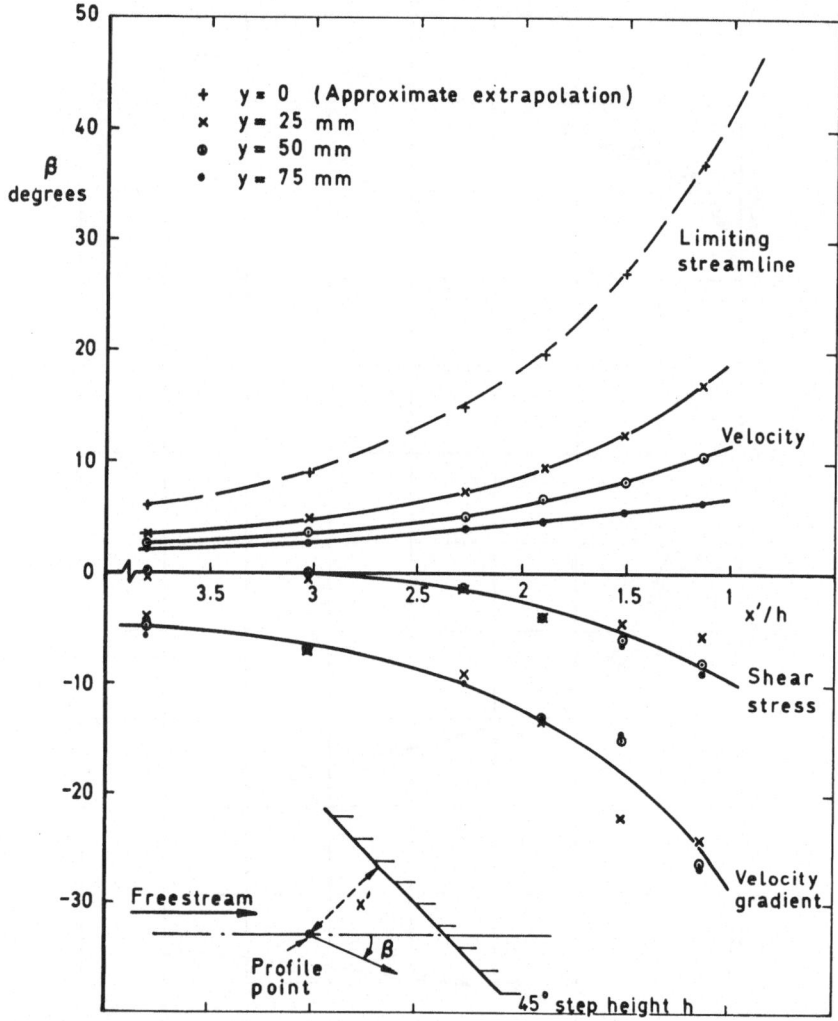

Figure 19. Streamwise distribution of the directions of velocity, shear stress and velocity gradient in flow approaching a swept step (East and Sawyer, 1979).

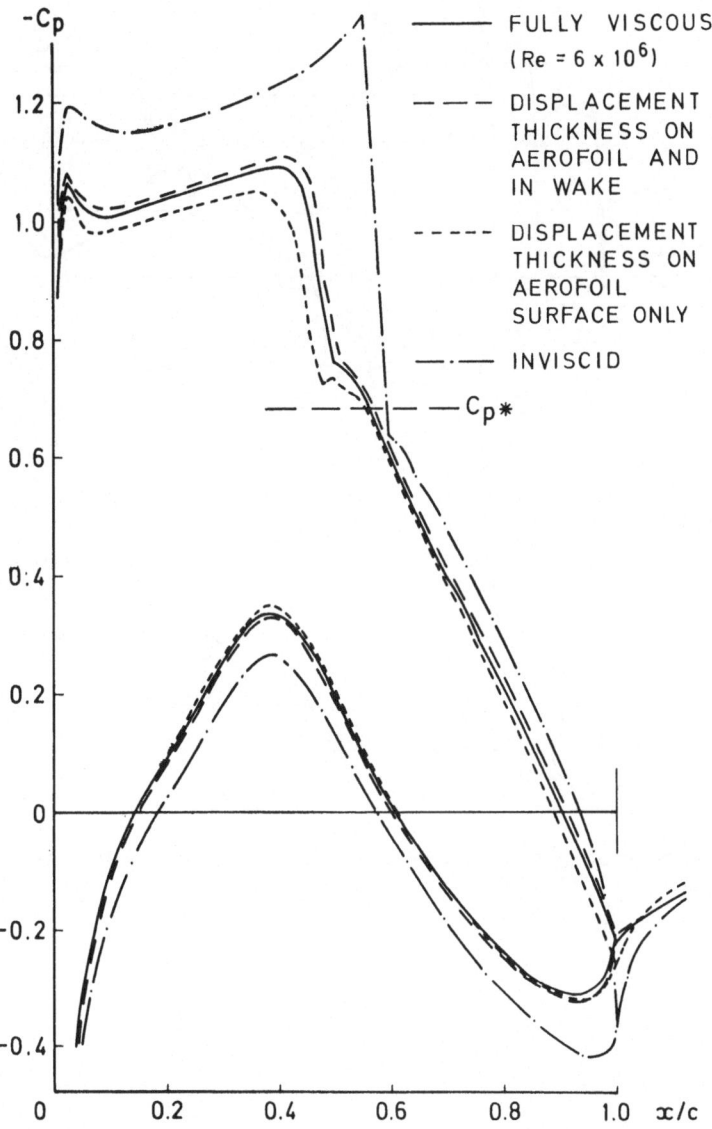

Figure 20. Comparison between viscous and inviscid solutions for RAE 2822 aerofoil at M_∞ = .725, α = 2.3° (Collyer 1977).

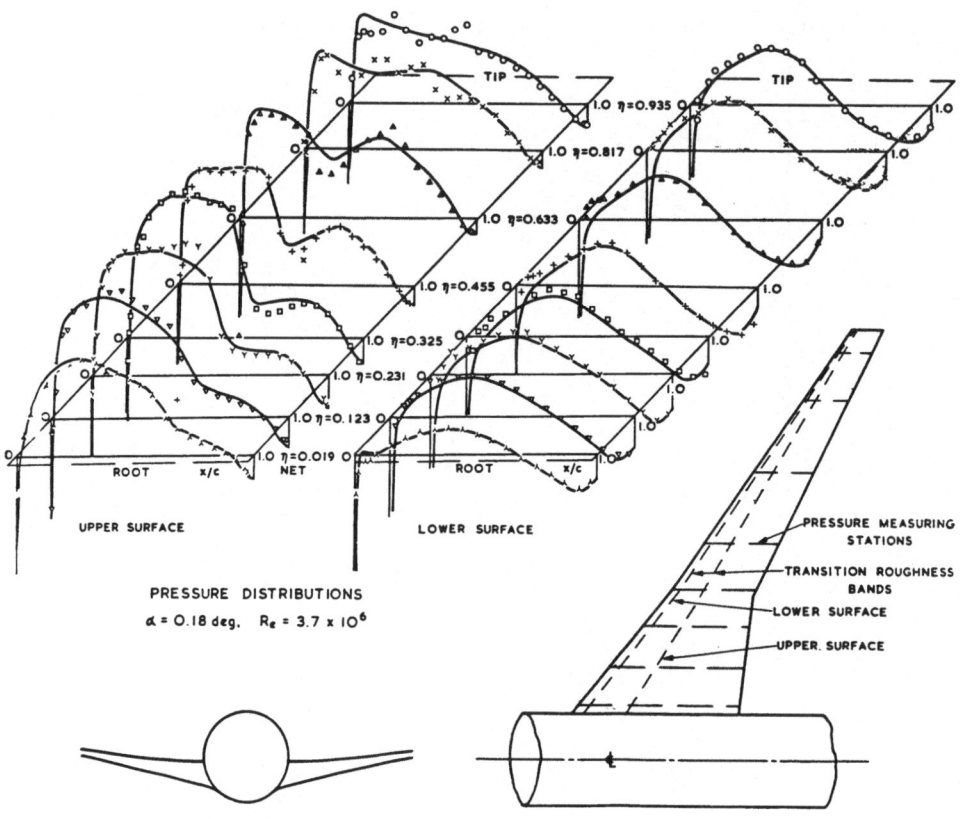

Figure 21. Prediction of pressure distribution over a swept wing with body, including wing viscous effects (Firmin 1980).

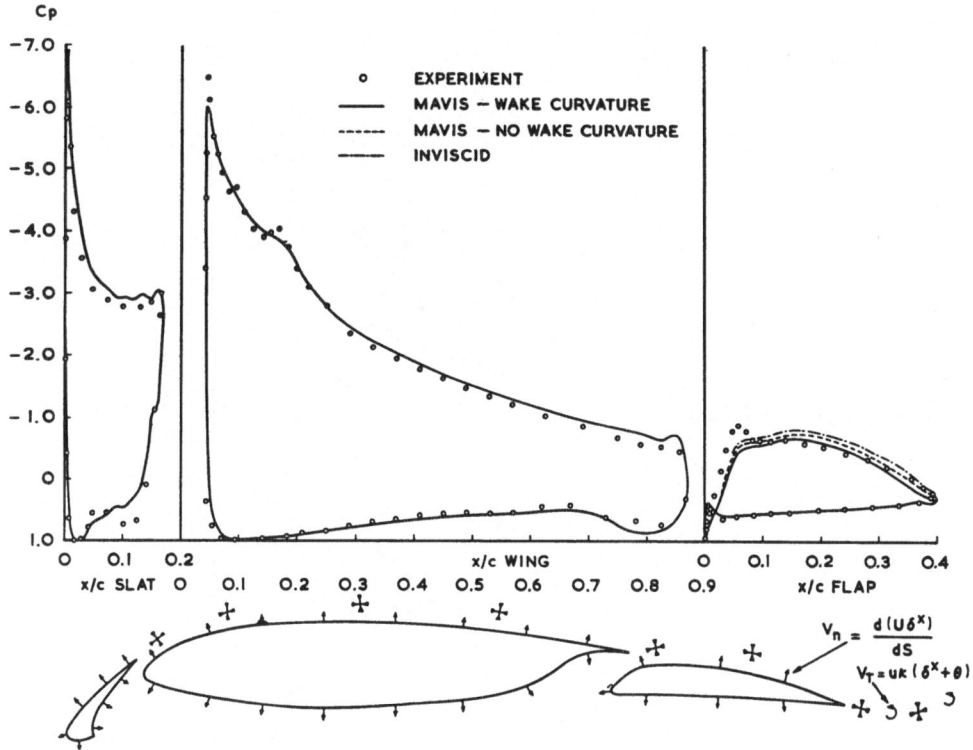

Figure 22. Prediction of pressure distribution on a three-element aerofoil, including viscous effects (Butter and Williams 1980).

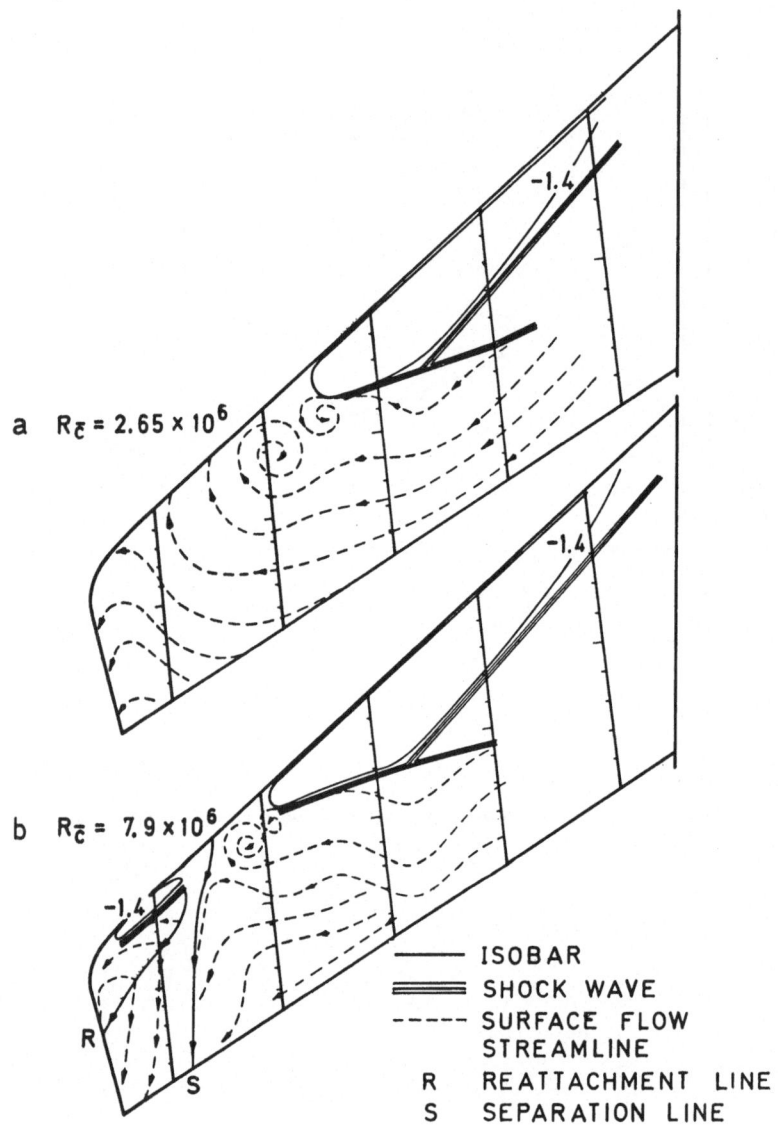

Figure 23. Effect of Reynolds number on upper surface flow patterns on a swept wing at transonic conditions and moderate angle of attack.

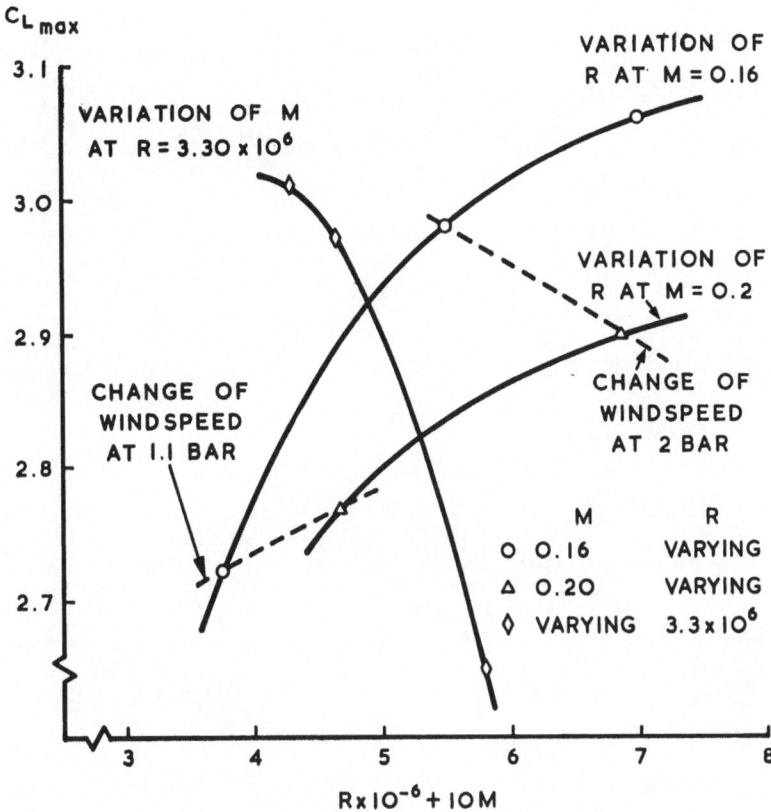

Figure 24. Tests in RAE 5m tunnel showing variation of CL_{max} with Reynolds number and Mach number on swept wing with slat and flap.

Figure 25. Aerial view of the RAE 5m pressurised low speed wind tunnel.

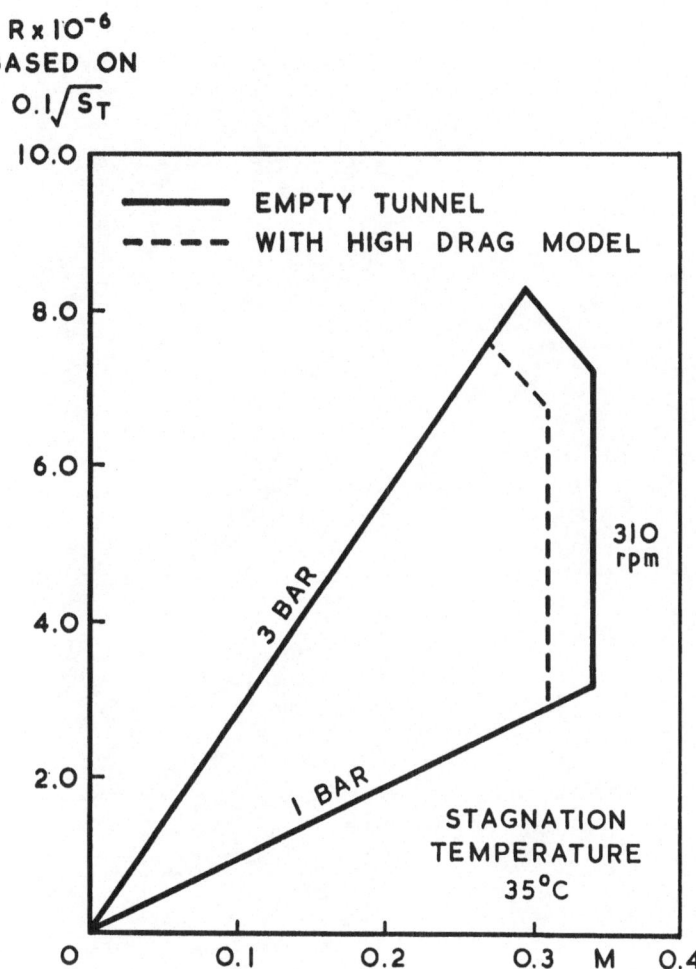

Figure 26. Performance envelope of RAE 5m tunnel.

Figure 27. A 300B model on sting rig in RAE 5m tunnel.

147

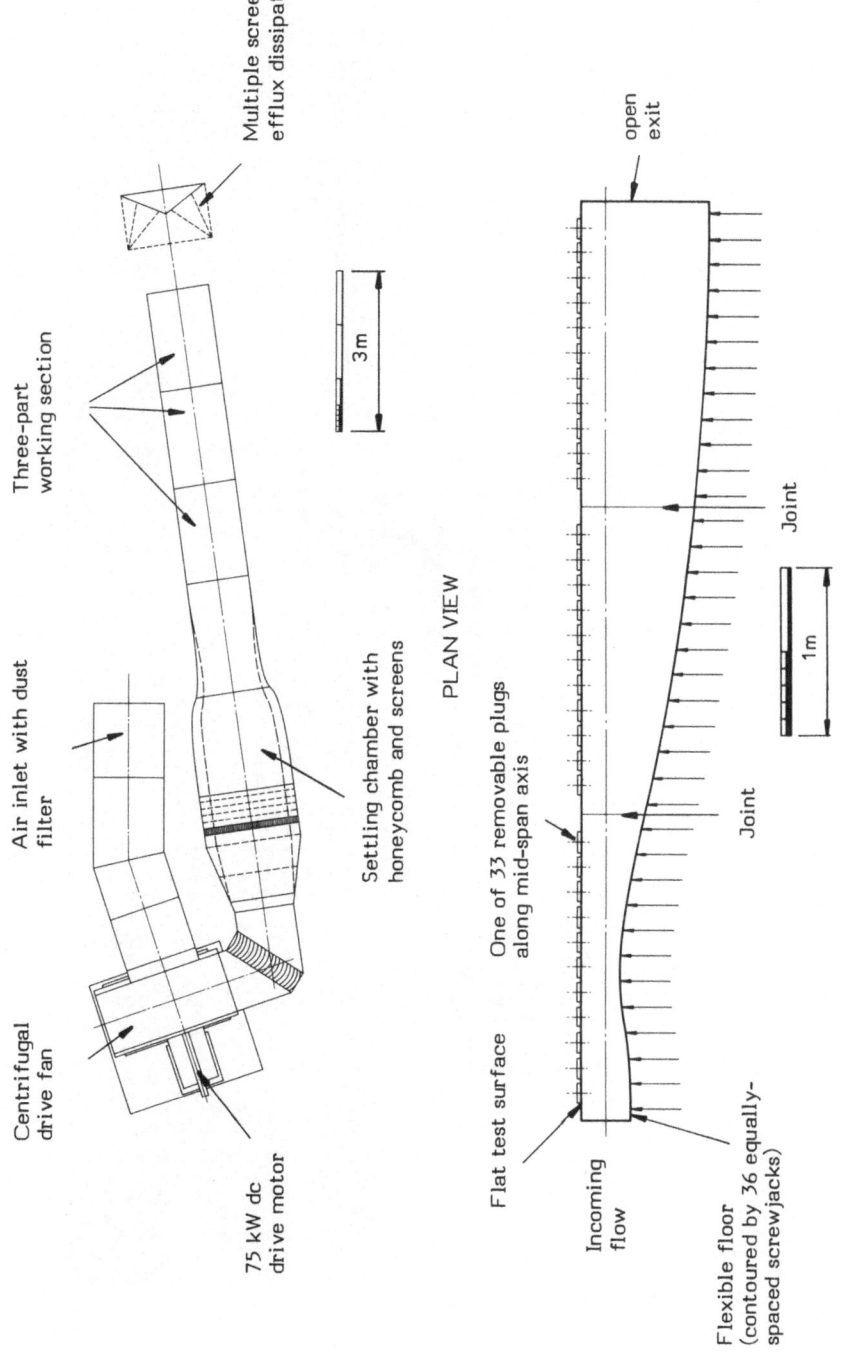

Multiple screen
efflux dissipator

Three-part
working section

Air inlet with dust
filter

Centrifugal
drive fan

75 kW dc
drive motor

Settling chamber with
honeycomb and screens

3 m

PLAN VIEW

Flat test surface

Incoming
flow

Flexible floor
(contoured by 36 equally-
spaced screwjacks)

One of 33 removable plugs
along mid-span axis

open
exit

Joint

Joint

1m

DIAGRAMMATIC SIDE VIEW OF WORKING SECTION

Figure 28. Boundary layer tunnel, RAE Bedford.

Figure 29. Artist's impression of the European Transonic Wind Tunnel.

High-Reynolds Number Boundary-Layer Shock-Wave Interaction in Transonic Flow

M. Sirieix and J. Délery
Chatillon
E. Stanewsky
Göttingen

1. Introduction

The phenomenon of viscous interactions plays an important role in transonic flow due to the presence of embedded shock waves and their interaction with the boundary layer which, in the case of interest, is generally turbulent. The interaction influences the entire flow field, especially when separation occurs provoked by the shock either directly or indirectly as a consequence of its destabilizing effect on the boundary layer.

The interaction phenomenon is present on many aerodynamic configurations such as airfoils and wings, cascades, helicopter rotor blades, air intakes and afterbodies, to name just a few. This explains the considerable effort put into experimental and theoretical studies of the problem by industry and research organizations. Concerning the development of resulting theoretical methods, one can distinguish the following approaches:

1.) A strong coupling of the external inviscid flow and the dissipative zone is carried out, with the latter being as a first approximation treated by boundary layer theory.

2.) The time-averaged Navier-Stokes equations are solved. This approach - although very time consuming - would be the best possible, especially in the presence of large sepa   rated regions.

3.) One might apply an intermediate procedure using a strong coupling concept outside certain domains or subdomains where a solution of the Navier-Stokes equations or approximations to this solution, such as provided by multiple deck theories, are applied.

Efficient and accurate numerical methods exist for all these different approaches; however, computational results are often deceiving and frequently do not agree well with experiment, especially when large separated regions are present. The main reason for this is that the effects of turbulence and viscosity are not accounted for correctly due to deficiencies in turbulence modelling. Also certain phenomena associated with large variations in Reynolds number on airfoils and wings in transonic flow are not well understood and cannot be predicted.

For these reasons the need for thorough and detailed experiments exists and a number of cooperative investigations have developed in this domain, of which the cooperation between

ONERA and DFVLR is one example. The present conference, held on the occasion of inaugurating the reconstructed building of the Aerodynamisches Institut of the RWTH Aachen, is deemed a good opportunity to present typical and complementary results of the cooperative work limiting the subject to viscous/inviscid interactions in steady two-dimensional transonic flow.

In order to describe the domains of research covered in this presentation a typical - and by now classical - example of the effect of viscous/inviscid interaction shall be given first, Fig. 1. Large differences in the upper surface shock location and in the flow structure and, as a consequence, in the pitching moment, which mainly depend on Reynolds number and mode of transition were observed on the Lockheed C-141 aircraft[1]. To explain these differences it is necessary to examine the development of certain dissipative regions.

At low Reynolds numbers with transition occuring close to the leading edge, separation is present at the trailing edge, covering a large part of the downstream region of the wing. As a consequence, the displacement thickness grows considerably subsequent to the jump caused by the shock. At high Reynolds numbers, the boundary layer is thinner upstream of the shock; the relative jump in displacement thickness across the shock is higher due to a higher Mach number upstream of the shock and the presence of a small separation bubble; however, the increase in boundary layer (displacement) thickness down to the trailing edge is more gradual due partly to the absence of rear separation. The results clearly lead to the main areas of research, viz.,

- the local and detailed study of shock-wave boundary-layer interaction to determine jump conditions for the dissipative layers and to determine the onset and development of separation,

- the coupling between the development of the viscous flow over the rearward part of an airfoil (wing) and the shock wave position due to changes in circulation with emphasis on Reynolds number and scaling effects.

The two topics constitute the essential part of this contribution, ONERA being mainly concerned with the first, DFVLR with the second topic. The paper is concluded by a brief discussion of some aspects of theoretical developments associated with the experimental research at both institutions.

2. Basic Studies

2.1 Test Facilities

The set-up implemented at ONERA for the experimental analysis of shock boundary-layer interaction phenomena has been conceived so as to be able to operate with facilities essentially devoted to fundamental research and hence presenting a great flexibility of use[2]. It was also necessary to obtain testing conditions providing a well established turbulent boundary layer, i.e., positioning the interaction far enough from the transition zone to obtain a turbulent flow in a situation of local quasi-equilibirum, the thickness δ being sufficient (several mm) for a detailed analysis of its structure.

It was also desirable to vary widely the Reynolds number $\mathcal{R}_\delta$ from moderate values close to usual conditions of wind tunnel tests on large models up to high values correspoding to conditions close to those in flight.

Lastly, an important effort has been made regarding instrumentation in view of the extreme sensitivity of transonic flow to the disturbing effects of solid probes, especially if separation occurs. It was decided to resort as far as possible to optical techniques for the measurement of both the main field quantities and turbulence levels, calling largely on holographic interferometry and laser doppler velocimetry (LDV).

Under these conditions, the experimental studies have been carried out in three typical installations, each comprising a specific domain of study.

2.1.1 Tests at Moderate Mach Numbers

The first of these installations, Fig. 2a, is the transonic tunnel S8A of the Fluid Mechanics Laboratory. The test section is 120 mm wide with a 100 mm height at its entrance. Its upper wall is solid and plane. A half-profile, or bump, is placed on the lower wall of the section, allowing the flow to accelerate up to slightly supersonic velocities ($M \le 1.4$).

A sonic throat of adjustable cross section is placed at the test section outlet, making it possible:

- on one hand to produce by blockage effect a shock wave whose position and hence intensity can be adjusted in a continuous and precise manner

- and on the other hand to free the flow under study from the pertubations that may be generated in the downstream ducts.

This wind tunnel is continuously fed with previously desiccated atmospheric air, the stagnation conditions being as follows: $p_i \simeq 0.95$ bar, $T_i \simeq 300$ K.

Under these conditions, the Reynolds number $\mathcal{R}_\delta$, calculated with the physical thickness δ of the boundary layer and the local Mach number immediately upstream of the shock wave, was between 0.5 and 0.8 x 10^5.

The bump technique in a blocked section allows the study of the shock boundary-layer interaction phenomenon either along the aerofoil wall or along the opposite plane wall and to widely vary, by changing the shape of the opposite wall, the local pressure gradient upstream of the shock as well as the curvature factors associated with the wall shape. This set up has been especially equipped for the systematic use of optical methods. Two measuring techniques have been developed, viz., holographic interferometry and two direction LDV [3, 4, 5].

The basic diagram of the interferometer used is given on Fig. 3. This single-crossing device comprises as the light source a helium-neon laser of 15 mW power, emitting continuously on the 0.6238 μm wave length. The field covered is a 110 x 140 mm retangle and photographs are taken with a 1/500 second exposure time.

The quantitative exploitation of the holograms and the restitution of the fringe pattern is made by the reference hologram method, the apparatus being adjusted either at zero order to obtain a visualization of the phenomenon, or with finite fringe mode of operation for the quantitative exploitation of the results. This exploitation has been subject of research to define semi-automatic conditions of picture analysis and for restituting the parameters of the viscous and the inviscid fields, i.e., density, local Mach number, etc.

The effort also concerned the implementation of the laser velocimeter developed at ONERA, Fig. 4. This is a two dimensional device allowing the simultaneous measurement of two components of the velocity vector. The source is an ionized argon laser emitting a maximum power of 15 W. The measuring volume has a 300 μm diameter and the inter-fringe is of the order of 25 μm.

In order to allow measurements in the regions where the velocity component changes sign (separation, for instance), the beams cross acousto-optical modulators (Bragg cells) inducing a fringe movement whose associated modulation frequency is 8 MHz.

The collecting optics are arranged so as to collect the forward scattered light. The whole velocimeter is fixed on a table allowing remote controlled and precise (0.01 mm pitch) dis-placements along three mutually perpendicular directions.

Particle seeding results essentially from the very fine dust emitted by the silicagel of the dryer, an enrichment by incense smoke being only necessary during probing of separated zones.

The signals delivered by the photo multipliers of the apparatus are subjected to a statistical processing on an HP 2100 mini-computer, so as to obtain the components $\bar{u}$, $\bar{v}$ of the mean velocity as well as the principal terms of the Reynolds tensor ($\overline{u'^2}$, $\overline{v'^2}$, $\overline{u'v'}$) and the various moments, i.e., skewness and flatness factors.

2.1.2 Tests at High Reynolds Numbers

The experiments on interaction at high Reynolds numbers were performed in the R1-CH blow-down wind tunnel, whose stagnation pressure is adjustable from 0.5 bar up to about 9 bar. The principle of the bump at the wall has also been retained in this set up, Fig. 2b.

The transonic section is a flat rectangle, 80 x 380 mm, the span being large relative to the section height in order to minimize end effects. As in the case of the S8A tunnel, flow blockage is ensured by a second throat located at the end of the test section.

The Reynolds numbers, $\mathcal{R}_\delta$, based on the boundary-layer thickness and local conditions upstream of the shock wave can be varied from 0.6×10^5 to 1.3×10^6, covering a range ensuring the link between wind tunnel tests and flight. In this set up conventional measuring techniques, including the measurement of wall pressure and the survey of the dissipative layers by classical probes, are employed.

2.1.3 Complementary Tests on Airfoils

In order to analyze some aspects of shock boundary-layer interaction phenomena specific to supercritical airfoils, a third type of mounting has been used. The tunnel employed is the pressurized injection-driven transonic wind tunnel T2 of ONERA CERT ($1 \leq p_i \leq 4$ bar), operated by DERAT in Toulouse.

In the square test section of 400 mm x 400 mm, a model of the airfoil LC 100D of 400 mm chord is installed close to the lower wall, Fig. 2c. This model can be set at an incidence relative to the upstream flow direction by means of adjustable struts; the upper wall is contoured in order to avoid flow blockage.

The objective is not to generate around the airfoil a flow corrected for the effects of the limitation by the upper and lower walls, but to realize pressure and velocity distributions on the profile upper surface similar to those encountered on a supercritical profile under normal operating conditions.

A downstream sonic throat ensures stable and reproducible working conditions.

2.1.4 Domain of Study of Shock Boundary Layer Interaction Phenomena

The Mach number range covered by the set ot these experimental studies is represented in Figs. 5a and 5b in the two diagrams $M_o = f(\mathcal{R}_{\delta o}^*)$ and $M_o = f(H_{io})$ which enable one, on the one hand, to ascertain the simulation conditions realized as regards Reynolds number and, on the other hand, to characterize the situations and states of the turbulent boundary layers to be subjected to the interaction with a quasi-normal shock wave, states resulting from the previous history of these boundary layers, i.e., in particular from the pressure gradients to which they have been subjected.

We shall first observe, Fig. 5a, the range of the simulation domain realized, extending from the usual conditions in wind tunnel tests to conditions close to those in flight, then, in the second diagram, Fig. 5b, the choice of the incompressible shape parameters for characterizing the velocity profile of the turbulent boundary layer. Here, the lowest values ($H_i = 1.15 - 1.20$) correspond to situations where the flow around the obstacle is highly accelerated upstream of the shock wave, which is the case of profiles used before the supercritical airfoil generation and of some turbomachine blades. The intermediate values ($H_i = 1.25 - 1.30$) concern the case of shock waves located towards the rear of modern profiles at high Reynolds numbers. The highest values ($H_i = 1.4 - 1.5$) correspond to a situation where the Reynolds number is low and/or where the flow is slowed down upstream of the shock with examples being air intakes and supercritical profiles with shock locations relatively close to the leading edge, especially in wind tunnel tests.

Having thus defined the domain covered by the experimental studies, we shall now examine in more detail the evolution of the phenomena related to the interaction between shock waves and turbulent boundary layers, first in the case where this interaction does not provoke any separation of the boundary layer.

2.2 Boundary-Layer Shock Wave Interaction Without Separation

2.2.1 Structure of the Flow in the Interaction Domain

A typical example of the evolution of the interaction phenomena for a bump-on-the-wall configuration [2] as the Mach number M_o upstream of the shock increases starting at a value of 1.08 - 1.10 is given in Fig. 6. This figure shows the development of the local Mach numbers, deduced from wall pressure readings with the assumption of an isentropic relation, as well as the quantitative visualization by interferometry of the corresponding flows. The variation of the Mach number upstream of the shock is obtained by the gradual opening of the second throat. The interaction phenomena are abserved on both walls, the one with the bump and the plane upper one.

The rapid pressure increase through the shock wave is followed by a more gradual recompression resulting from the profile curvature and the boundary layer displacement effect.

Note the continous structure of compression at the foot of the shock, marked by the focusing of interference fringes, the corresponding domain widening as M_o increases.

A more detailed investigation of this phenomenon is provided by the enlargement of the interferometric picture for the configuration corresponding to the highest upstream Mach number M_o, Fig. 7. This picture brings to light two domains, characteristic of the interaction phenomenon:

- In domain I there occurs a continuous and rapid compression of supersonic nature and of simple wave type down to an almost sonic value of the velocity.

- In domain II a noticeable less rapid evolution of the velocity field takes place resulting from a much slower variation of the boundary-layer displacement effects in the absence of separation.

It is clear from these considerations that the general and specific properties of the shock boundary layer interaction phenomena can only be searched within domain I, of supersonic nature, while in domain II, of subsonic nature, the pressure distribution results from the integration of effects extending far downstream and taking into account the complete flow and, in particular, the shape and curvature of the wall.

That is why we shall concentrate on the analysis of phenomena within in Domain I, of length L_s, whose downstream end is defined by the critical pressure p^*.

2.2.2 Factors of Influence Acting on the Extent of the Domain of Rapid Interaction

Among the factors likely to influence the domain of rapid interaction, the most commonly investigated concern the effects of the Mach number, M_o, and the Reynolds number, $R_{\delta_o}^*$, immediately ahead of the interaction domain, the Reynolds number being calculated with the displacement thickness.

An analysis of these effects has been performed in the R1-CH wind tunnel on a bump-on-the-wall configuration, the wall Mach number distributions being practically insensitive to changes of the free stream stagnation pressure, which governs the variations of $R_{\delta_o}^*$. Thus, a quasi-similitude of the states of the boundary layer upstream of the interaction domain has been realized, the action of the pressure gradient on the evolution of the shape parameter, for instance, being preponderant relative to that of the Reynolds number.

The results obtained are presented in Fig. 8a in the form of a diagram giving the evolution of the reduced scale L_s/δ_o^* as a function of Mach number M_o $(1.09 \leq M_o \leq 1.3)$ with the variations of $R_{\delta_o}^*$ being between 0.5×10^5 and 1.3×10^6 and the value of H_i for the whole set of results being close to 1.2.

First, we observe an excellent overall grouping with a moderate dispersion, the latter being mainly due to the difficulty of defining precisely the length L_s. The influence of the Reynolds number, very marked on the thickness and the physical extent L_s of the domain, disappears when these two variables are reduced one by the other.

Moreover, it appears that L_s is not very sensitive to the effect of the upstream Mach number M_o; the dispersion, occurring when M_o comes close to 1.3, corresponds, in fact, to a situation close to separation, if not actually slightly separated, so that, as a first approximation, incipient separation can be deduced by using this property.

When this reduction L_s/δ_o^* is applied to the complete set of results, obtained for the various experimental devices and corresponding to very different situations as regards the Mach number development upstream of the interaction domain, and hence of the state of the turbulent boundary layer, a very pronounced dispersion of results appears, Fig. 8b. However, the experimental points are regularly spaced as a function of the value of the shape parameter H_i of the upstream boundary layer.

For instance, we observe that the value L_s/δ_o^* is practically increased twofold when H_i passes from 1.2 to 1.4.

To take this effect into account, a correlation of the results has been looked for within the actual domain of variation of the parameters involved $(1.15 \leq H_i \leq 1.50, 1.10 \leq M_o \leq 1.30)$.

The empirical law shown in Fig. 9 leads to a rather satisfactory grouping of the results and makes it possible to define with a reasonable precision the characteristic scale of the interaction phenomena with all effects being accounted for (H_{io}, $R_{\delta_o}*$, etc.).

2.2.3 Definition and Calculation of the "Viscous Ramp" Associated with the Domain of Rapid Interaction

The above empirical law may be used for determining the displacement effect associated with the strong viscous interaction taking place at the foot of the shock. The evolution $\delta*(x)$ during the recompression between the upstream Mach number M_o and the sonic state defines a "viscous ramp" which represents, in a schematic way, the complex phenomena related to the formation of the shock wave; taking them strictly into account would require either a particular treatment (e. g. by triple deck method) or a very severe local tightening of the mesh pattern in the calculation of the inviscid fluid.

Considering the limited number of parameters acting on L_s (at least as a first approximation), it is possible to carry out a priori a calculation of the viscous ramp which would be associated with

- an upstream Mach number M_o and

- a boundary layer state, characterized by the local Reynolds number and the incompressible shape parameter H_{io}.

A simple method has been devised to compute the shape of the "viscous ramp" associated with strong interaction phenomena occurring in the previously defined Domain I. This method relies on some experimental evidences.

The basic hypothesis is to assume that the fast recompression from the upstream Mach number M_o to the sonic value is a simple wave process leading to a quasi-linear variation $M(x)$ over the extent of L_s of Domain I (Hypothesis a). Under such circumstances, the coupling relation between inviscid and dissipative flow writes:

$$\frac{d\delta*}{dx} = P(M) - P(M_o)$$

where $P(M)$ is Busemann's pressure number.

This equation allows one to compute the development $\delta*(x)$ of the dissipative layer displacement thickness along the distance L_s from a given initial value δ_o* when the law $P(x)$ is known (Hypothesis a). One is now able to define the geometrical shape of the ramp.

Knowing $\delta*(x)$, the von Kármán integral equation allows us to compute the momentum thickness Θ, especially its value Θ_s at the end of the viscous ramp. Fig. 10 presents the shape of viscous ramps, calculated for $H_{io} = 1.3$.

The calculation of the ramp provides not only the effect of viscous interaction (expressed, for the inviscid fluid calculation, in the form of a displacement effect or a wall injection effect), but also the state of the boundary layer at the level of the sonic point, in particular its integral thicknesses and its shape parameter. From there, it is thus possible to continue the calculation of this boundary layer by a classical method (integral or finite-difference method) within the framework of a coupling procedure such as that presented in [9] and [10].

Two examples of application of such a procedure are presented in Fig. 11, viz.,

- for the Dornier supercritical profile CAST 10-2/DO A2 at $M_\infty = 0.765$ and $\alpha = 2.4^o$, Fig. 11a, and

- for the Aérospatiale profile RA 16 at $M_\infty = 0.680$ and $\alpha = 0^o$.

In both cases, the Mach number M_o upstream of the shock is between 1.2 and 1.3.

The calculation performed for the upper surface of both profiles leads to a rather good prediction of the measured characteristics of the boundary layer.

The calculation of ramps also makes it possible to predict the limit of the separation onset, i.e., to find, for a given state of the upstream boundary layer, the Mach number M_o for which a separated bubble first forms at the foot of the shock wave. To define this limit, it is postulated that the point of incipient separation appears necessarily at the end of the viscous ramp, i.e., for a sonic local external state. The separation limit is then defined by the condition of coincidence of the points where $C_f = 0$ and $M = 1$. This very intuitive approach, based on a number of experimental observations, is well confirmed by the comparison presented on Fig. 12, which shows a very good agreement between the theoretical limit and the experimental limit determined from very careful measurements.

It should also be noted that the Reynolds number effect is to a great part included in this representation due to the Reynolds number dependency of the evolution of H_{io}.

We shall now pass to the analysis of a clearly separated configuration, appearing when the Mach number M_o is higher than the values defined by the above law. Our analysis will mainly concern the structure of the turbulent field.

2.3. Boundary-Layer Shock Wave Interaction with Separation

2.3.1 Configuration Considered

The general characteristics of the flow under study (after [7])are reproduced in Fig. 13, where we present

- in (a) the Mach number distribution deduced from wall pressure measurements and the interferometric visualization of the flow,

- in (b) the structure of the field outside the dissipative layers, given in the form of lines of equal Mach number with the structure resulting from both the exploitation of interferometric pictures and a verification by laser velocimetry.

The conjunction of a rather strong shock occuring for an upstream Mach number of 1.37 and the corner effect at the trailing edge of the obstacle leads to a rather wide-spread, almost isobaric separation whose lengthwise development between separation S and re-attachment R is of the order of 12 δ_o, δ_o being the boundary layer thickness at the origin of the interaction.

The lambda structure of this configuration is classical, and we shall only point out the following main elements:

- The separation process is a rapid process of supersonic nature and of the "free interaction" type; the resulting shock C_1, formed by the focusing of compression waves, is a weak shock.

- Downstream of shock C_2, the flow is subsonic. This situation is not general and is due to the fact that the upstream Mach number M_o is not very high (M_o = 1.37); for higher values of the upstream Mach number (M_o > 1.4), a local supersonic zone may develop downstream of the shock (supersonic tongue), whose extent is related to the particular conditions of the coupling associated with the deflexion of the dissipative layer towards the wall which leads to reattachment.

As regards the dissipative layer, Fig. 14, the field of mean streamline which in particular defines the reverse flow bubble results from the integration of the flow rate calculated from the profiles of mean velocity, obtained with a two-dimensional laser velocimeter in a number of sections normal to the wall.

The scale of the phenomena relative to the displacement thickness δ_o^* of the initial boundary layer is indicated. One may observe the important thickening of the dissipative layers, related to the development of the separation bubble.

Let us now examine in more detail the structure of these dissipative layers, in particular at the edge of the separation bubble.

2.3.2 Structure of the Separated Zone

As regards the mean velocity field, Fig. 15a, the choice of adequate reduced variables - the same as those used for the treatment of mixing layers - emphasizes interesting properties concerning the flow structure [7].

The difference of reduced velocity, $(\bar{u} - \bar{u}_m)/(\bar{u}_e - \bar{u}_m)$, is plotted here as a function of the transverse scale, taken equal to unity for a variation of the reduced velocity from 0.05 to 0.95.

We observe a good correlation of the set of measurements taken within the separation bubble, showing that a structure of mixing types is reached very rapidly, whose profile is similar to that of an incompressible isobaric mixing.

As regards turbulence, Fig. 15b, the example presented here, which concerns the fluctuation of longitudinal velocity, confirms this similarity which extends to the various statistical moments, the influence of the wall entailing only a very localized distortion [7].

This mixing layer type flow structure has a number of consequences as regards the properties of the turbulent field:

The first is the onset and development of large turbulent structures, more or less, organized which are brought out by short exposure time (a few micro-seconds) interferometric viszualizations, Fig. 16a.

The second one, which is a consequnce of the first, is a great unsteadiness of the separation bubble caused by the interaction between those large structures and the reattachment process.

In Fig. 16b the curves of equal probability of negative value of the longitudinal component u of the velocity are presented. A zero probability means that the velocity is always directed downstream; a probability equal to one means that, on the contrary, u is always negative.

We can see that levels of probability close to one are only apparent within a narrow domain along the wall at the trailing edge of the obstacle. Such a situation is indicative of large scale unsteady motions [7].

2.3.3 Analysis of the Turbulent Field

Let us now survey the statistical properties of the turbulent field as a whole, including the part behind the reattachment point [8].

An example of results on turbulent shear stress, $\bar{\tau} = -\overline{u'v'}/\overline{u}_e^2$, is presented in Fig. 17a. It is limited to a few typical profiles among the many probings performed, the exploration extending far downstream of reattachment.

From these results it has been possible, Fig. 17b, to build the diagram of lines of constant shear stress. We observe, in a general way,

- the high values of $\bar{\tau}$ created by the mixing process at the bubble edge, the maximum values of $\bar{\tau}$ (>2.5 %) being found around the reattachment point;

- the slow and gradual decrease of this maximum value of $\bar{\tau}$ downstream of reattachment within a domain where the pressure gradient is slightly positive or zero; the maximum of $\bar{\tau}$ is also at a distance from the wall, and its level is much higher than that of a turbulent boudary layer in a state of quasi-equilibrium and closer to a wake type situation.

Two important parameters as regards the structure of the turbulent field have also been analyzed. They are, Fig.18,

- the turbulent kinetic energy $\bar{k}$, estimated by assuming the equality of $\overline{v'^2}$ and $\overline{w'^2}$, Fig. 18a, and

- the term of shear stress production proportional to the product $\overline{u'v'} \, \partial\bar{u}/\partial y$, Fig. 18b.

The lines of equal $\bar{k}$ present a general shape very similar to those of equal $\bar{k}$, the ratio $\bar{\tau}/\bar{k}$ varying, however, within the field. The maximum of $\bar{k}$ lies also near reattachment at about the same position as that of $\bar{\tau}$.

As regards the term of shear stress production, we observe that the zone of highest production corresponds to the onset and development of the mixing layer, which is associated with high values of transverse velocity gradients and of shear stress.

Downstream of reattachment, the term of production within the dissipative layer decreases rather rapidly, while it develops near the wall in a way which marks the return to an equilibrium state.

To close this analysis of the turbulent field it seems interesting to compare experimental results with a number of simple algebraic models of turbulence used in calculation models. This comparison will concern the term of eddy viscosity $\nu_t = -\overline{u'v'}/(\frac{\partial\bar{u}}{\partial y})$. We shall distinguish two regions:

- The separated region, where the compared models are those of Alber and Levy.

- The region downstream of reattachment where the compared model is that of Michel.

We first remark that the values of ν_t deduced from experiments are not very precise near the edge of the dissipative layers because they are provided by the ratio between two very small quantities. This situation is not very important for the conclusions drawn from this comparison.

In a general way, we observe:

- For the separated zone, Fig. 19a, generally significant discrepancies between experimental values of ν_t and those provided by the two models occur, especially in the lower part of the dissipative layers; the values of ν_t are generally highly underestimated, except at reattachment in the case of the Levy model.

- Downstream of reattachment, Fig. 19b, the discrepancies are very severe, displaying a very important difference to the situation of a turbulent boundary layer in a state of quasi equilibrium where much smaller values of ν_t are to be found.

The effect of turbulence re-structuring associated with the presence of large structures involves a strong local non-equilibrium, which slowly disappears. The associated relaxation phenomenon develops downstream of reattachment over a range corresponding to about ten times the thickness δ_R of the dissipative layer at reattachment.

Such a situation must necessarily be taken into consideration in the turbulence modelling for calculation methods of separated regions.

We should also note that the first attempts made by using more sophisticated but still classical models based on the transport of the turbulent quantities (models with 2 or 3 transport equations) are not able to provide a good representation of the phenomena related to the formation, development, and re-structuring of large turbulent structures within a separated zone and downstream of it.

3. Global Investigation of Viscous/Inviscid Interactions in Transonic Airfoil Flow

Transonic airfoil flow is, as was already outlined in the Introduction, to a considerable degree affected by viscous/inviscid interactions. Here, the interaction of the upper surface shock wave and the interaction of the sustained adverse pressure gradient close to the trailing edge with the boundary layer are the most important; they are closely linked through their influence, directly or indirectly, on circulation. Detailed experimental knowledge of the flow phenomena related to the interaction and their effect on the overall flow development is essential in support of any theoretical treatment of the problem and, of course, to the improvement of airfoil and ultimately transonic aircraft performance. What led to the investigation of viscous/ inviscid interactions at DFVLR, however, was the concern about

Reynolds number and scale effects and the simulation and/or determination of high Reynolds number flow behavior in wind tunnel tests at Reynolds numbers of less than 10×10^6 with flight Reynolds numbers being of the order of 30×10^6 to 100×10^6.

3.1 Program and Test Set-Up

The main parameters affecting the viscous/inviscid interaction are

- the airfoil shape, i.e., essentially the type of pressure distribution occuring on the airfoil dependent on freestream Mach number and angle of attack, and

- the initial boundary- layer condition represented in the case of airfoil flow, e. g., by the displacement thickness and the shape factor upstream of the begin of strong adverse pressure gradients.

The initial boundary-layer condition depends mainly on the Reynolds number - for a given airfoil to a lesser degree on the forward pressure distribution - but can also be altered artificially by adding a transition strip to the model surface and changing its size and/or location.

Test program: According to objectives and main parameters of influence, boundary layer, wake and flow field measurements were carried out at DFVLR on supercritical airfoils having different characteristics in the pressure distribution. The free stream conditions were such that the local shock upstream Mach number (representative of the shock strength) on the upper surface varied between $M_1 \approx 1.2$ and $M_1 \approx 1.4$ at Reynolds numbers between $Re \approx 2 \times 10^6$ and $Re \approx 3.5 \times 10^6$. The initial boundary-layer condition was, in addition, varied by changing the tripping device location.

Besides the boundary layer measurements, surface pressure and wake measurements were conducted on a large number of supercritical airfoils testing one, viz., CAST 10-2/DO A 2, at Reynolds numbers up to $Re = 31 \times 10^6$ in the Lockheed CFWT transonic tunnel.

The airfoil sections most thoroughly investigated are the DORNIER supercritical designs CAST 10-2/DO A 2 and CAST 7/DO A 1 [11, 12]. They are shown in Fig. 20 together with their design pressure distributions. Note the somewhat stronger adverse pressure gradient in the case of CAST 10-2/DO A 2 on the upper surface at chord stations around $x/c = 0.90$ and the larger trailing edge angle of this airfoil. These features have a pronounced negative effect on the boundary layer development and on the magnitude of scale effects, respectively.

Test set-up and experimental techniques: The boundary layer and flow field measurements were carried out in the 1 x 1 Meter Transonic Wind Tunnel of the DFVLR-AVA Göttingen [13]. A front view of one of the models tested and the boundary-layer probe installed in the test section are shown in Fig. 21. The model chords were generally c = 200 mm with the span being b = 1000 mm. The relatively large span/chord ratios essentially eliminated side wall effects up to free stream conditions for total separation. Note that in Fig. 21 the boundary-layer probe is mounted for measurements on the lower surface. Probe and probe drive unit were attached to a sting, independent of the model, with the sting, in turn, fastened to the support usually employed to change the angle of attack of sting-mounted complete aircraft models. This allowed the probe to be always adjusted tangential to the model surface and to traverse normal to it.

The boundary layer probe employed here is comprised of a 0.15 mm fish-mouth type pitot probe, a cone-cylinder static probe and a directional probe consisting of two tubes cut-off under 45°. The use of a conventional probe was deemed sufficient for the present investigation since the main objective, as stated, was to determine the overall boundary layer and flow development as affected by viscous/inviscid interaction. Details of certain experimental techniques concerning, e. g., the calibration of the probe with respect to the flow direction, and the correction of the position due to probe deflection, as well as data reduction procedures are given in Ref. [14].

The test set-up for the surface pressure and wake measurements was similar to the one shown in Fig. 21 with the probe and probe support replaced by a wake rake.

3.2 Experimental Results

The conditions for which the most complete boundary-layer surveys were made on the upper surface of the airfoil CAST 10-2/DO A 2 (and similarly on the airfoil CAST 7/DO A 1) are shown in Fig. 22. The test parameters were at a constant free stream Mach number of $M_\infty = 0.765$

- The angle of attack - and hence the shock upstream Mach number - in a range centered about the (vaguely known) shock upstream Mach number for the onset of shock induced separation, i.e., $M_1 = 1.3$, covering part of the linear and non-linear range of the lift curves,

- the initial boundary-layer condition by changing the transition-strip location[1] at a constant Reynolds number of $Re_c = 2 \times 10^6$ and

- the Reynolds number at a constant transition-strip location of $(x/c)_{TS} = 0.07$.

The lift curves of Fig. 22 demonstrate, in addition, quite nicely the effect of changing the initial boundary-layer condition on the flow development like, for instance, the onset of flow break-down. The latter is indicated by the lift divergence which seems to occur for the more rearward transition-strip location at a higher angle of attack. Note, however, that there is already a strong influence in the linear range of the lift curves. It is now to be examined how viscous/inviscid interactions lead to these differences.

3.2.1 Effect of Increasing Angle of Attack (Shock Upstream Mach Number)

As the angle of attack and hence the shock upstream Mach number is increased, the load on the boundary layer becomes more severe. The effect is shown in Fig. 23 where the displacement-thickness distribution on the upper surface together with the "generating" pressure (Mach number) distribution are plotted for a Reynolds number of $Re = 3.5 \times 10^6$ and transition fixed at 7 % c.

Note, Fig. 23b,

- the strong increase (jump) in displacement thickness across the shock and the even stronger (absolute) increase due to the rear adverse pressure gradient.

The jump due to the interaction with the shock becomes more severe as the shock upstream Mach number is raised from about $M_1 = 1.25$ to $M_1 = 1.35$ with the interaction region increasing in size. The rear development starts at the end of the plateau downstream of the

---

[1] Identification of transition strips in all figures: 107 BA = glass spheres of 0.097 to 0.107 mm diameter; 124 BA = glass spheres of 0.107 to 0.124 mm diameter; 220 K = carborundum grit with an average height of 0.08 mm. The width of the transition strip was 1 % C.

shock; the existence of the plateau itself and its extent seem to indicate that only a minor recovery of the boundary layer occurs before the rear adverse pressure gradients take effect.

The lift curve divergence, Fig. 24, begins at $\alpha > 3.2^{\circ}$ as a result of the rapid increase in displacement thickness at the trailing edge which reaches, for instance at $\alpha = 3.5^{\circ}$, already a height corresponding to 3 % of the airfoil chord. As the shock upstream Mach number is raised (lower diagram of Fig. 24), one can observe a sudden strong increase in displacement thickness downstream of the shock, δ_2^*. The onset of this sudden increase coincides closely with the one for the trailing-edge displacement thickness, δ_3^*. One may conclude that the shock boundary-layer interaction strongly influences conditions at the trailing edge - at least at the higher shock upstream Mach numbers - which is most likely due to the fact that the boundary layer lacks time to recover sufficiently downstream of the interaction. The close coupling of conditions at the trailing edge with the shock can be seen by going back to Fig. 23a where at $\alpha = 3.5^{\circ}$, i.e., after the δ_3^*-divergence, the downstream movement of the shock has halted. It should be kept in mind that trailing edge conditions are, of course, also affected by the severity of rear adverse pressure gradients and curvature, i.e., by the airfoil geometry.

The development of the velocity profiles for angles of attack of $\alpha \geq 2.85^{\circ}$ ($M_1 \geq 1.274$) is shown in Fig. 25. The retardation of the profiles due to the shock, the subsequent (minor) recovery as the severity of the pressure gradient relaxes and the renewed decay of the profiles due to the rear adverse pressure gradients are clearly indicated in all cases shown. Before proceeding with analyzing the profiles with respect to separation, one should note: The reading of the pitot probe is in the presence of small separation bubbles in the vicinity of a wall generally erroneous. This is, firstly, due to disturbances introduced by the probe itself and, secondly, due to pressure fluctuations in the separated region. An extension of the quasi-linear part of the velocity profiles to the surface was, therefore, used, where applicable, (also see Ref. [15]) to determine separation in a very crude way.

Examining Fig. 25 in that regard it seems that

- neither shock induced nor rear separation are present at angles of attack of $\alpha = 2.85^{\circ}$ and $\alpha = 3.2^{\circ}$, i.e., at shock upstream Mach numbers of $M_1 = 1.274$ and $M_1 = 1.318$, respectively (however, at the latter a small shock induced bubble could exist with its extent being less than 5 % c),

- shock induced separation is present at shock upstream Mach numbers of $M_1 = 1.329$ ($\alpha = 3.5^{\circ}$) and $M_1 = 1.345$ ($\alpha = 3.9^{\circ}$), respectively but rear separation occurs only at $M_1 > 1.33$ although the rapid increase in δ^* at the trailing edge and the drop in lift curve slope, Fig 24, commence already at $\alpha \approx 1.32$. The delayed trailing edge separation is most likely a consequence of the viscous/inviscid interaction where the rear adverse pressure gradients are softened by the thickenning of the boundary layer.

3.2.2 Reynolds Number and Scale Effects

Comparing the chord-wise development of the displacement thickness at about the same shock upstream Mach number ($M_1 \approx 1.307$) but different initial boundary-layer conditions, one can observe, Fig. 26b,

- that the initially thinner boundary layer stays thinner across the shock and down to the trailing edge, an observation not as trivial as it seems since it is the displacement thickness that the outer inviscid flow sees and reacts to, and

- that the Reynolds number has as far as the interaction with the shock is concerned only a minor effect - the initial displacement thickness being nearly the same - , however, noticeably influences the development between shock and trailing edge.

The interaction pressure (Mach number) distributions, Fig. 26a, show the effects of the differences in displacement-thickness distribution downstream of the shock, viz., for increasing displacement thickness a reduction in pressure recovery between shock and trailing edge and as a consequence a more forward shock position. (Also note the increase in lower surface pressures due to the reduction in displacement thickness.) The large change in circulation due to the (slight) change in the initial δ^* is best demonstrated by the fact that in the present comparison a much lower angle of attack had to be selected for the thinner boundary layer in order to allow the comparison to be made at about the same shock upstream Mach number.

The curves in Fig. 27 indicate that the small change in the initial displacement thickness, caused by moving the transition strip from 7 % c to 30 % c, is significantly amplified by the shock and again by the rear adverse pressure gradient. A numerical example shall underline this (also see vertical bars in Fig. 27): At M_1 = 1.346 the change in δ_1^* due to relocating the transition strip corresponds to 0.075 % of the chord (c = 200 mm). This change is boosted by the shock to 0.375 % c and by the total interaction, i.e., across the shock and down to the trailing edge, to 2.2 % c, resulting in "amplification" factors of 5 and almost 30, respectively. It is this "amplification" of an initially small difference that results in the large Reynolds number and scale effects observed on many modern airfoils.

Investigations with the airfoil CAST 7/DO A 1 have shown this airfoil to be less sensitive to scale effects than CAST 10-2/DO A 2 which can partly be attributed to the less severe rear adverse pressure gradients. The reduction of the boundary layer data obtained for this airfoil has, unfortunately, not proceeded far enough for results to be included here.

It was shown, see for instance Fig. 24, that the sudden increase in δ^* across the shock, as the shock upstream Mach number is raised, nearly coincides with the δ^*-divergence at the trailing edge. This emphasizes the importance of the shock boundary-layer interaction on the overall flow development and the knowledge of the shock upstream Mach number for the onset of the rapid increase dependent on the initial boundary-layer condition seems essential. Fig. 28 indicates that for the range of conditions investigated this Mach number is only weakly affected to M_1 = 1.318. The insert to Fig. 28 shows that the same holds for the shock upstream Mach number for incipient separation, the latter determined by ONERA (and discussed earlier).

The two curves of the insert to Fig. 28 necessitate one more comment: There is a marked difference in the curves for incipient separation and the onset of the rapid increase in displacement thickness (e. g., ΔM = 0.045 at H_i = 1.5). It is most likely that the "kink" in the curves $\Delta \delta_{12}^{*}$ = $f(M_1)$ indicates the begin of rapid shock-induced bubble growth which occurs after the bubble - with increasing shock strength - has reached a certain size. Essential is that the "kink" Mach number signals the beginning of the flow break-down while the (lower) Mach number corresponding to incipient separation leaves a certain safety margin to the designer.

3.2.3 High Reynolds Number Flow

The maximum Reynolds number for which boundary-layer data were obtained was Re = 3.5 x 10^6. Considering the effect of the initial displacement thickness just discussed and the difference in lift resulting from small differences in the initial δ^{*}, one must assume that increasing the Reynolds number further will have a further strong positive effect on lift. Surface pressure measurements with the CAST 10-2/DO A 2 airfoil in the Lockheed transonic wind tunnel [16] have confirmed this. Fig. 29, lower curve, shows that a sustained increase in lift coefficient exists up to the highest Reynolds number investigated, viz., Re = 31 x 10^6.

The upper curve of Fig. 29 is obtained by moving the transition strip downstream keeping the Reynolds number constant at Re = 2.4 x 10^6. Since the pressure gradient upstream of the shock is favourable, transition always occurs at the strip (not before) with the most aft transition location being the one for free transition, the latter being caused by the shock. The downstream shift in transition results, as was demonstrated by the boundary-layer measurements, in a reduction in the initial displacement thickness and, as a consequence, in an increase in lift. Fig. 29, upper curve, shows that this trend is continued until the transition point reaches the shock.

It is quite obvious from this figure that high Reynolds number lift coefficients can, within limits, be duplicated at low Reynolds numbers by selecting the proper transition strip location. However, how close do corresponding pressure distributions under these circumstances agree? Fig. 30, comparing the c_p-distributions corresponding to Point ⑤ of the lower and Point ④ of the upper curve of Fig. 29, shows that the pressure distributions obtained under such different test conditions, viz.,

- a Reynolds number of Re = 2.4 x 10^6 and transition fixed at 45 % of the chord and

- a Reynolds number of Re = 31 x 10^6 with transition occuring close to the leading edge

are, for all practical purposes, identical. This comparison demonstrates, as have the boundary-layer measurements discussed earlier, the crucial role of the initial boundary-layer (displacement) thickness in viscous/inviscid interactions in transonic flow; it also indicates a method - or better confirms a method that has up to now been applied by DFVLR-AVA without much proof - to simulate high Reynolds number flow in conventional wind tunnels at low Reynolds numbers.

Considering the latter, there exist certain restrictions when using rear transition strip locations to reduce the initial displacement thickness. Here, this method covers only free stream conditions for which the shock - or the onset of other severe adverse pressure gradients - is located sufficiently far downstream, as is usually the case in the vicinity of the design point and the drag rise and buffet boundaries. However, an extension of this approach to include all essential free stream conditions is possible by applying, for instance, suction to obtain the desired displacement thickness upstream of strong interactions.

4. Theoretical Approaches to the Viscous/Inviscid Interaction Phenomena

4.1 A Computational Procedure Including a Special Solution for Shock Wave Boundary-Layer Interaction Developed at DFVLR

From the experimental results described in Section 3 one can deduce that small differences in the displacement thickness downstream of the shock will be amplified by the subsequent rear adverse pressure gradients resulting in large differences at the trailing edge and, as a consequence, in the overall pressure distribution and lift. Considering theory, this means that the change in boundary-layer properties due to the interaction with the shock must be determined with a high degree of accuracy. In shock boundary-layer interaction the physics of the flow are not well represented by boundary-layer theory and it seems very unlikely that conventional boundary-layer theory will meet the high accuracy requirement. In the present approach to transonic airfoil flow analysis, limited to non-separating turbulent flow and modelled after concept (3) given in the Introducion, an analytical solution for near-normal shock wave boundary-layer interaction was, therefore, included as a module in a state-of-the-art viscous/ inviscid computation code [17].

To solve the transonic shock boundary-layer interaction problem for non-separating turbulent flow (local Mach number upstream of shock ≤ 1.3 in the Reynolds number range $Re \approx 10^6$ to 10^8), a non-asymptotic triple deck disturbance flow model is employed [18]. The model, Fig. 31, is comprised of an upper mixed flow region outside the boundary layer consisting of an incoming potential supersonic flow in Region ① and a subsonic potential flow in Region ③ separated by a given shock discontinuity. Underlying these regions is a double-infinite non-uniform boundary-layer region (Region ②) that contains a highly rotational, mixed transonic linear disturbance flow. Near the wall a viscous disturbance sublayer exists that contains the upstream influence and the skin-friction pertubation. The equations describing the flow model are solved by operational methods to obtain the interactive pressure rise, the displacement thickness growth and the local skin-friction solution [18].

The viscous flow analysis is carried out by adding the displacement thickness to the geometric airfoil contour and performing an inviscid computation for the effective airfoil shape. The displacement thickness is obtained outside the shock boundary-layer interaction domain, Fig. 32, by Rotta's integral dissipation method [19] providing at the interface between Regions (A) and (B) , Fig. 32, the boundary-layer parameters needed to start the interaction code, viz., the local Reynolds number based on δ_1^* and the shape factor H. Shock upstream Mach number and shock location which are further inputs to the interaction code are taken from the pressure distribution given either by experiment of inviscid theory. The interaction module then determines the displacement thickness, the shape factor and the skin-friction coefficient together with the surface-pressure distribution due to shock impingment, supplying the parameters necessary to reinitiate the boundary-layer method downstrem of the interaction.

The inviscid computation utilizes Jameson's version [20] of the relaxation method of [21]. The computation for the airfoil plus displacement surface starts with an initially assumed analytically modelled displacement-thickness distribution. Subsequently, the pressure distribution calculated by inviscid theory is used by the boundary-layer/shock boundary-layer interaction code to determine a new displacement thickness. The latter is approximated analytically, smoothing over the local more sudden change in displacement thickness due to shock impingment, the approximation added to the actual airfoil contour and the inviscid computation repeated. This procedure is performed iteratively until the desired agreement between added and calculated displacement thickness is reached.

Numerical examples of computations performed by the boundary-layer/ shock boundary-layer interaction code with the measured pressure distribution as input and the complete viscous/ inviscid computation method, using the non-conservative scheme for the inviscid computation and a comparison with corresponding experimental data are given in Figs. 33 and 34, respectively. The agreement between experiment and theory is in both instances quite good. It can be shown [17] that the accurate prediction of the experimentally determined boundary layer (δ^*) development downstream of the interaction, Fig. 33, is mainly due to the interaction code providing the "correct" input for the subsequent boundary-layer computation.

4.2 Coupling Method Developed at ONERA for the Treatment of Viscous/Inviscid Phenomena

The calculation of strong viscous interaction effects has been developed at ONERA along two complementary approaches:

The first consists in adopting traditional methods used in weak viscous interaction regimes; it is based on the simplified introduction of the effect of boundary-layer displacement [22].

The second approach concerns the development of more refined methods based, on one hand, on a generalization of the boundary-layer concept and, on the other hand, allowing one, thanks to strict numerical methods for the coupling, to treat any problem of strong viscous interaction including separations by conjugating inverse calculation methods for the dissipative layers and a semi-inverse method for coupling interactions [9,10, 23].

In the first case, the principles which governed the implementation of a practical calculation method are essentially the same as those used at DFVLR and presented previously. However, the "ingredients" are somewhat different.

The ideal fluid problem is also treated by a non-conservative method, in this case that of Garabedian and Korn so as to obtain jump conditions across the shock wave leading to a state close to critical conditions.

The wall under consideration is displaced by the boundary-layer displacement thickness, calculated in direct mode from the ideal fluid solution and appropriately smoothed if necessary. The viscous ramp associated with the presence of the shock wave is, however, not defined in a theoretical manner but in a semi-empirical manner according to the above mentioned principles.

An iterative procedure makes it possible to tie together the series of calculations without the convergence of the process having to be automatically ensured.

As long as no separation occurs, such a method provides a usually satistactory working tool, as shown by the two examples given in Fig. 35 for the CAST-7/DO A 1 airfoil.

As already seen, this profile has been experimentally studied at DFVLR within a wide Reynolds number range. The examples presented here are from tests performed at a Reynolds number of 6 million and a Mach number of 0.76. On the profile polar for this Mach number, the corresponding points of these examples are located on either side of the drag rise. The values of C_L obtained are 0.57 and 0.63, respectively.

We observe a rather satisfactory agreement in the shock wave position and also in the pressure distributions on both the upper and lower surfaces down to about 80 % of the chord. However, the prediction of the pressure at the trailing edge is not satisfactory, as the wake effects are not taken into account.

As regards the second method, it is, by its very principle, able to automatically capture the phenomena of rapid interaction related to the presence of shock waves or of separations whatever their number. This means in the case of a shock wave, for instance, that the extent of the numerical interaction domain related to shock capturing or to the mesh pattern must be noticeably smaller than the length L_s of the physical interaction domain [9, 10, 23].

This condition necessarily entails a local mesh subpattern of the flow within the domains where rapid interaction phenomena occur.

In the present state of affairs and for flows at high Reynolds numbers for which the dissipative layers are very thin, such an operation appears very difficult, if not impossible, to realize by means of current potential ideal fluid techniques. Indeed, the generally used relaxation methods negotiate rather badly the pattern distorsions or densities that they are required to use imposing difficulties for the calculation, slow to converge.

Research for overcoming these difficulties is under way. However, to illustrate the possibilities of such a method, we shall give an example in which, although the adaption of the mesh pattern to the scale of the viscous effects in shock boundary-layer interaction is just sufficient, the coupling in strong interactions has been strictly realized at every node of the calculation, the numbers of nodes along the wall and the wake being the same for viscous and inviscid calculations.

This example concerns the RAE 2822 supercritical profile investigated in the 8 by 6 foot wind tunnel of RAE (Fig. 36).

The calculation of viscous effects is performed by an integral method governing the normal velocity of the ideal fluid on the wall as well as a discontinuity in normal velocity located along the median line of the viscous wake, whose position is actually calculated.

The strong interaction at the trailing edge is fully accounted for, including the case of sepa-ration with reattachment in the wake. We solve simultaneously the viscous differential equations acting as boundary conditions and the potential equation by the factorized fully conservative relaxation method.

The calculation example presented here corresponds to a Mach number $M_\infty = 0.732$ and an angle of attack $\alpha = 2.85^o$ and a Reynolds number of 6.5 million with a triggered transition at 3 % of the chord on both upper and lower surfaces. The experimental angle of attack, $\alpha = 3.22^o$, has been corrected to match approximately the experimental lift coefficient of $C_L = 0.79$. The calculated drag, $C_D = 0.0185$, is slightly higher than the experimental one, $C_D = 0.0175$. The comparison of calculated and measured pressures is satisfactory as a whole, although the Mach number behind the shock is somewhat low because of the choice of a conservative ideal fluid method and a rather wide mesh pattern relative to the shock boundary-layer interaction domain. One should mention the very good numerical behaviour of the method which does not include smoothing around the trailing edge, and, in partiular, the disappearance of stagnation points whatever the profile, as well as a good agreement with experiments. In this example, the Mach number before the shock was about 1.25 so that the boundary layer remainded attached everywhere, conditions close to separation being only reached near the trailing edge.

5. Conclusion

This account of strong viscous interaction phenomena in 2-D transonic flow, although not complete, nevertheless shows that a certain progress has been made in the understanding of the mechanism of this interaction occuring on supercritical airfoils. It has been possible to characterize the conditions for incipient separation due to the shock and conditions downstream of the interaction. In this context the influence of Reynolds number and associated scale effects have been discussed together with the proper experimental means for the simulation of these phenomena.

It has been shown that current computing methods give a good description of the flow past such airfoils when there is no separation. Nevertheless, there still exists a number of difficulties for the numerical computation when shock induced separation occurs and develops downstream. These difficulties are strongly related to

1) computing the fine details associated with the rapid development characteristic of supersonic flow when shock waves form,

and

2) the correct description of the turbulent phenomena with strong local non-equilibrium and associated relaxation effects, important to the boundary layer development downstream.

These two topics constitute subjects of research which seem to us among the most important. If solved, they would lead in the near future to the creation of very efficient computing tools.

The existence in the experimental domain of means for analyzing turbulent fields, such as LDV, can contribute substantially in improving our knowledge and consequently our methods of prediction.

Restricted to 2-D flows, this analysis, in fact, constitutes only a small part of our needs in the transonic domain. The most important problems as far as applications are concerned arise in 3-D flows for which the same research effort as in the 2-D case has to be undertaken.

Fortunately the need appears at a time of convergence of favourable factors: First, there are numerical methods under development allowing the computation of inviscid 3-D flows and 3-D boundary layers which reasonably approximate the real physics. Secondly, the development of high speed computers and computer architecture allows large computations in a relatively short time, simulating realistic shapes of wings and wing-fuselage combinations. Lastly, prospects for the realization of facilities able to simulate sufficiently high Reynolds numbers are now very good with the realization in some instances already in progress.

It is in this respect that our cooperative work has now to be defined.

6. References

[1]　S t a n e w s k y , E. and　L i t t l e , B.H. : Studies of separation and reattachment in transonic flow. AIAA Atmospheric Flight Mechanics Conference, Tullahoma, Tennessee, May 13-15, (1970), Paper No. 70-541.

[2]　D e l e r y , H. : Recherches sur l'interaction onde de choccouche limite turbulent. T.P. ONERA No. 1976-135.

[3]　D e l e r y , J., S u r g e t , J. and L a c h a r m e , J.P.: Interferometrie holographique quantitative en écoulement transsonique bidimensionnel. La Recherche Aérospatiale No. 1977-2, pp. 89-101.

[4]   S u r g e t , J., D e l e r y , J. and L a c h a r m e ,J.P. : Holographic Interferometry applied to the metrology of gaseous flows. 1er Congres Europeen sur l'optique appliquee a la Metrologie, Strasbourg Octobre 1977. T.P. ONERA no. 1977-169.

[5]   B o u t i e r ,   A. :Étude et réalisation d'un vélocimètre compact. Application a des mesures de vitesse en écoulement supersonic et transsonique très turbulents. Note Technique ONERA no. 237   (1974).

[6]   M i c h e l , R., M i g n o s i , A. and Q u e m a r d , C. : The induction driven tunnel T_2 at ONERA-CERT. Flow qualities, testing techniques and examples of results. 10th Testing techniques conference San Diego 1978, AIAA Paper No. 78-767.

[7]   D e l e r y , J. : Analyse du décollement résultant d'un interaction choc-chouche limite turbulent en transsonique. La Recherche Aérospatiale No. 1978-6, pp. 305-320.

[8]   D e l e r y , J. and L e D i u z e t , P.: Décollement résultant d'une interaction onde de choc-couche limite turbulente. XVIe colloque d'Aérodynamique Appliquée, Lille 1979. T.P. ONERA No 1979-146.

[9]   L e B a l l e u r , J.C. : Couplage visqueux - non visqueux: analyse de problème incluant dècollements et ondes de choc. La Recherche Aerospatiale No. 1977-6, pp. 349-358. English translation ESA TT 476 (1978).

[10] L e B a l l e u r , J.C. : Couplage visqueux - non visqueux: methodes numerique et applications aux écoulements bidimensionnels transsoniques et supersoniques. La Recherche Aèrospatiale No. 1978-2, pp. 65-76.

[11]K ü h l , P. and Z i m m e r , H. : Design of airfoils for transport aircraft with improved high speed characteristics. DORNIER GmbH, Report 74/16B, (1974).

[12] S t a n e w s k y , E. and Z i m m e r , H. : Development and wind tunnel tests of three supercritical airfoils for transport aircraft. Zeitschrift für Flugwissenschaften 23, (1975), Heft 7/8.

[13] H o t t n e r , Th. and L o r e n z M e y e r , W. : The Transonic wind tunnel of the Aerodynamische Versuchsanstalt Göttingen. DGLR-Jahrbuch (1968), pp. 235-244.

[14] S t a n e w s k y , E. and T h i e d e , P. : Boundary layer and wake measurements in the trailing edge region of a rear-loaded transonic airfoil. BMVg-FBWT 79-31, pp. 3135 (4th US-FGR-Meeting on Viscous and Interacting Flow Field Effects).

[15] S e d d o n , J. : The flow produced by interaction of a turbulent boundary layer with a normal shock wave of strength sufficient to cause separation. RAE Technical Memorandum No. Aero 667, (1960).

[16] S t a n e w s k y , E., P u f f e r t , W. and M ü l l e r , R.: Comparison of wind tunnel wall interferences in several transonic tunnels. Meeting of the DGLR-Fachausschuß 3.1 Hydro-, Aero- and Gasdynamics on Effect of Wind Tunnel Flow on Experimental Results, Göttingen, Februar 6 - 7, (1979), Paper 1-3.

[17] N a n d a n a n , M., S t a e w s k y , E. and I n g e r , G.R. : A computational procedure for transonic airfoil flow including a special solution for shock boundary layer interaction. AIAA 13th Fluid and Plasma Dynamics Conference, Snowmass, Colorado, July 14 - 16, (1980), Paper No. 80-1389.

[18] I n g e r , G.R. and M a s o n , W.H. : Analytical theory of transonic normal shock-turbulent boundary layer interactions. AIAA Journal, 14 (1976), pp. 1266-1272.

[19] R o t t a , J.C. : Turbulent boundary layer calculations with the integral dissipation method. Computation of Turbulent Boundary Layers. AFOSR-IFP Stanford Conference, I (1968), p.177.

[20] J a m e s o n, A. : Numerical computation of transonic flow with shock waves. Symposium Transsonicum II, Springer Verlag, Berlin, Heidelberg, New York, (1975).

[21] B a u e r, F., G a r a b e d i a n, P. K o r n, D. and J a m e s o n, A. : Supercritical wing sections II. Springer Verlag, Berlin, Heidelberg, New York, (1975).

[22] B o u s q u e t, J. : Calculs bidimensionnels transsoniques avec couche limite. 11e colloque AAAF (1974).

[23] L e B a l l e u r, J.C., P e y r e t, R. and V i v i a n d, H. : Numerical studies in high Reynolds number aerodynamics. Computers and Fluids, 8,(1980), pp. 1-30.

7. Nomenclature

a	speed of sound
c	airfoil chord
C_D	drag coefficient
C_L	lift coefficient
C_M	pitching moment
c_p	pressure coefficient
H	shape factor, $\delta*/\Theta$
$\bar{k}$	turbulent kinetic energy, Fig. 18
L_s	interaction length, Fig. 7
M	Mach number
P	probability of $\bar{u} \leq 0$
$\bar{P}r$	shear stress production, Fig. 18
p	static pressure
p*	critical pressure corresponding to M = 1.0
q	dynamic pressure
q_m	flow rate, Fig. 14
$\mathcal{R}e$, Re	Reynolds number based on chord
$\mathcal{R}_\delta*$, $Re_\delta*$	local Reynolds number based on $\delta *$
T	Temperature
$\bar{u}, \bar{v}$	components of mean velocity in x and y, resprectively
$\bar{u}_m$	maximum reversed velocity, Fig. 15
x, y	rectangular coordinates
x/c	non-dimensional chord

α	angle of attack
δ	boundary layer thickness
$\delta*$	displacement thickness
η	$0 \leq \eta \leq 1.0$ for $0.5 \leq (\bar{u}-\bar{u}_m)/(\bar{u}_e-\bar{u}_m) \leq 0.95$
Θ	momentum thickness
ν_t	eddy viscosity term
ρ	density
$\bar{\tau}$	turbulent shear stress

Subscripts

$\left.\begin{array}{l} 0 \\ 1 \end{array}\right\}$	initial conditions upstream of shock
∞	free stream conditions
B	boundary layer
e	edge of boundary layer
i	stagnation conditions
i	incompressible flow

Superscripts

	denotes fluctuating quantities

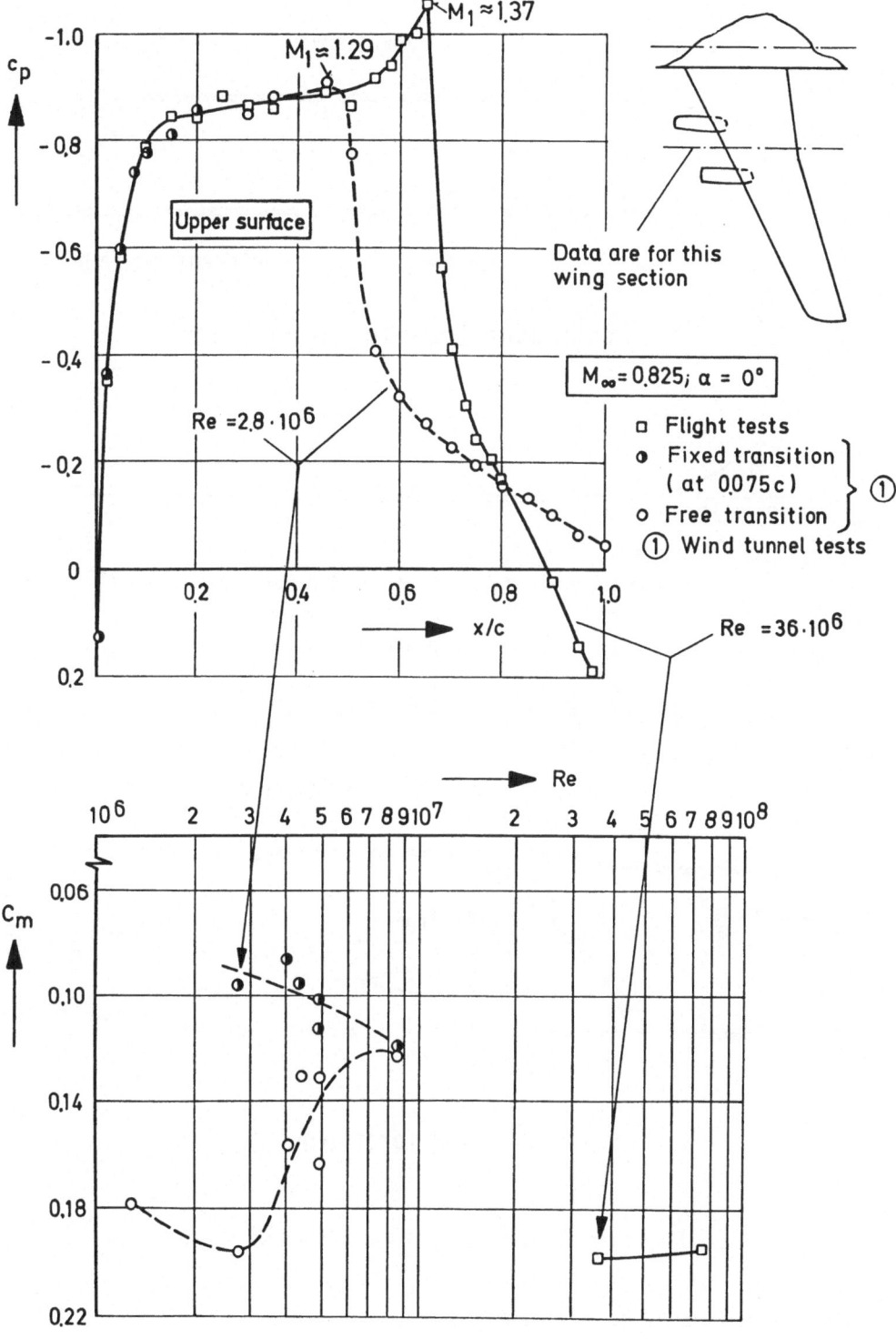

Fig. 1: Scale affects observed on C-141 aircraft.

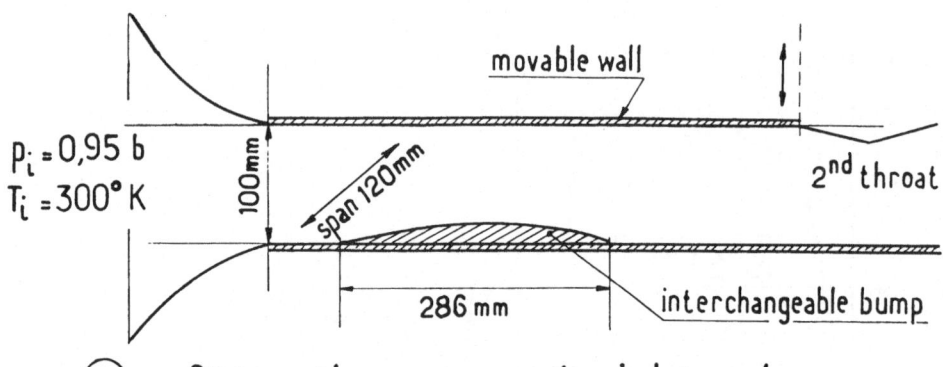

$p_i = 0.95\,b$
$T_i = 300°\,K$

100mm

span 120mm

movable wall

2nd throat

286 mm

interchangeable bump

(a)   S8A continuous transonic wind tunnel

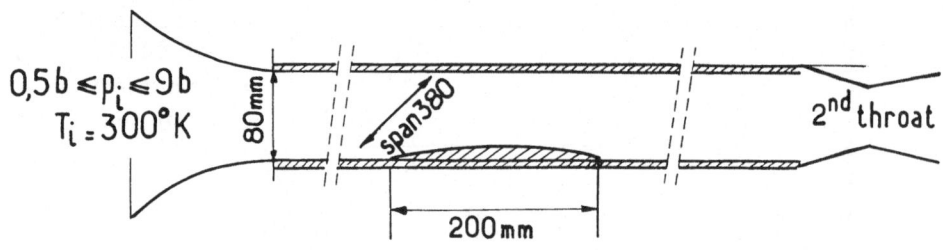

$0.5b \leqslant p_i \leqslant 9b$
$T_i = 300°\,K$

80mm

span 380

2nd throat

200 mm

(b) R_1Ch intermittent transonic wind tunnel

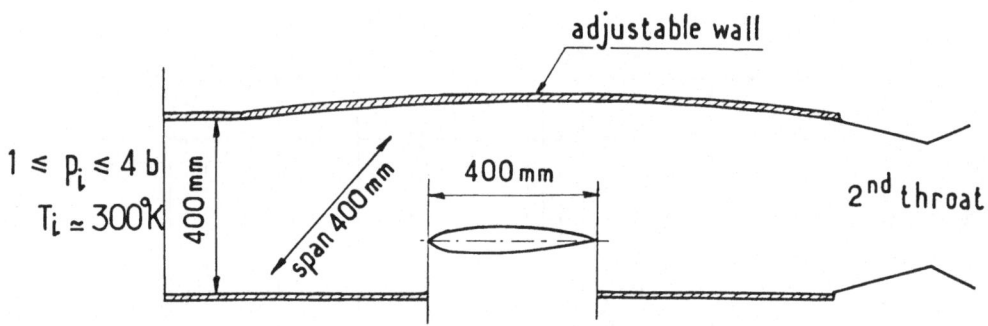

adjustable wall

$1 \leqslant p_i \leqslant 4\,b$
$T_i \simeq 300°K$

400 mm

span 400 mm

400 mm

2nd throat

(c) T_2 intermittent transonic wind tunnel

Fig. 2: Test Set-up for basic investigations (ONERA).

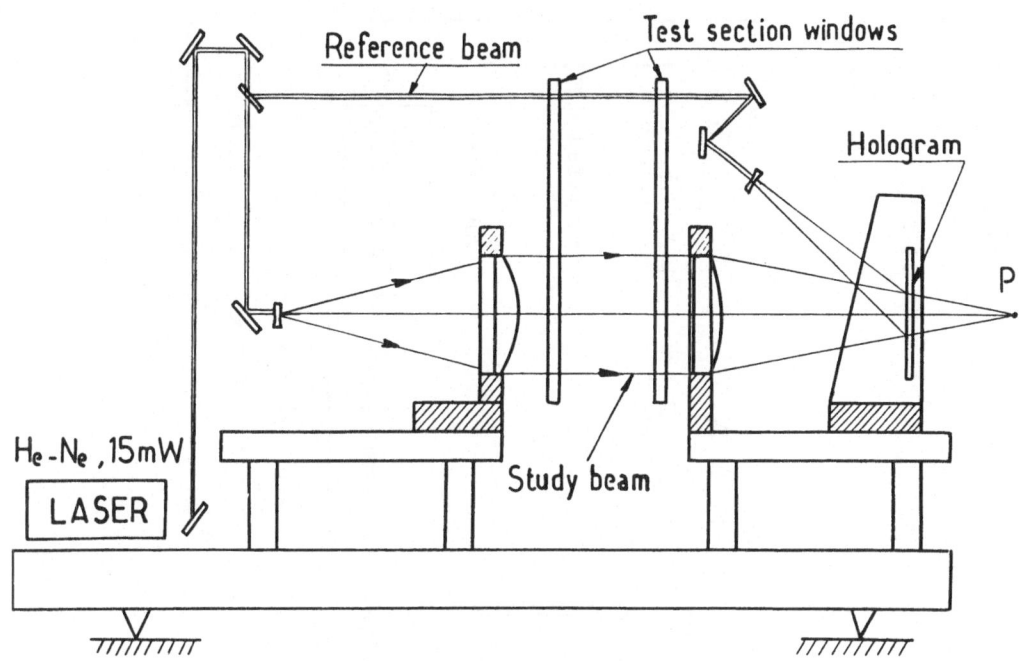

a_ Layout of the holography bench

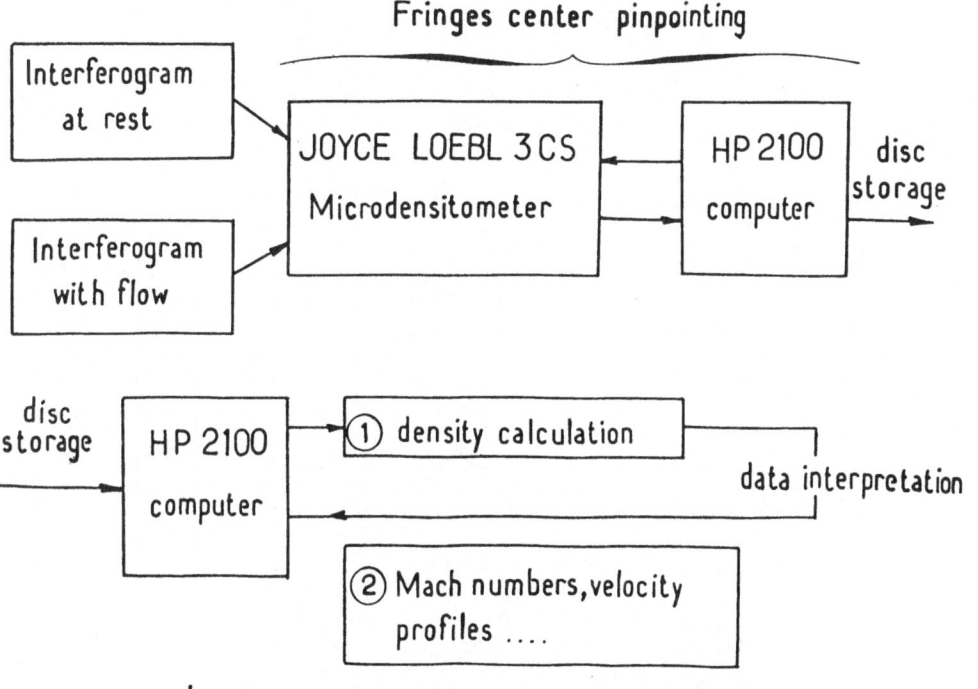

b_ Interferogram processing method

Fig. 3: Holographic interferometry system.

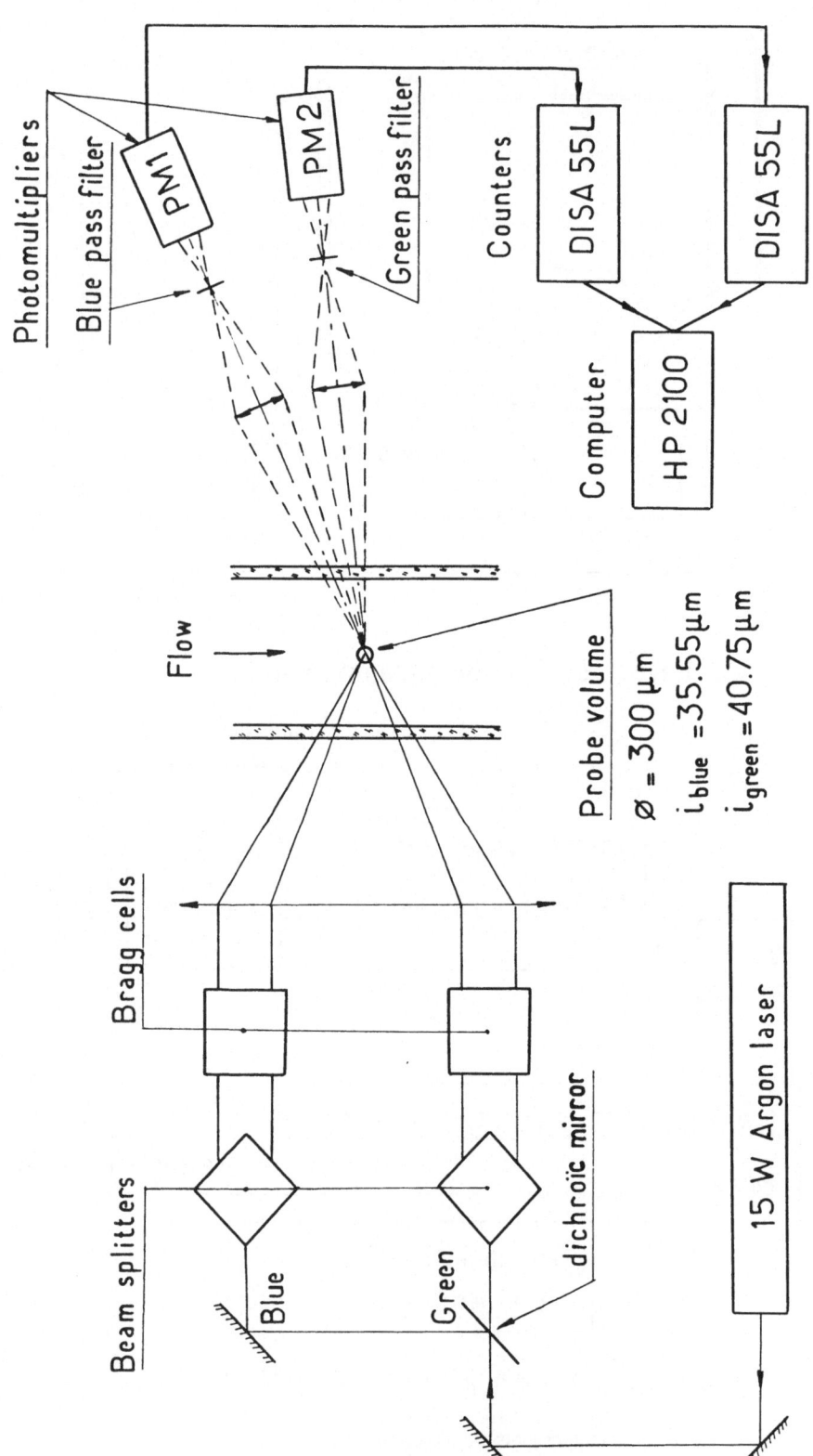

Fig. 4: Two colour laser velocimeter system.

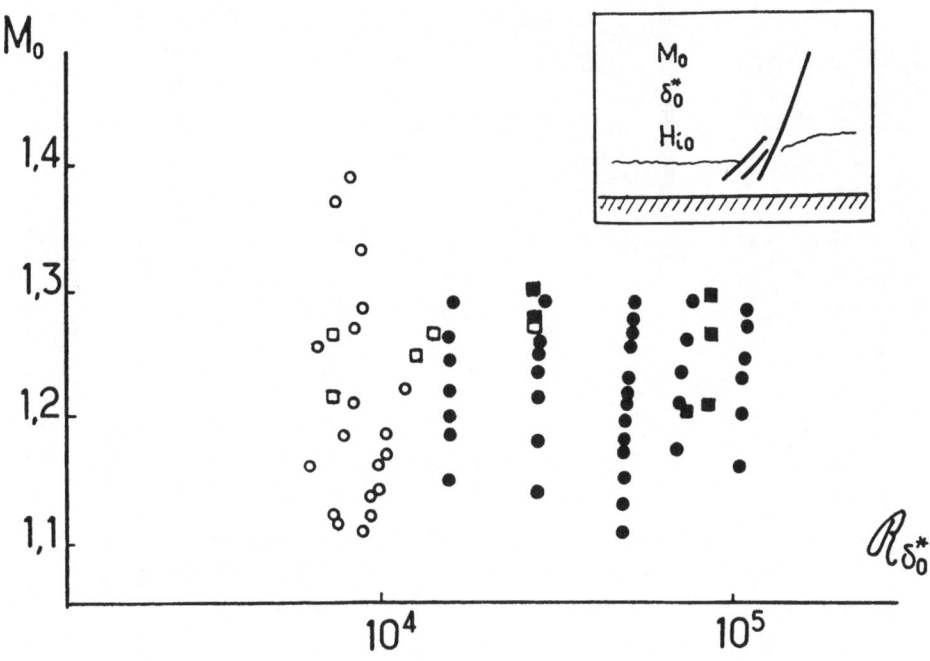

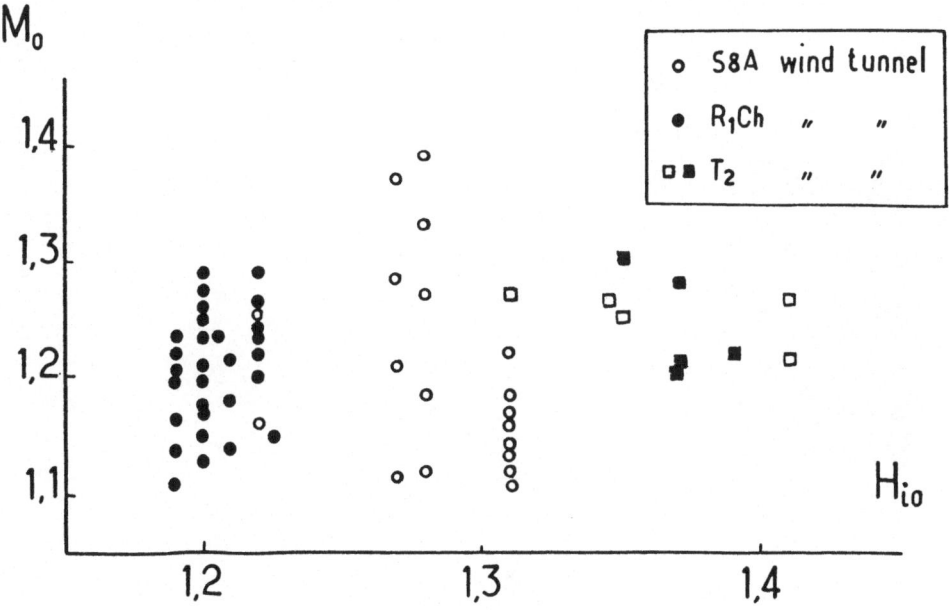

Fig. 5: Initial BL condition of the basic investigation.

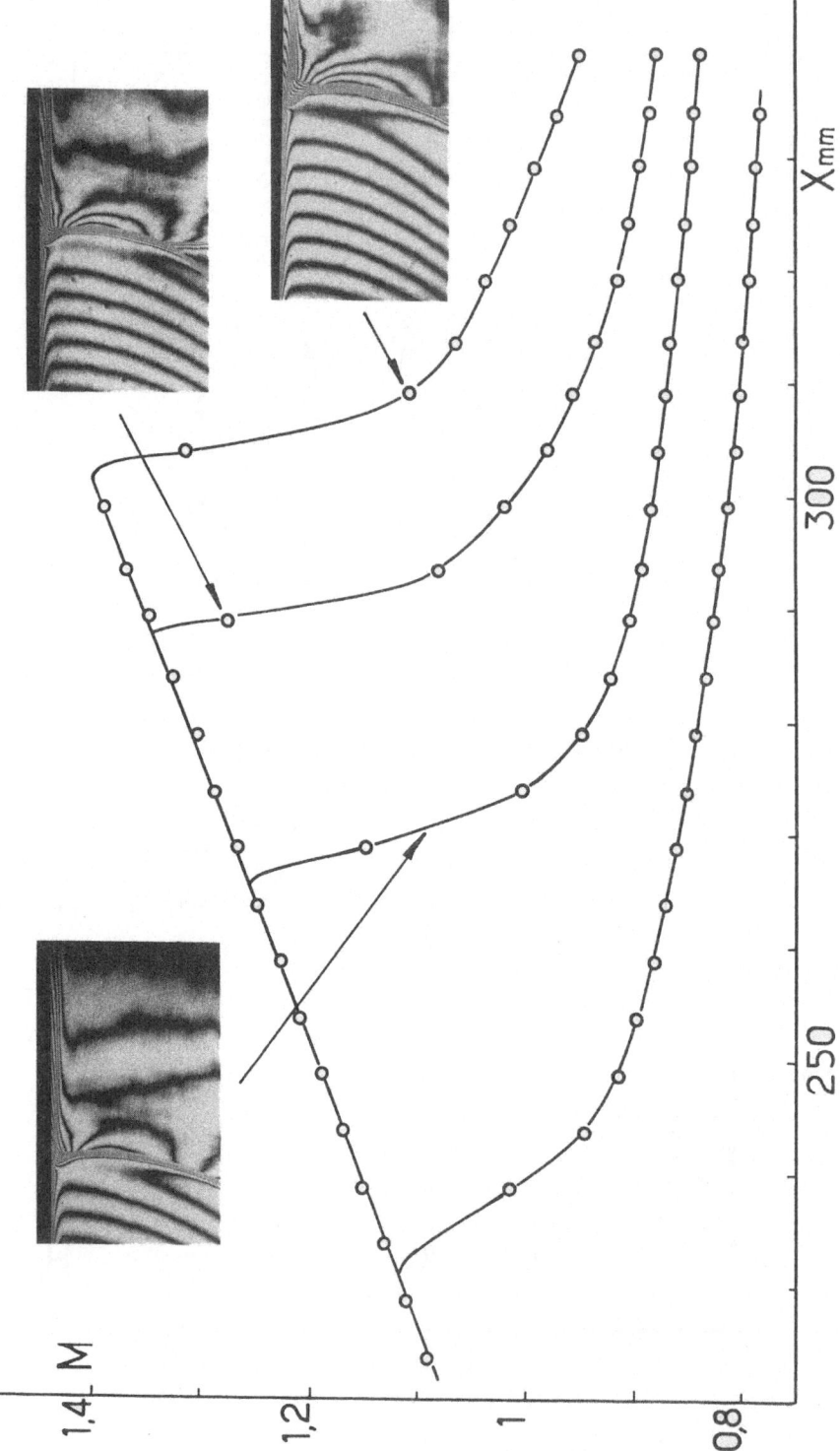

Fig. 6: Typical wall Mach number distributions investigated S8A transonic wind tunnel.
$H_{io} = 1,28$ $Re_{\delta o}^* \simeq 0,8 \times 10^4$.

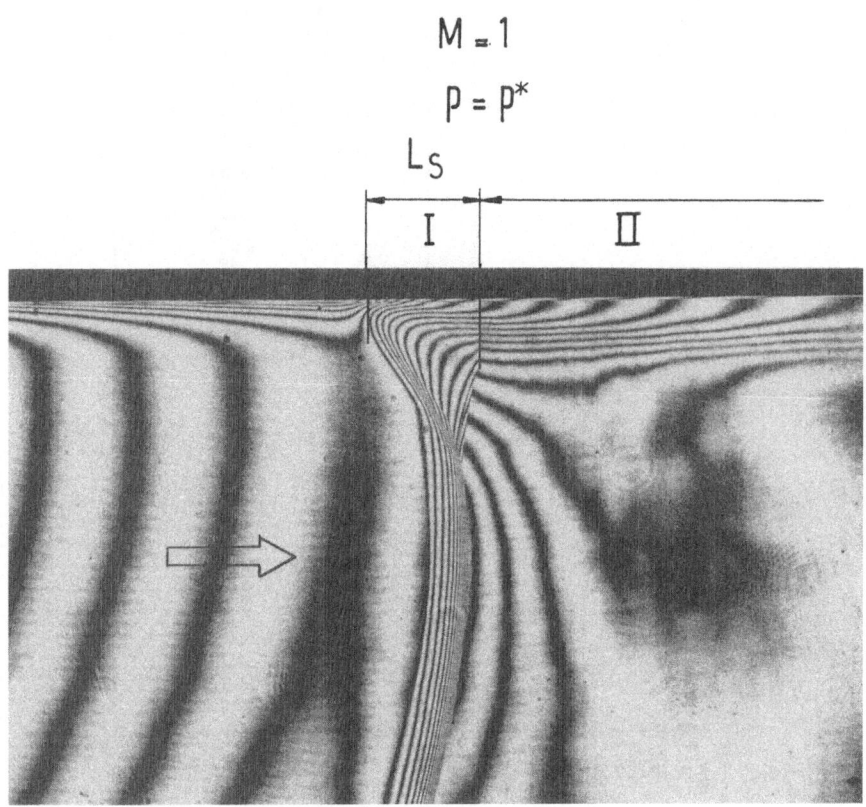

Fig. 7: Enlarged picture of interaction region.

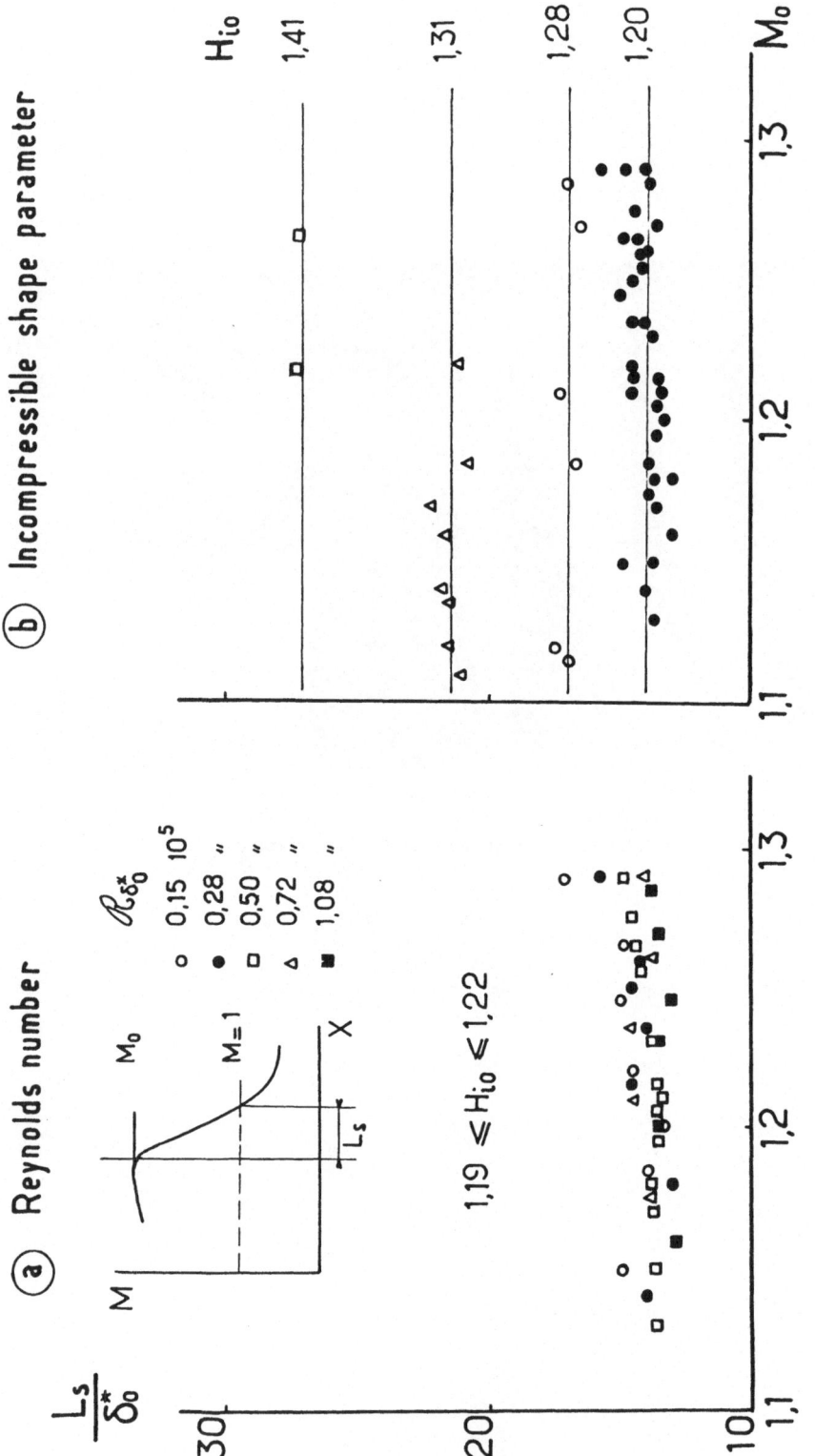

Fig. 8: Effect of the initial conditions on the interaction length.

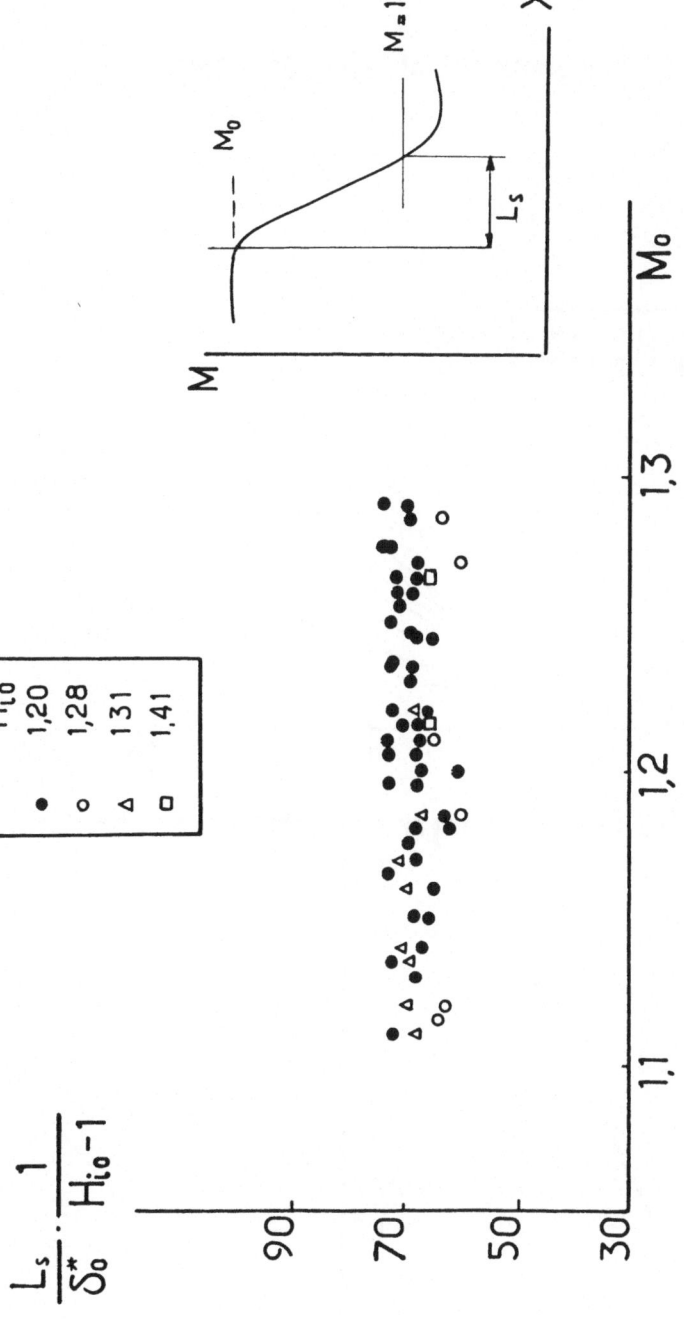

Fig. 9: Correlation of interaction length.

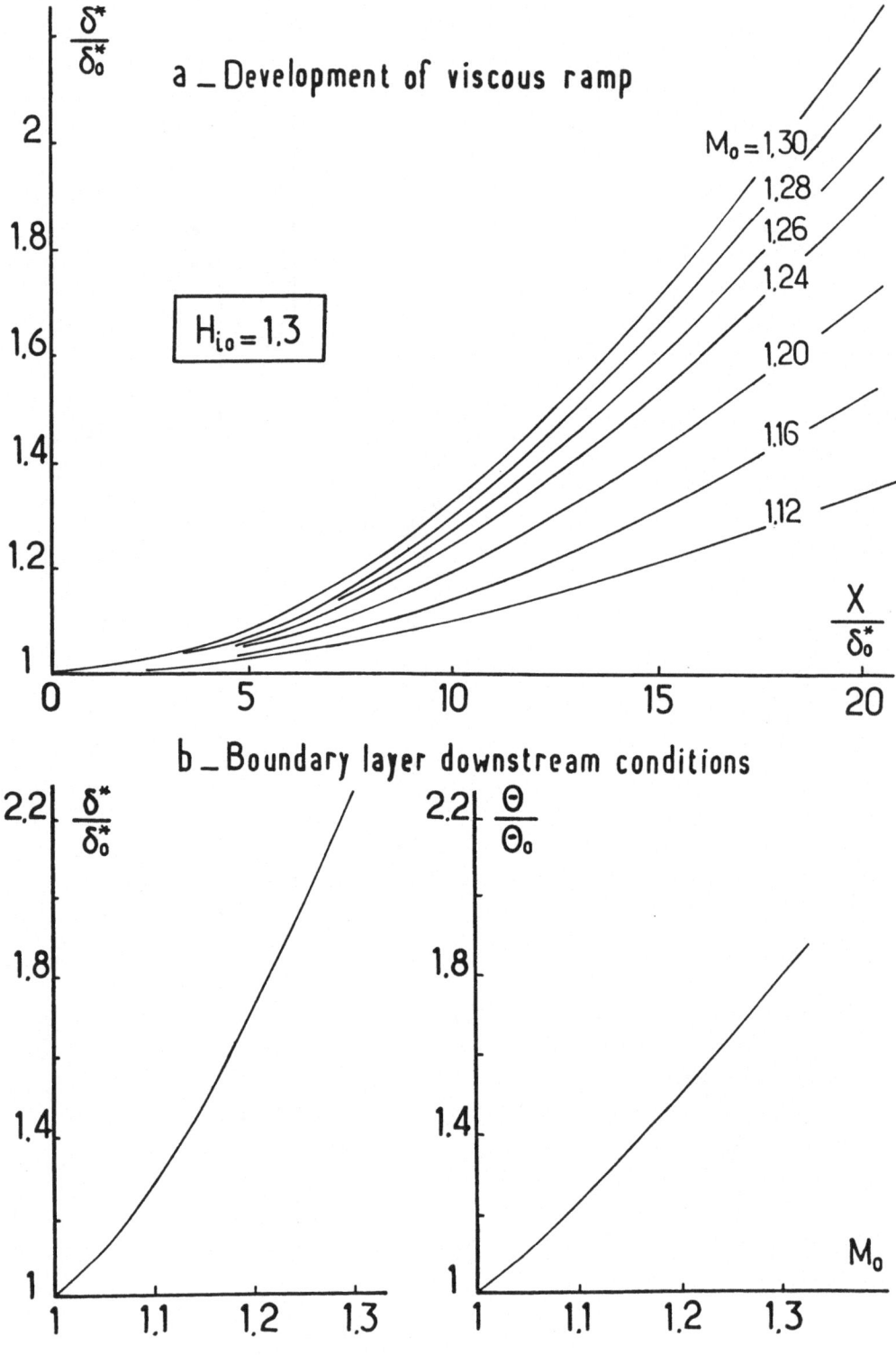

Fig. 10: Example of viscous ramp calculations.

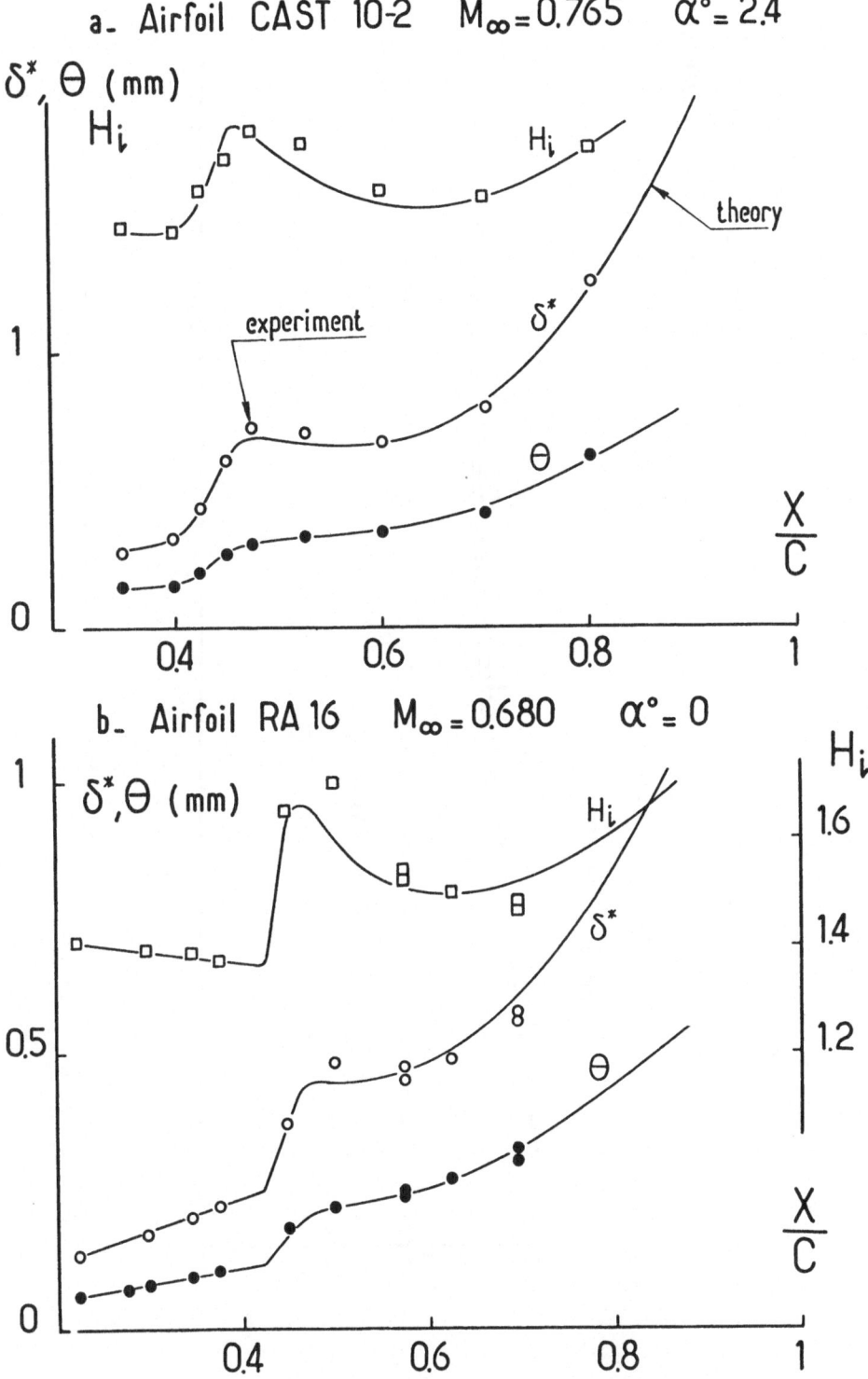

Fig. 11: Examples of boundary layer calculations.

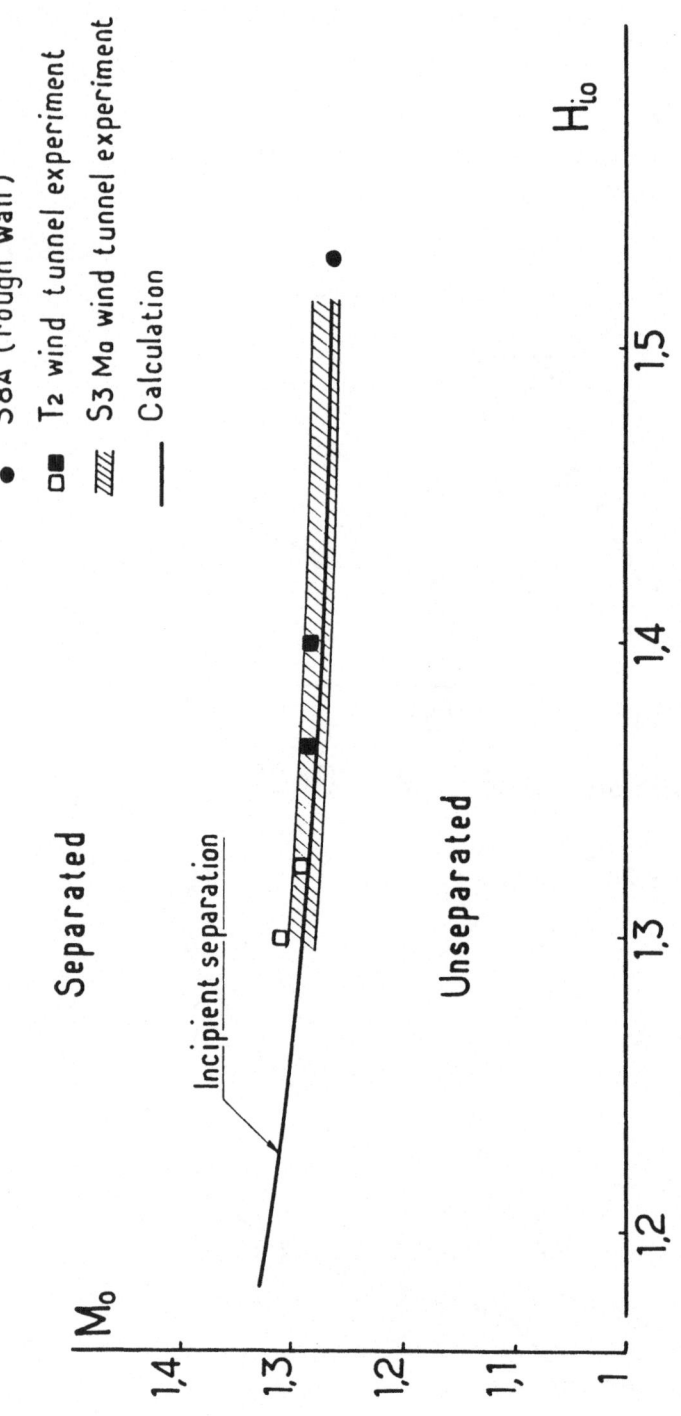

Fig. 12: Limit of shock induced separation.

a _ wall Mach number distribution

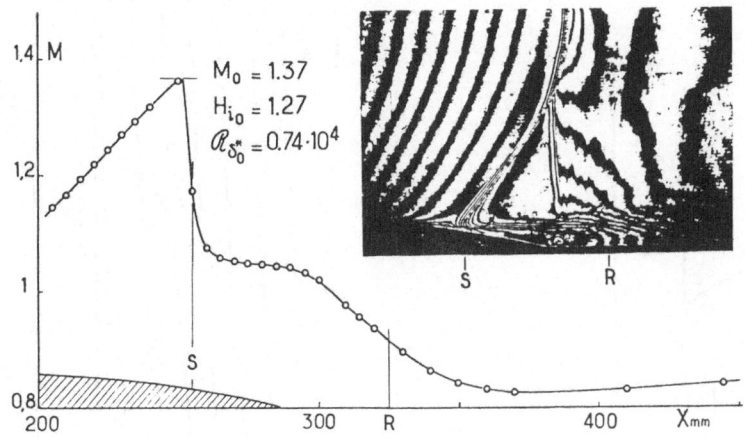

$M_0 = 1.37$

$H_{i_0} = 1.27$

$\mathcal{R}_{s_0^*} = 0.74 \cdot 10^4$

b _ inviscid flow field structure

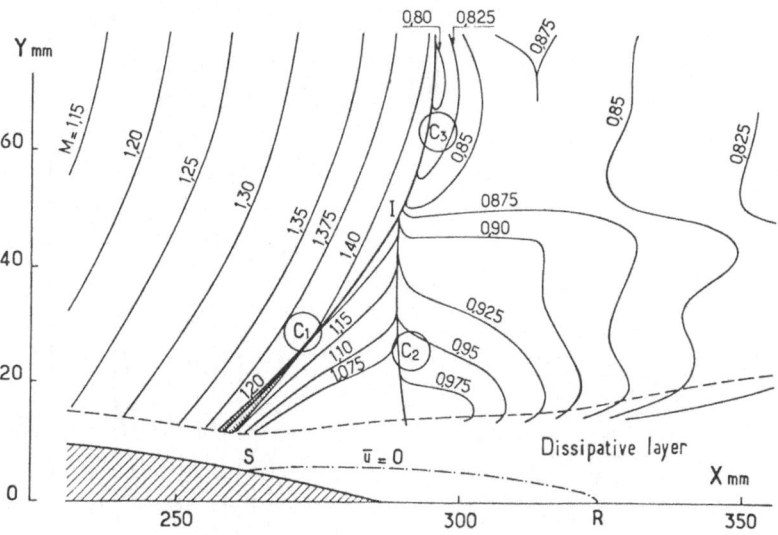

Fig. 13: General features of the flow analysed.

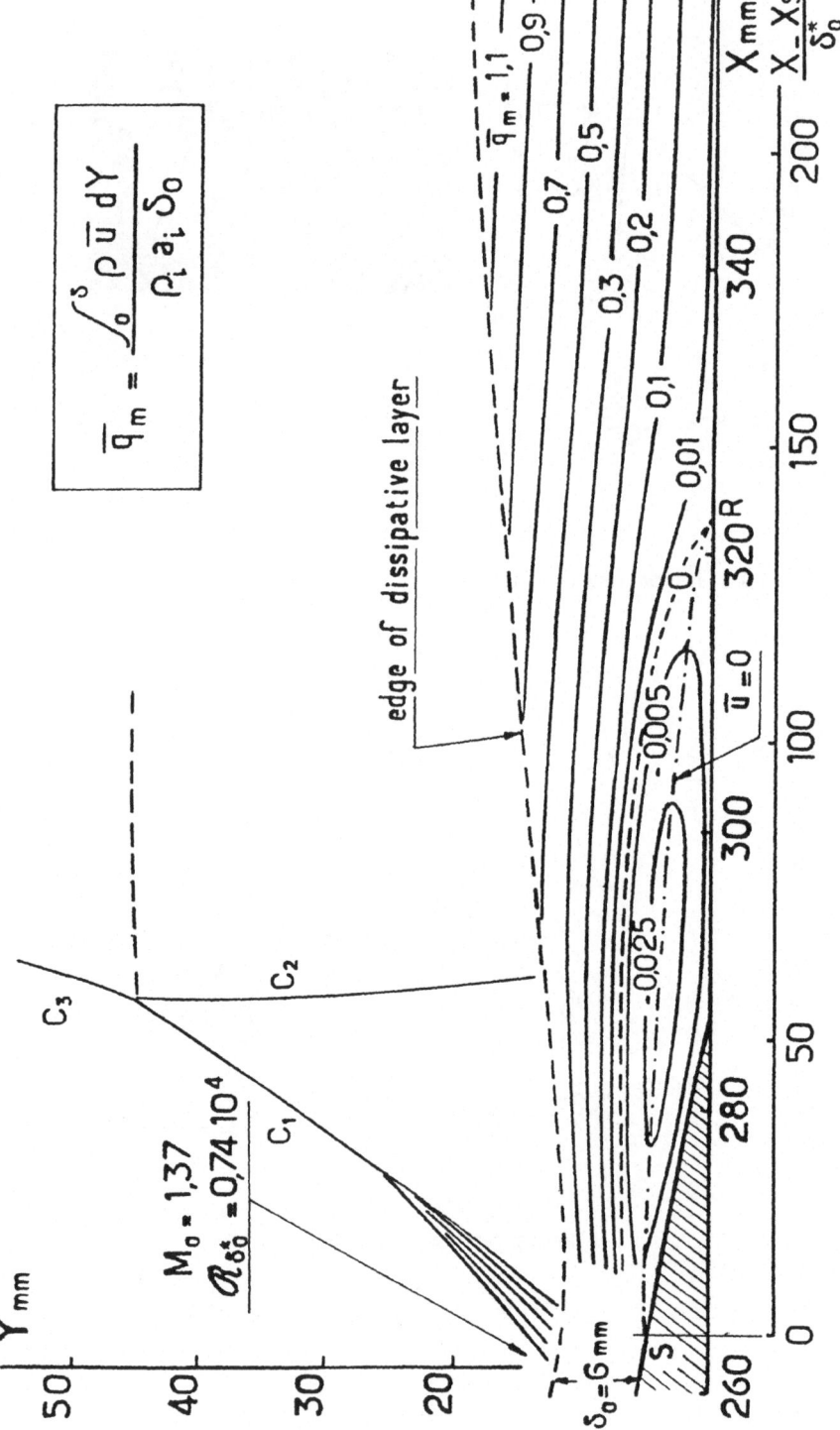

Fig. 14: Mean flow streamlines.

189

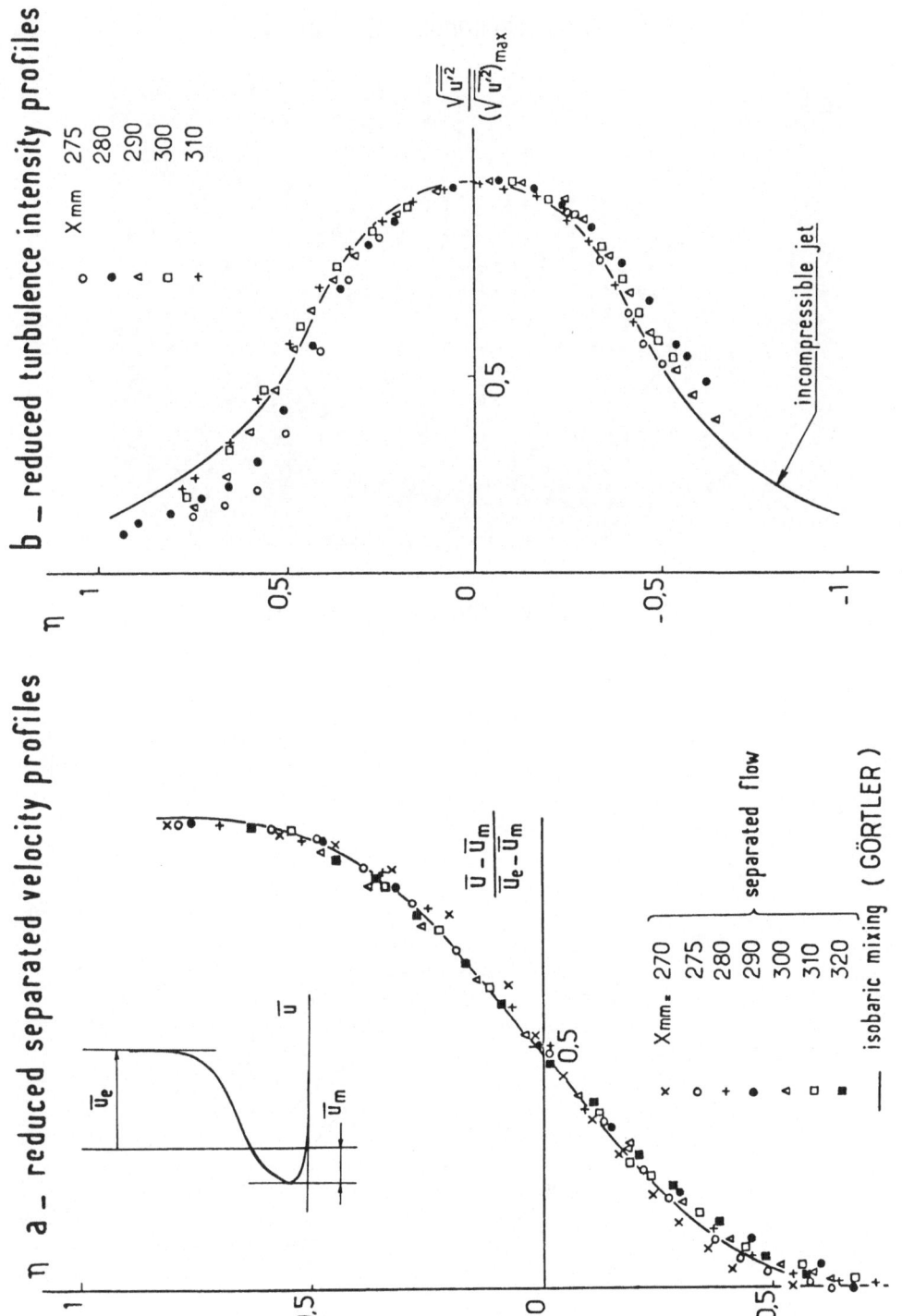

Fig. 15: Turbulence quantities.

a _ short exposure interferogram of flowfield

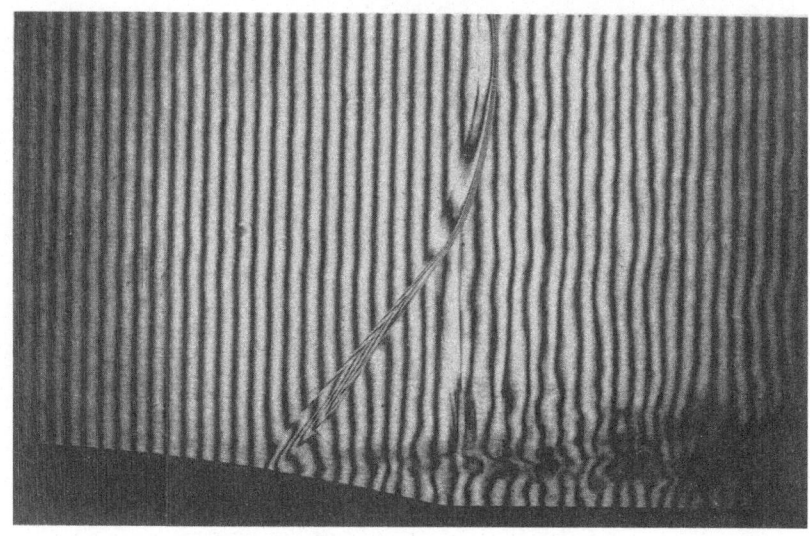

b _ lines of constant values for the probability $P(u \leq 0)$

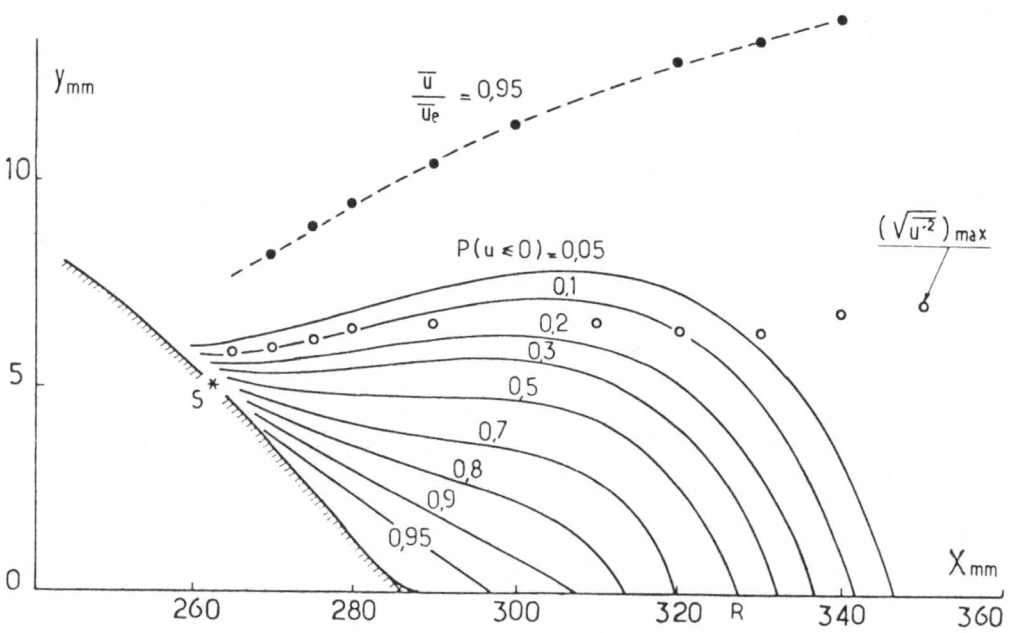

Fig. 16: Turbulence quantities.

a _ shear stress profiles

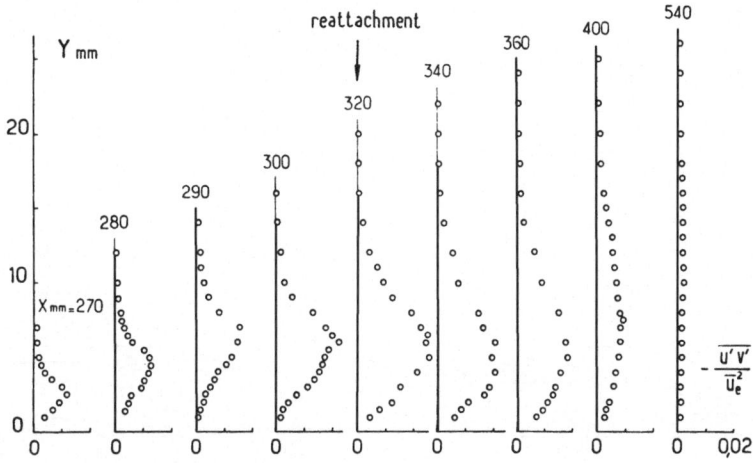

b _ lines of constant shear stress

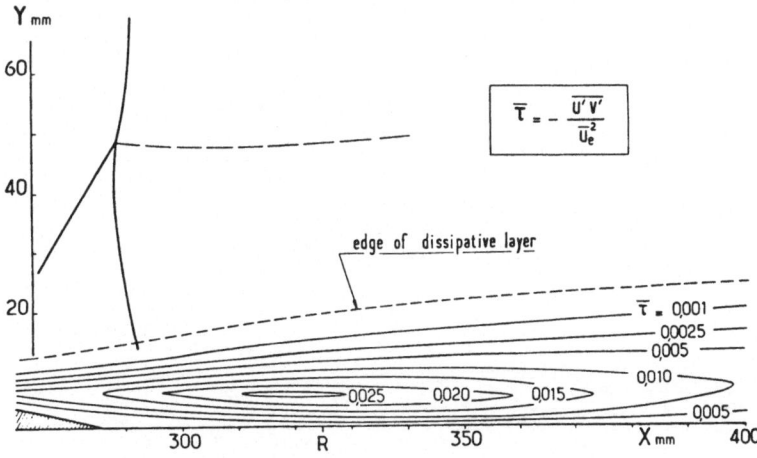

Fig. 17: Turbulence quantities.

192

a – lines of constant kinetic energy

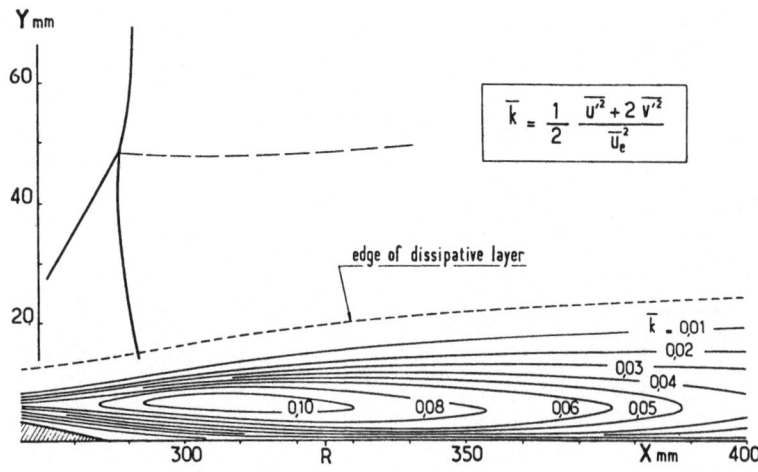

b – lines of constant shear stress production

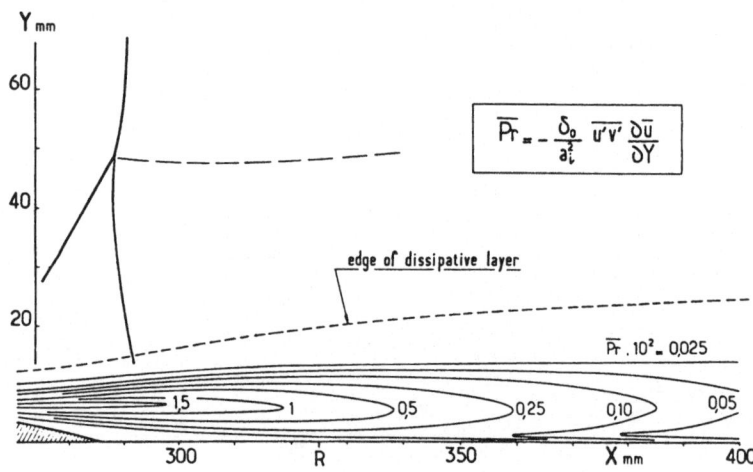

Fig. 18: Turbulence quantities.

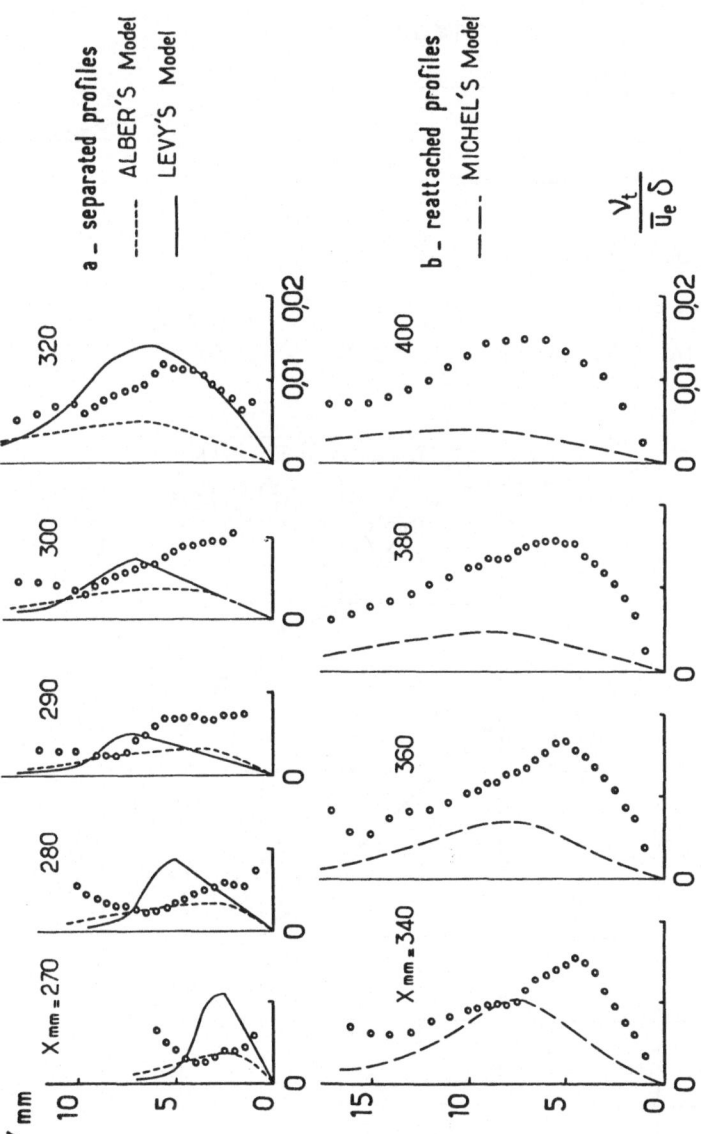

Fig. 19: Turbulent eddy viscosity. Comparison with different models.

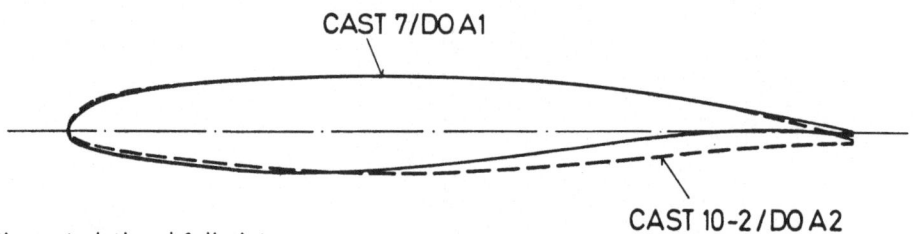

CAST 7/DO A1

CAST 10-2/DO A2

Characteristic airfoil data

CAST 7/DOA1 : t/c =11.8% at 35% c ; $(t/c)_{TE}$ =0.5%   Design M_∞=0.76, C_L=0.573, α=0°
CAST 10-2/DOA2: t/c=12.1% at 45%c; $(t/c)_{TE}$ =0.5%   Design M_∞=0.76, C_L=0.595, α=3°

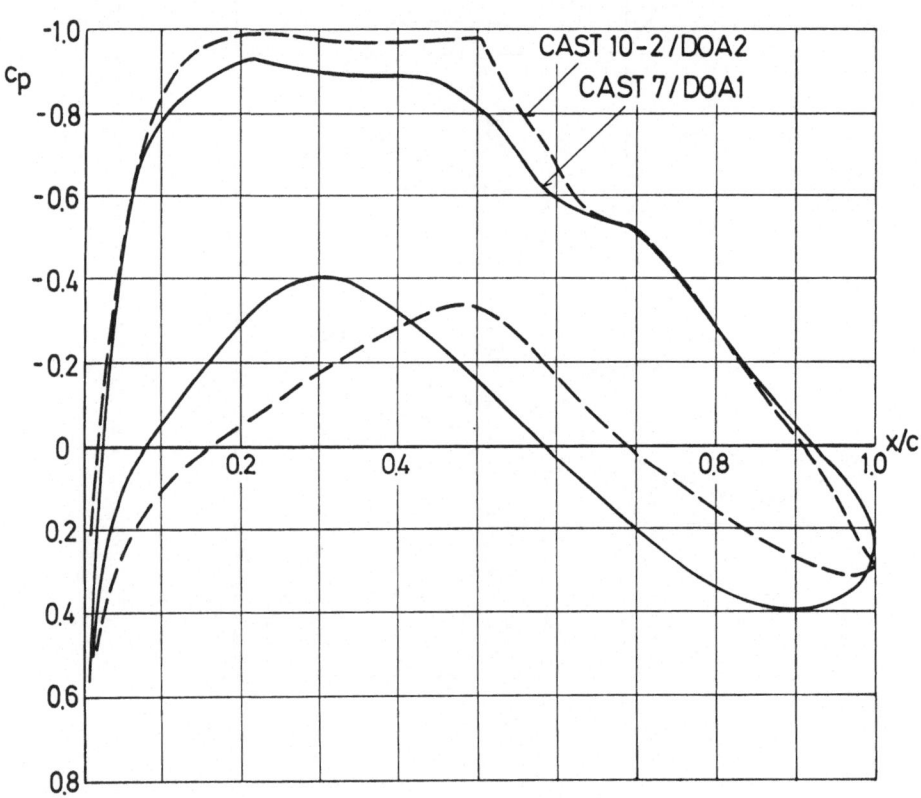

Fig. 20: Airfoil sections used in main investigation.

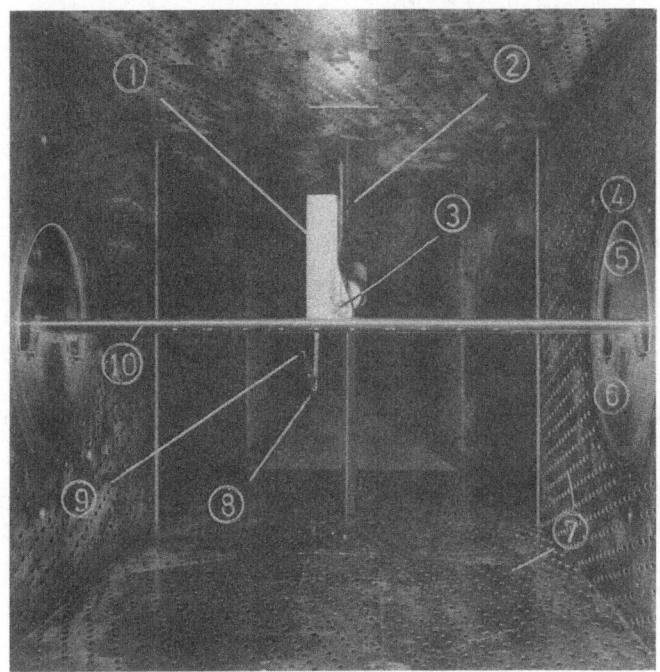

(1) Drive unit encasement (2) Strut (3) Sting (4) Turn-table
(5) Schlieren window (6) Steel insert to carry model loads
(7) Perforated (6% open) test section walls (8) Probe support
(9) Probe (10) Airfoil model

Fig. 21: Test set-up in the 1x1 meter transonic wind tunnel of the DFVLR-AVA Göttingen.

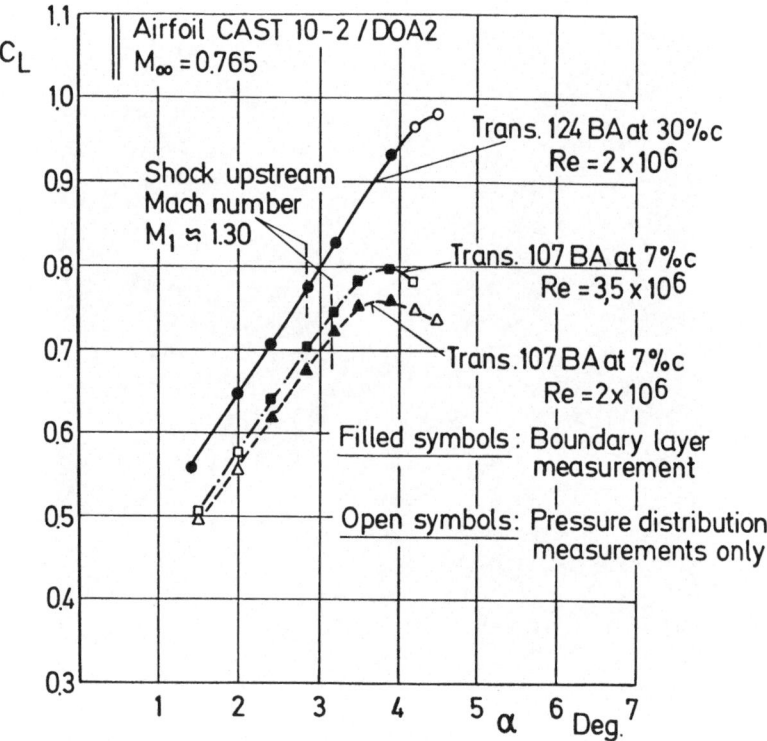

Note: Boundary layer measurements on the airfoil CAST 7/DO A 1
were carried out under similar conditions

Fig. 22: Conditions for boundary layer measurements on the supercritical airfoil CAST 10-2
(also: Effect of tripping device location and Reynolds number on lift coefficient).

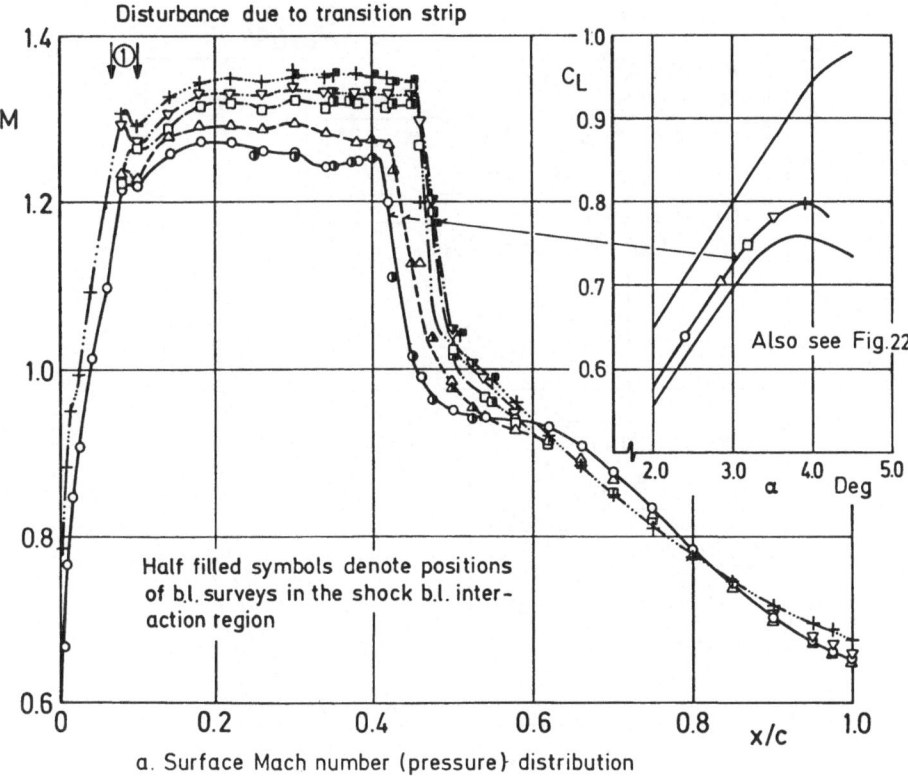

a. Surface Mach number (pressure) distribution

Fig. 23: Viscous/inviscid interaction at increasing angles of attack (shock strengths).

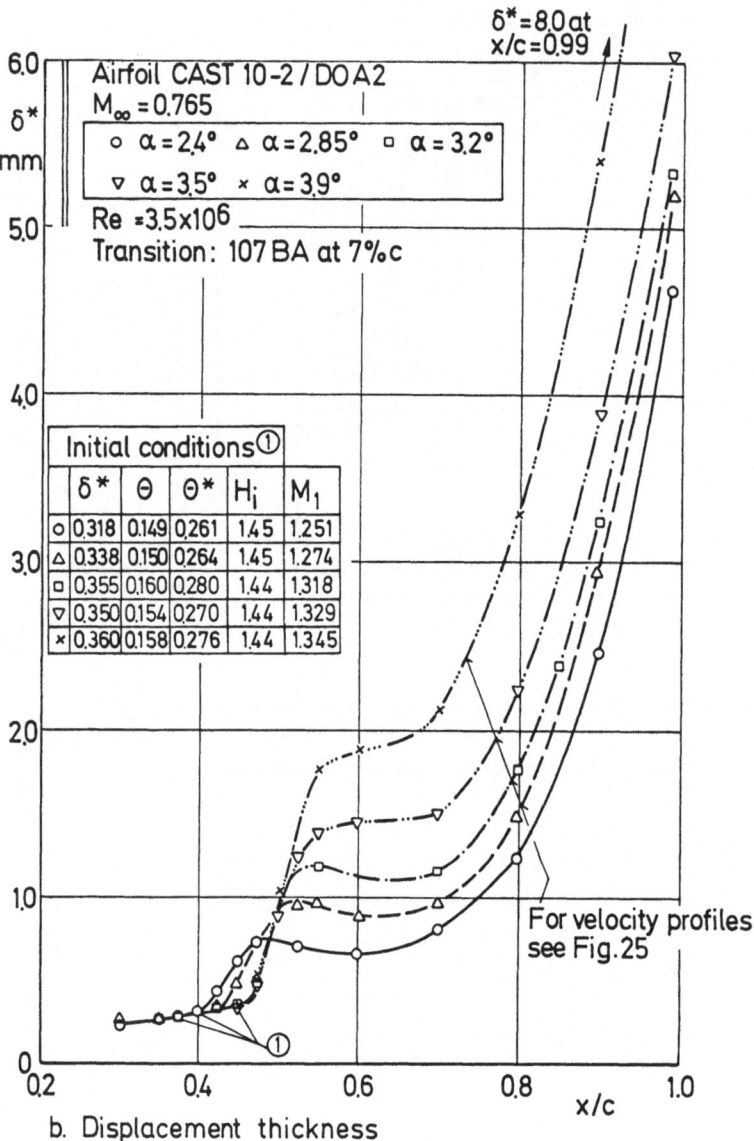

The graph contains the following labels and data:

δ* = 8.0 at
x/c = 0.99

6.0

δ*
mm

5.0

4.0

3.0

2.0

1.0

0

Airfoil CAST 10-2 / DO A2
M∞ = 0.765

o	$\alpha = 2.4°$	△ $\alpha = 2.85°$	□ $\alpha = 3.2°$
▽	$\alpha = 3.5°$	× $\alpha = 3.9°$	

Re = 3.5×10^6
Transition: 107 BA at 7% c

Initial conditions①

	δ*	Θ	Θ*	H_i	M_1
o	0.318	0.149	0.261	1.45	1.251
△	0.338	0.150	0.264	1.45	1.274
□	0.355	0.160	0.280	1.44	1.318
▽	0.350	0.154	0.270	1.44	1.329
×	0.360	0.158	0.276	1.44	1.345

For velocity profiles
see Fig. 25

0.2   0.4   0.6   0.8   1.0

x/c

b. Displacement thickness

Fig. 23: Concluded

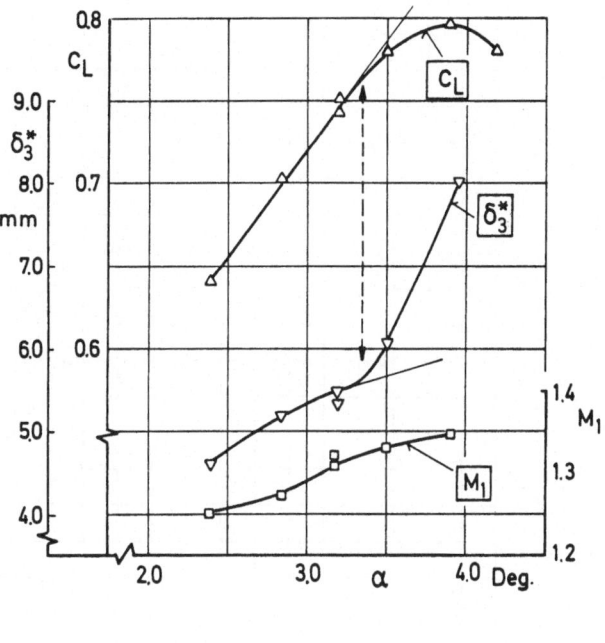

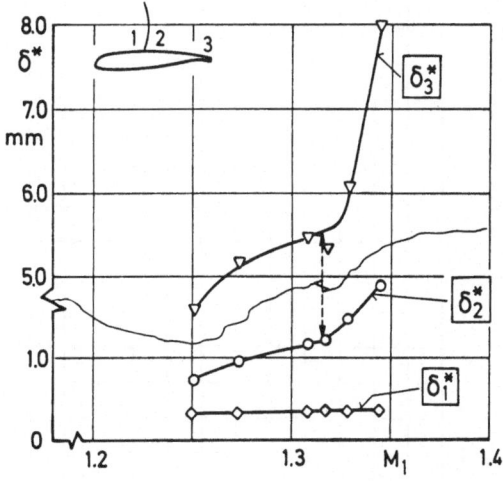

Fig. 24: Displacement thickness and lift curve . (Test cases of Fig. 23: Airfoil CAST 10-2/ DO A2, $M_\infty = 0.765$, Re = 3.5 x 10^6, transition: 107 BA at 7% c).

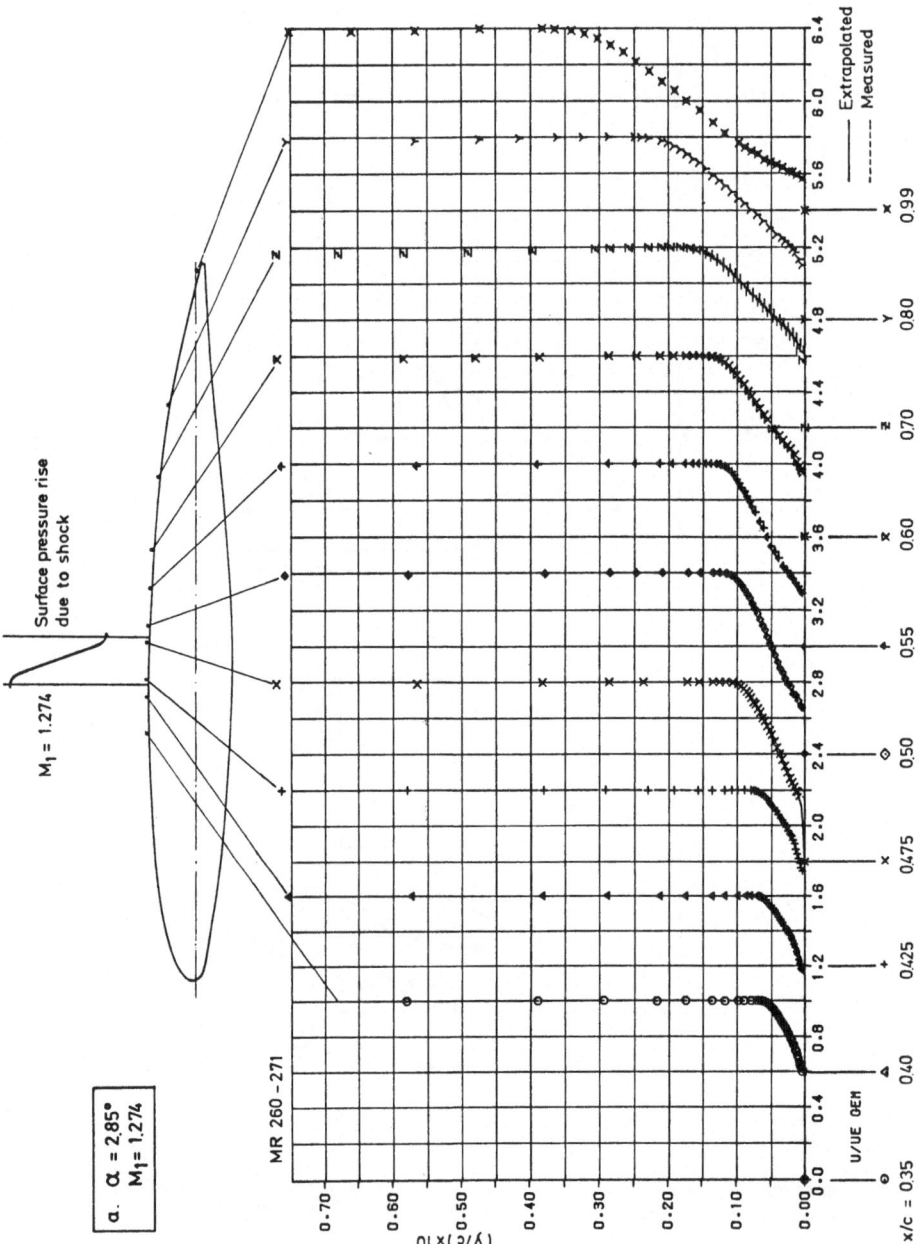

Fig. 25: Velocity profiles corresponding to the test cases of Fig. 23: Airfoil CAST 10-2/DO A 2, $M_\infty = 0.765$, Re = 3.5×10^6, transition: 107 BA at 7 % c.

201

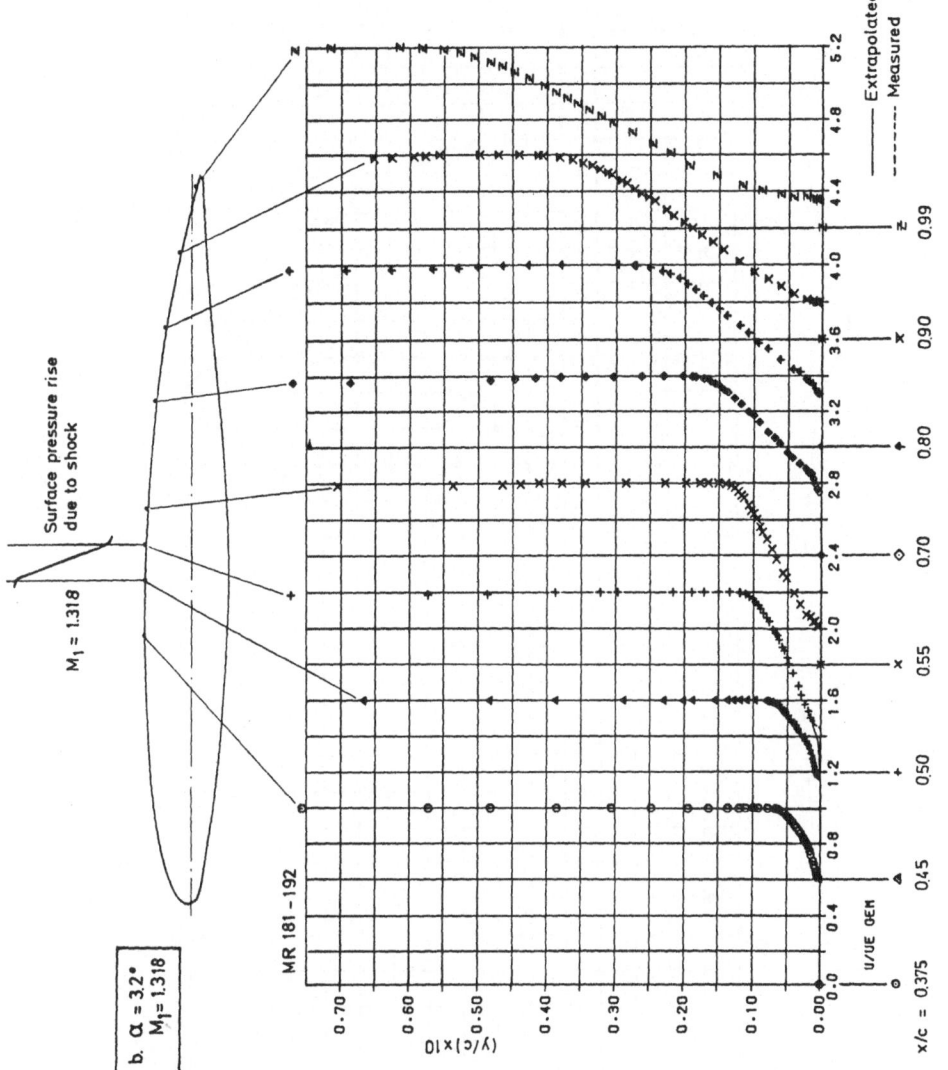

Fig. 25: Concluded

b. α = 3.2°
M₁= 1.318

Surface pressure rise
due to shock

M₁ = 1.318

MR 181 -192

——— Extrapolated
------- Measured

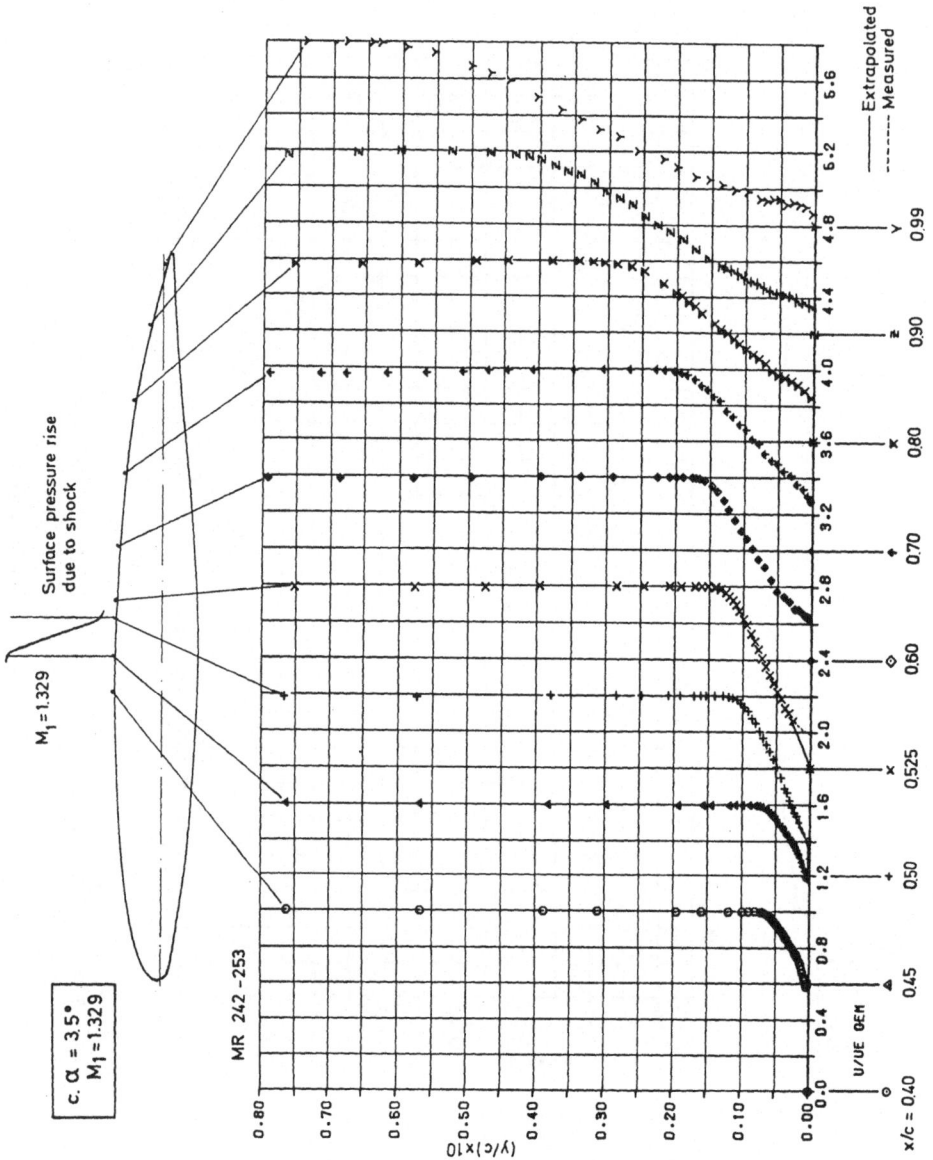

Fig. 25: Concluded

203

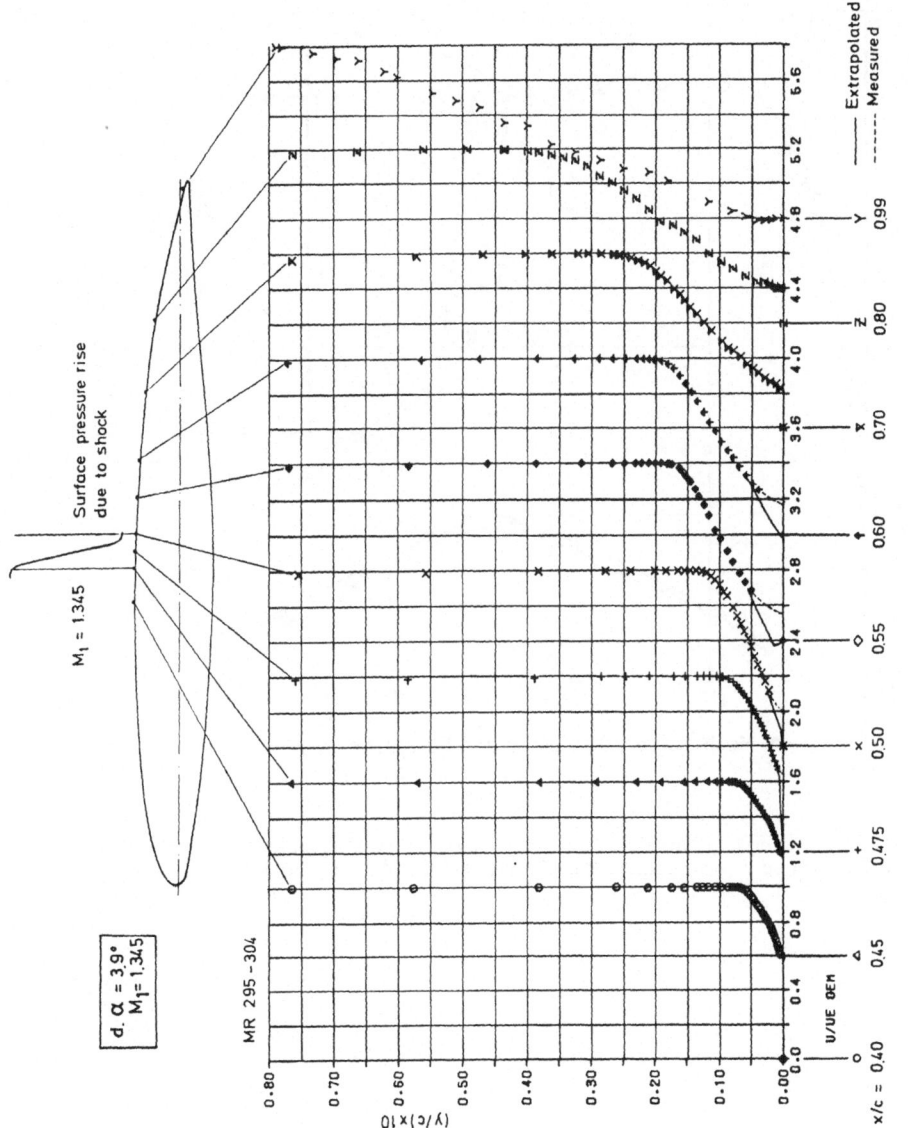

Fig. 25: Concluded

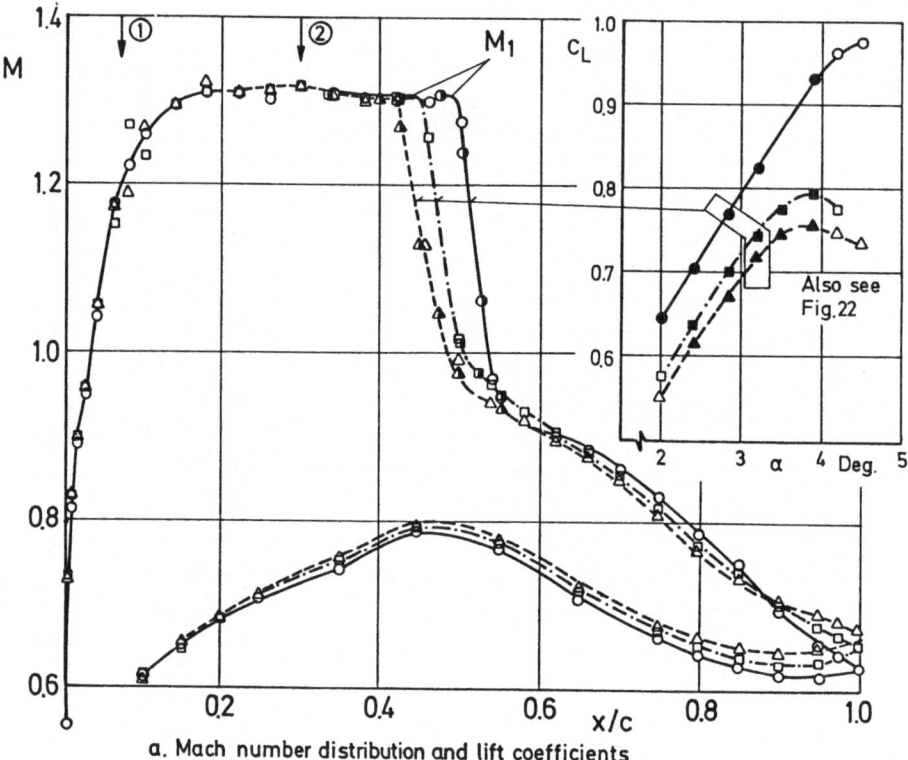

a. Mach number distribution and lift coefficients

Fig. 26: Effect of initial boundary layer condition. (Transition location and Reynolds number) on viscous/inviscid interaction. Airfoil CAST 10-2/DO A2, $M_\infty = 0.765$.

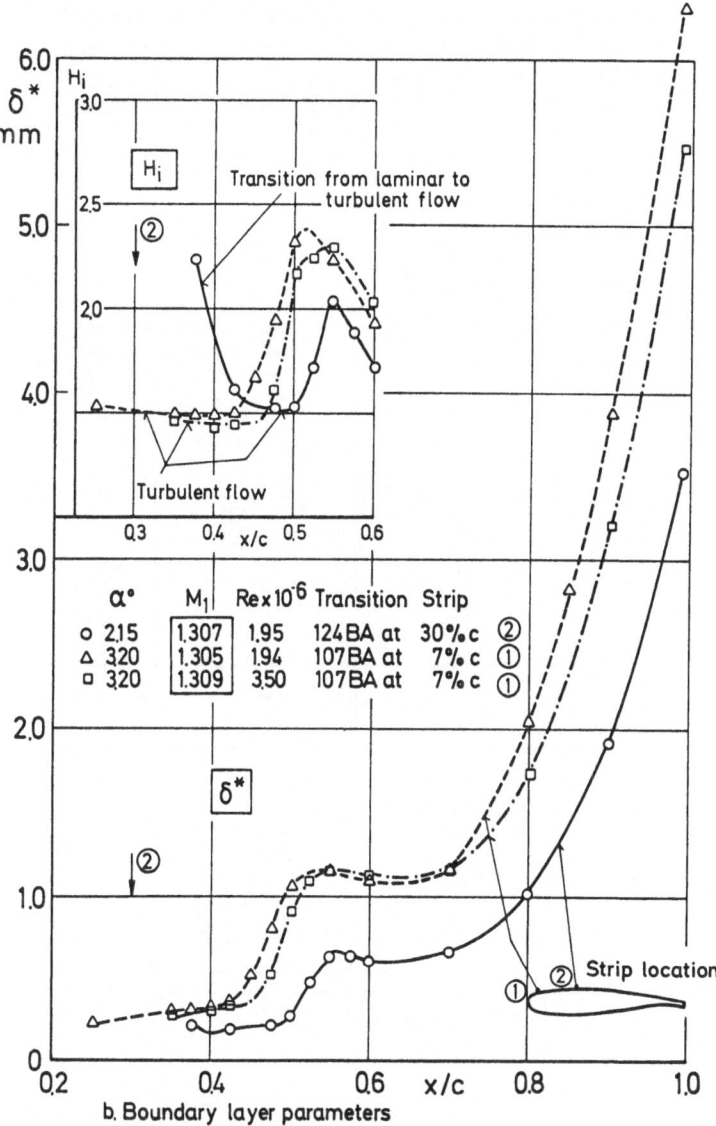

Fig. 26: Concluded

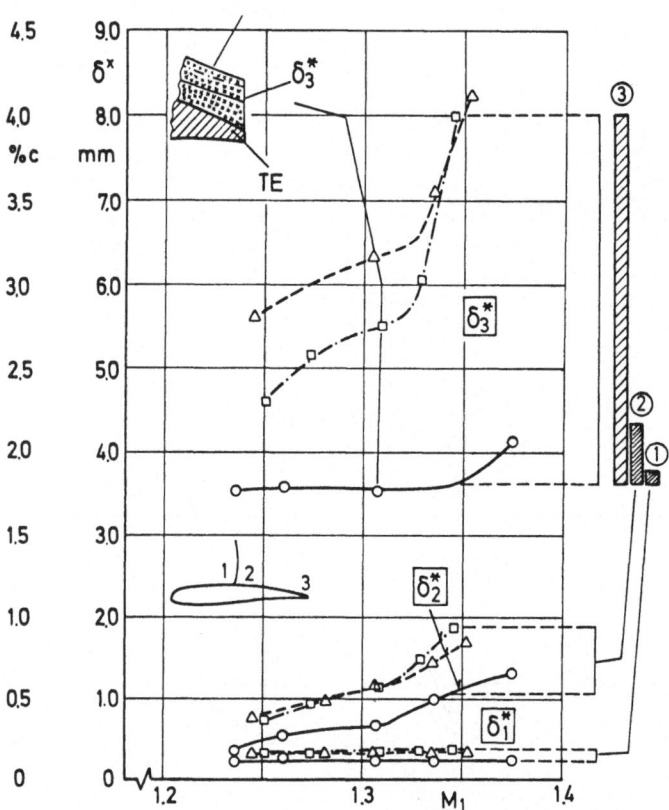

Re × 10⁻⁶	Transition	Initial difference (I.d.)

<table>
<tr><td>○</td><td>1.95</td><td>124 BA at 30% c</td></tr>
<tr><td>△</td><td>1.94</td><td>107 BA at 7% c</td></tr>
<tr><td>□</td><td>3.50</td><td>107 BA at 7% c</td></tr>
</table>

① I.d. amplified by shock
② I.d. amplified by total viscous /
③ inviscid interaction

Scaled TE conditions

Fig. 27: Amplification of initial difference in displacement thickness due to shock and rear adverse pressure gradient, respectively. Airfoil CAST 10-2/DO A2, $M_\infty = 0.765$.

207

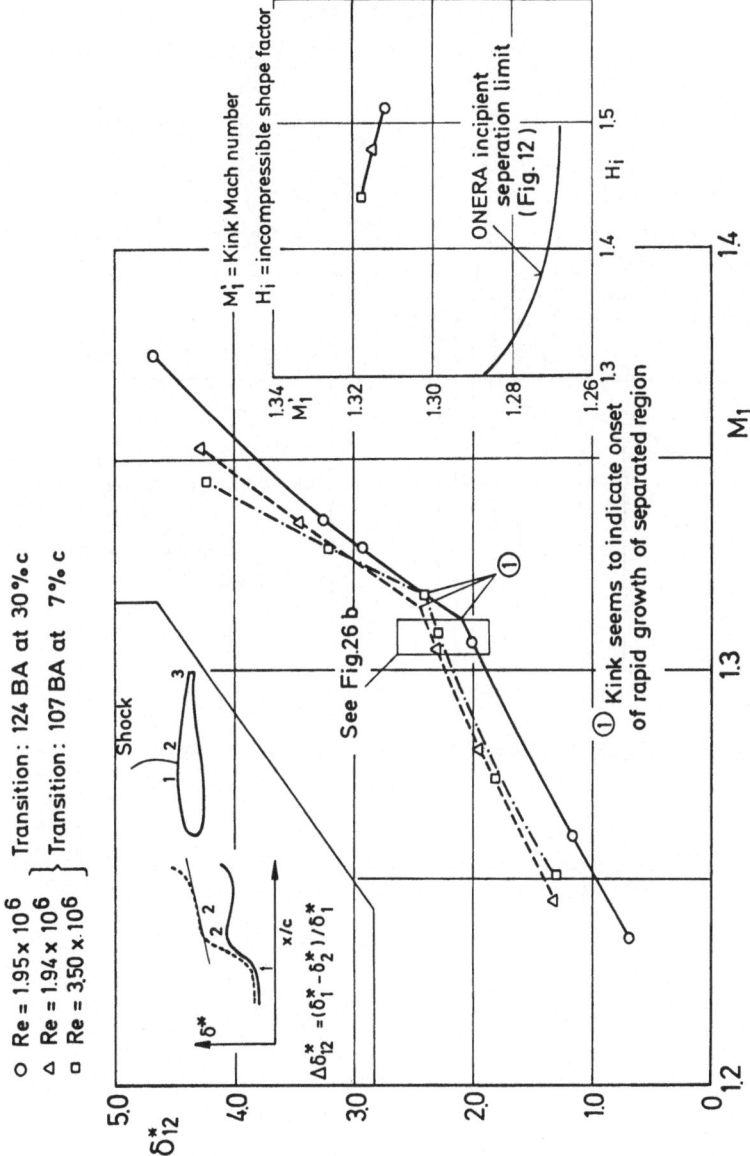

Fig. 28: Change in displacement thickness on airfoil upper surface dependent on shick-upstream Mach number and initial boundary layer condition. Airfoil CAST 10-2/DO A2, $M_\infty = 0.765$.

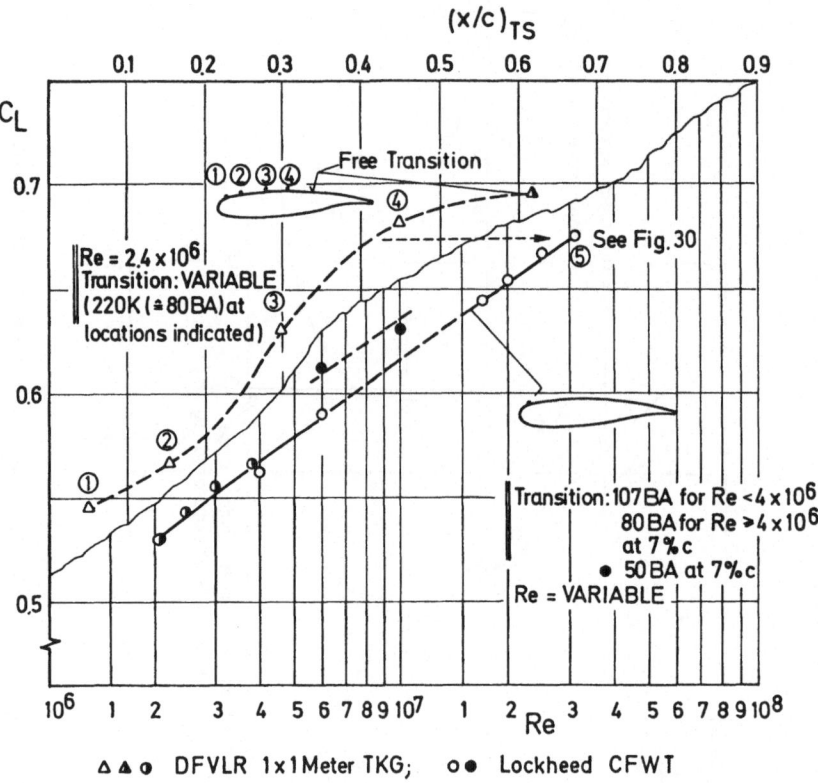

Fig. 29: Effect of Reynolds number and transition strip location on viscous/inviscid interaction. Airfoil CAST 10-2/DO A2, $M_\infty = 0.765$, $\alpha = 2^o$.

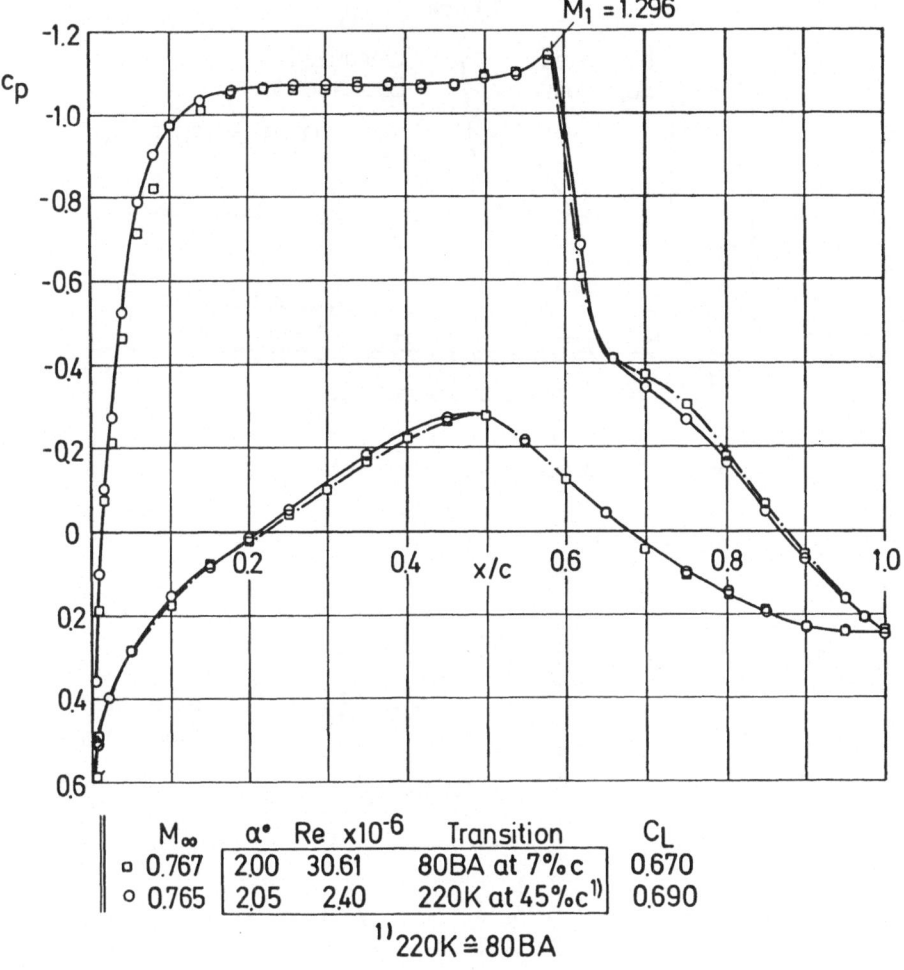

	M_∞	α°	Re $\times 10^{-6}$	Transition	C_L
□	0.767	2.00	30.61	80BA at 7%c	0.670
○	0.765	2.05	2.40	220K at 45%c[1)]	0.690

[1)] 220K ≅ 80BA

Fig. 30: Surface pressures corresponding to points ④ and ⑤ of Fig. 29. Airfoil CAST 10-2/DO A2.

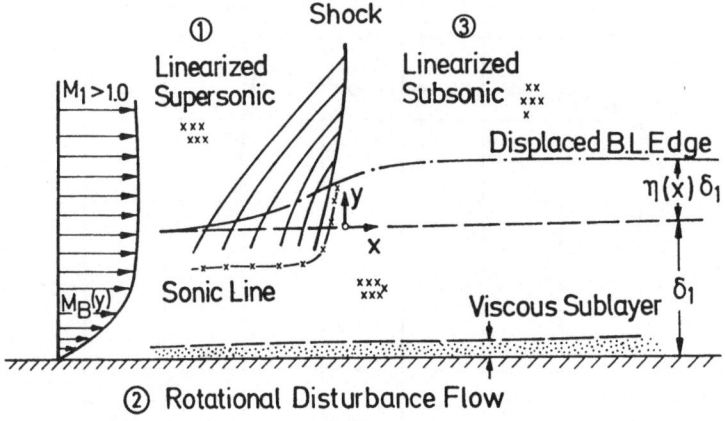

Fig. 31: Triple deck model of shock boundary layer interaction.

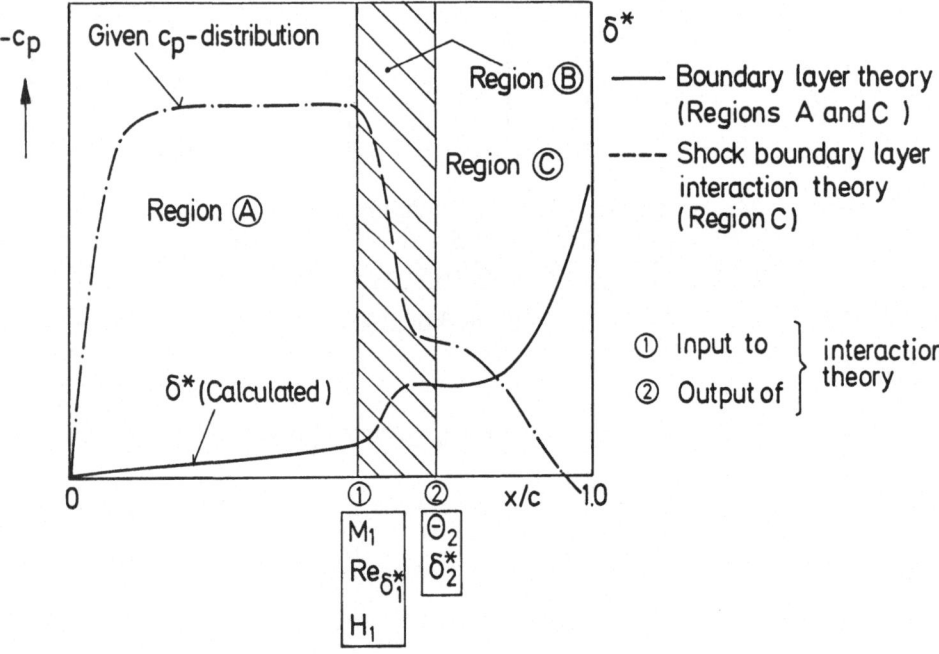

Fig. 32: Schematic representation of coupling of boundary layer/shock b.l. interaction theory.

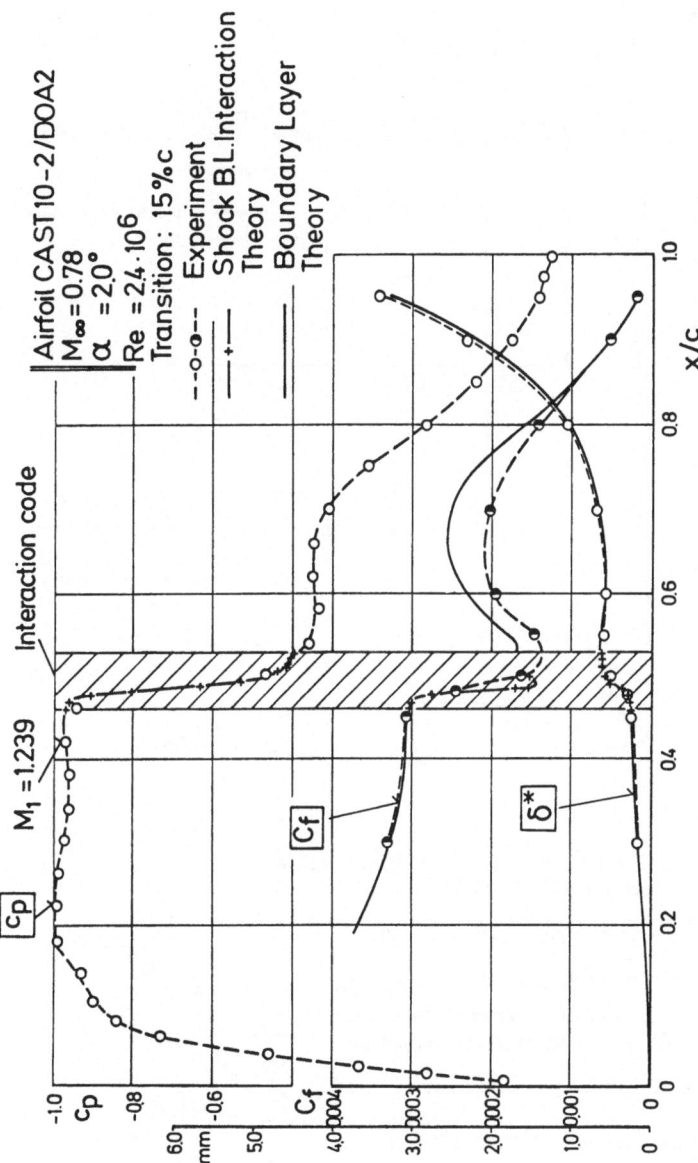

Fig. 33: Comparison of boundary layer/shock boundary layer interaction theory with experiment.

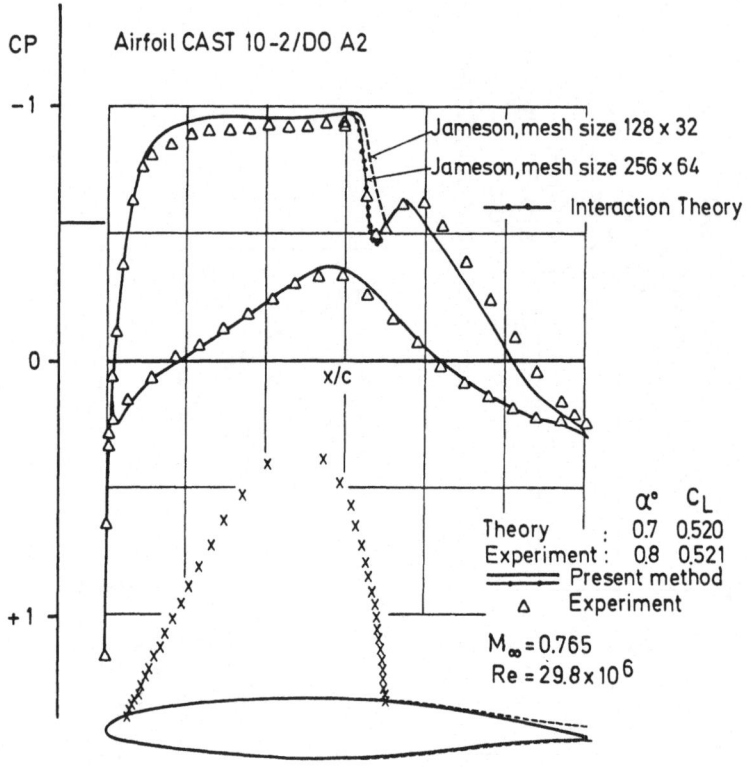

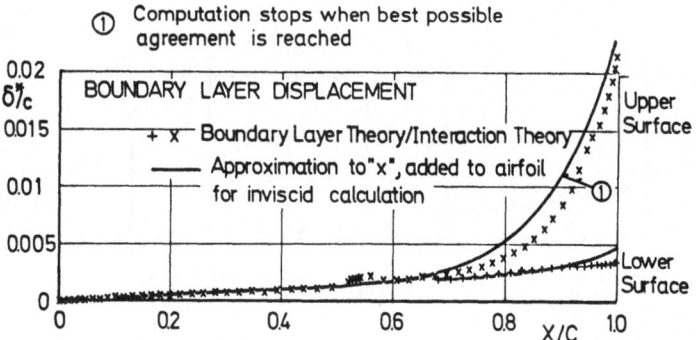

Fig. 34: Comparison of results of complete method with experiment.

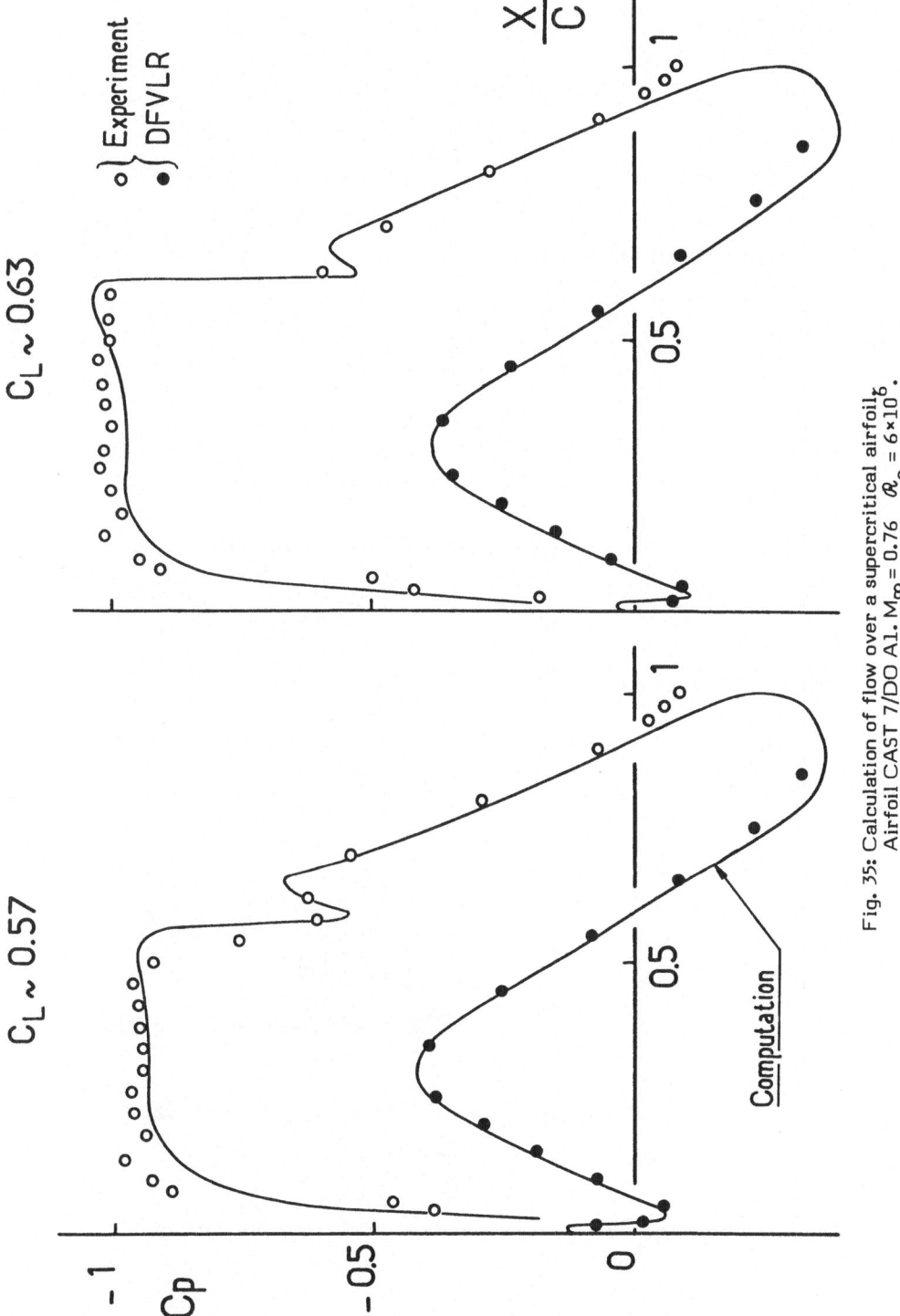

Fig. 35: Calculation of flow over a supercritical airfoil.
Airfoil CAST 7/DO A1. $M_\infty = 0.76$ $R_e = 6\times10^6$.

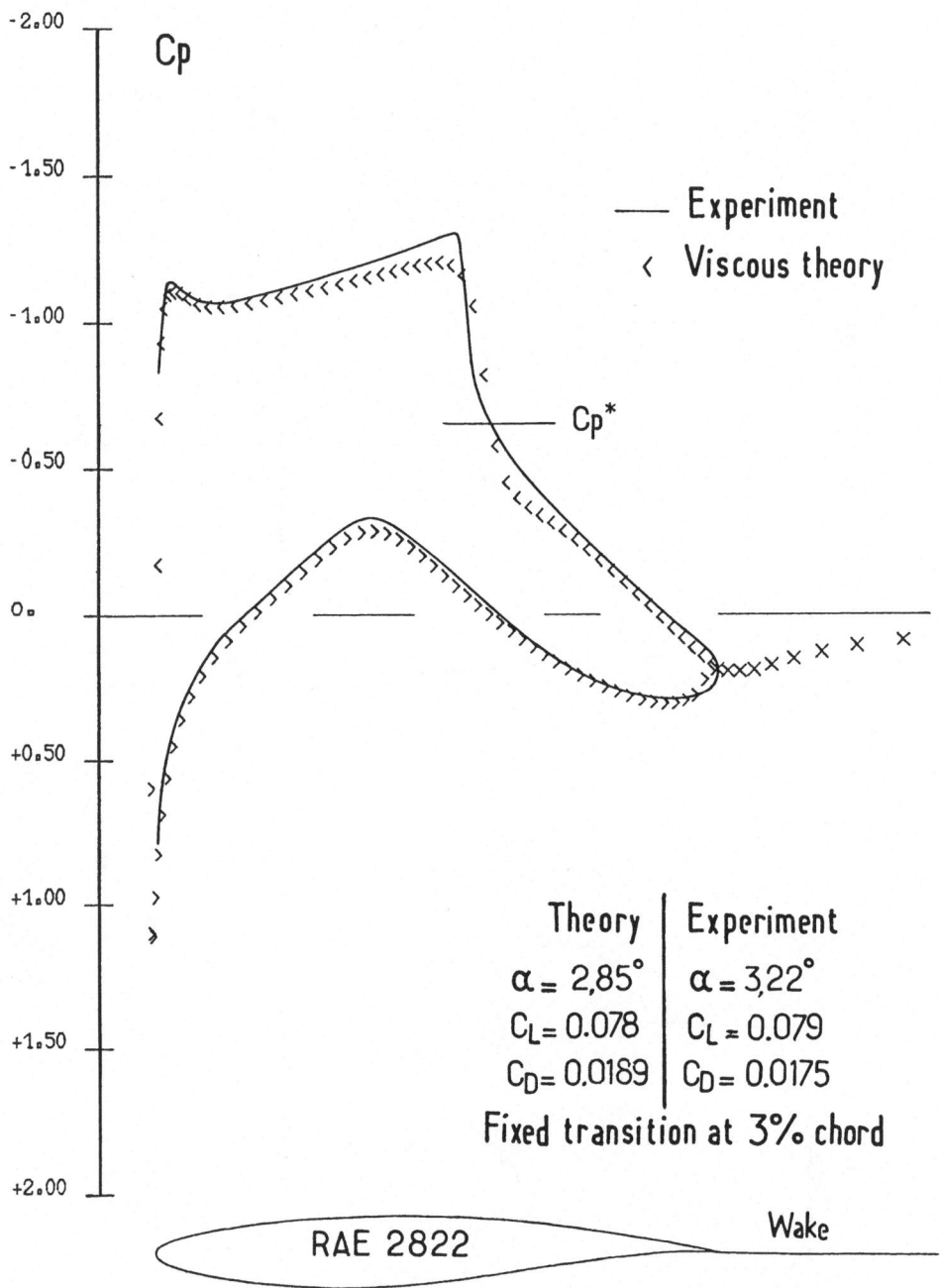

Fig. 36: Calculation of flow over a supercritical airfoil. (14 % RAE 2822 profile).
$M_o \simeq 0.73$ $\mathcal{R}_e = 6.5 \times 10^6$.

Flow in Multiply-Connected Domains

W. J. Prosnak
Warsaw

1. Introduction

The difficulties in solving theoretical problems of fluid mechanics stem from the mathematical properties of the differential equations involved as well as from boundary conditions.

The first kind of these difficulties is rather well-known, and doesn't need do be discussed here.

The second kind follows from the fact, that boundaries appearing in realistic problems are rahter complicated both geometrically and kinematically, which makes their mathematical description - the prerequisite for mathematical formulation of the boundary conditions - not too easy a task. On the other hand, the solution of a particular problem may depend even qualitatively on slight differences in the shape of the boundaries, and the flow around laminar profiles may be mentioned here as an example of such a sensitivity. Although no mathematical discontinuity of the solution with respect to boundary condition appears, nevertheless, the importance of describing the boundaries with proper accuracy becomes only too obvious. Sometimes, the requirements imposed on accuracy may concern even higher derivatives.

The difficulties conncected with boundary conditions were for a long time rather overshadowed by those stemming from the system of equations. Only during the last decade, and with the advance of electronic computers, they began to be properly appreciated.

The present paper deals with computation of flow in a domain represented by plane infinite exterior of a finite system of closed, nonintersecting lines, which will be further referred to as profiles. The system may be regarded as a section of a multi-element airfoil, Fig. 1, or of a hydraulic machine, Fig. 2 .

Such a domain is rather complicated geometrically. Moreover, its multiply-connectivity brings specific difficulties to the mathematical problem to be solved, e.g. nonuniqueness of the solution, which has to be overcome by means of certain complementary conditions. Finally, it should be stressed, that the domain considered in the paper may be depending on time.

The flow in the domain just described will be assumed as plane, inviscid, incompressible, irrotational, and attached to the boundaries, i.e. the profiles. The last two assumptions may be, however, properly modified, in order to enable modelling of flows with wakes of separation regions.

Three following kinds of problems will be considered within the frame of the paper:

- steady flow around a given system of profiles;
- unsteady flow in time-depending domains;
- design of a system of profiles possessing certain hydronamical properties given in advance.

Just the main ideas will be presented, sparsely illustrated by simple examples. No flood of numerical results is to be feared.

There are several methods of dealing with complicated boundaries, and domains in theoretical fluid mechanics. One of the most general and effective methods of this kind consists in transformation of such boundaries, and domains into simpler ones. Fluid mechanics is very rich in various transformations concerning independent variables as well as the unknown functions. In plane and axisymmetric domains a specially convenient and powerful transformation may be applied, known already for a rather long time, and called conformal mapping. In spite of this long time of application it was confined essentially to doubly- connected domains, with very few exceptions.

All considerations presented in the paper will be founded on conformal mapping. No comparison with other methods of approach is intended, however, some distinct merits of the method of conformal mapping should be pointed out. It allows:

- very high and easily controlled accuracy in mathematical description of the shape of profiles as well as in formulation and fulfilling of boundary conditions;
- entire independence of the transformation from the system of equations governing the flow;
- entire independence from the number and shape of profiles;
- obtainment of general semi-analytic solution to the whole class of problems defined by the set of assumptions introduced formerly.

The possibility of using the method of conformal mapping in combination with discrete methods of solutions exemplified by the method of finite differences, or the method of finite elements should also be reminded here, although it remains beyond the scope of the present paper. The obvious advantage of such combination consists in dealing with canonical domains bounded by regular lines.

2. Steady Flow Around a Given System of Profiles

Let us consider any mathematical problem defined by a system of partial differential equations together with proper boundary and initial conditions. The solution to this problem sought in an <u>original domain</u> by means of the method of conformal mapping, can be thought of as a sequence of solutions to the following "subproblems":

1. determination of the functions, mapping conformally the original domain onto an appropriate auxiliary one;

2. solution of the transformed problem;

3. transformation of the solution onto the original domain.

In case of the steady, plane, irrotational flow of inviscid and incompressible fluid these sub-problems are mutually almost independent, and will be considered separately.

2.1 The Conformal Mapping

The original domain characteristic for the class of problems under consideration consists of the plane, infinite exterior of a given system of profiles. It will be assumed, that the profiles are represented by Jordan curves. In the majority of realistic configurations the profiles have continuous slope and curvature everywhere, except at one point, which will be usually referred to as the trailing edge.

The auxiliary domain will be assumed correspondingly as plane, infinite exterior of circles, the number K of the circles being equal to the number of the given profiles. Such a domain is illustrated schematically in the Fig. 3 for the particular case K = 3. The selected form of the auxiliary domain has the obvious advantage of being independent on the number of profiles.

Denoting by z a complex plane containing the original domain, and by ζ - another complex plane containing the auxiliary domain, one can formulate the subproblem to be considered in this section as seeking a complex function:

$$z = z(\zeta),\tag{2.1}$$

transforming conformally the auxiliary domain into the original one.

The existence and uniqueness of the mapping function (2.1) in case of multiply-connected domains follows directly from the Koebe Theorem. In the formulation given by Gaier [1], and slightly reformulated for the present purpose the Theorem ascertains that every K-connected domain can be transformed conformally onto an infinite domain bounded by K circles, the transformation being unique when certain normalizing conditions are satisfied.

The problem of conformal mapping posed for the formerly defined system of profiles meets the fundamental requirements of the Koebe Theorem, and in view of hydrodynamical applications the following complementary conditions will be assumed:

1. the points $z = \infty$ and $\zeta = \infty$ correspond mutually;

2. the derivative of the mapping function (2.1) equals one at the lastmentioned point:

$$\left.\frac{dz}{d\zeta}\right|_{\zeta=\infty} = 1;\tag{2.2}$$

3. the mapping function (2.1) is regular everywhere in the auxiliary domain, except at isolated points of the circles.

The mapping function satisfying alle these conditions can be assumed as the following series of rational fractions:

$$z = \zeta + C_o + \sum_{k=1}^{K} \sum_{n=1}^{n} C_{kn} \left(\frac{a_k}{\zeta - \zeta_k} \right)^n , \qquad (2.3)$$

where the new symbols denote:

C_o, C_{kn} - complex constant coefficients;

a_k - radius of the k-th circle;

ζ_k - a complex number such that its real and imaginary part are equal to the respective coordinates of the centre of the k-th circle.

The adopted form (2.3) does not mean any restriction imposed on the mapping functions. In case $K = 1$ it reduces to the Laurent series.

The infinite number of terms in the inner sums of the series (2.3) takes care even of the formerly mentioned "edges" at the profiles. However, only a finite number N of the terms can be retained in practical computations, which is equivalent to "rounding off" the edges, and restrictions - in fact - to profiles with continuous slope and curvature. It is possible, however, to account exactly for the existing edges even in case of truncated series - by introduction of special factors in (2.3), representing the singularities involved, as it will be shown in Chapter 4.

In view of (2.3) the problem of determination of the mapping functions (2.1) reduces to computation of the constants:

$$C_o, C_{kn}, a_k, \zeta_k, \qquad k = 1,2,\ldots K; \ n = 1,2\ldots \qquad (2.4)$$

either C_o or ζ_1 being assumed arbitrarily. Computation of the remaining constants (2.4) can be executed for any given system of profiles by means of an algorithm designed and described in our former paper [2]. Also the corresponding computer program will be published soon [3]. It will suffice therefore at the present place to mention only, that the numerical values of the constants (2.4) are arrived at in a convergent process of consecutive approximations.

A numerical example is presented below, borrowed from [3].

The given system of three profiles to be transformed is shown in Fig. 4, and the corresponding system of circles - in Fig. 5. The constants determining the circles are gathered in Table 1.

Table 1

k	a_k	$\xi_k = \text{Re}(\zeta_k)$	$\eta_k = \text{Im}(\zeta_k)$	ε_k
1	0.042 963 59	-0.363 667 9	0.135 175 6	57.710 099°
2	0.279 401 05	0	0	-1.204 287°
3	0.079 662 75	0.528 027 3	-0.234 150 4	-6.584 600°

The last column in the Table contains values of an angle ε_k, which determines on the k-th circle the image of the trailing edge of the corresponding profile. The values of the remaining constants (2.4) of the mapping function (2.3), i.e. of C_o, C_{kn} can be found in [3] as well as data determining the system of profiles shown in Fig. 4.

The process of computing the constants (2.4) of the mapping function may be quite time-consuming, if a high accuracy of the transformation is demanded. In order to cut-down this time it is advisable to take advantage - if possible - of some special properties of the system to be transformed, such as symmetry or periodicity.

A symmetric system of profiles which appears in considerations regarding influence of the "ground" on flow ground the system [4], is shown in Fig. 6. Assuming, that the system is symmetric with respect to the real axis, one can easily arrive at the corresponding special form of the mapping function:

$$z = \zeta + C_o + \sum_{k=1}^{K} \sum_{n=1}^{\infty} \left[C_{kn} \left(\frac{a_k}{\zeta - \zeta_k} \right)^n + \bar{C}_{kn} \left(\frac{a_k}{\zeta - \bar{\zeta}_k} \right)^n \right]. \tag{2.5}$$

the dashes denoting complex conjugates, and K referring to the number of profiles situated just at one side of the symmetry line. In consequence, although the system considered contains 2K profiles, the time needed for computation of the constants (2.4) in the function (2.5) is not larger than the one corresponding to a system of K profiles only.

The same is true as far as a periodic system is concerned. Such a system is presented in Fig.2, the period being determined by an angle:

$$\lambda = \frac{2\pi}{K}, \tag{2.6}$$

and K denoting, as previously, the number of profiles. The corresponding special form of the mapping function (2.3) can be written as:

$$z = \zeta + \sum_{k=1}^{K} \sum_{n=1}^{\infty} C_{1n} \, e^{i(n+1)(k-1)\lambda} \left[\frac{a}{\zeta - \zeta_1 \, e^{i(k-1)\lambda}} \right]^n \tag{2.7}$$

the index 1 being prescribed to an arbitrarily selected profile of the system. Due to the special form (2.7) of the mapping function the time needed for computing the constants is not much higher than for just one comparable profile.

The function (2.7) corresponds to a single circular cascade of profiles shown in Fig. 2. It can, however, be easily generalized for a system of such cascades having the same number of profiles.

Finally, a system of K profiles situated between two parallel lines should be considered (Fig. 7), as important from the viewpoint of aerodynamical applications. The lines represent here walls of a test section of a wind tunnel. Assuming that the domain of flow bounded by the said two lines occupies part of a complex w-plane, the distance of the lines being $\tau/2$, as it is shown in

Fig. 7., one can immediately transform this original domain to a symmetric and infinite one, by using the auxiliary mapping function:

$$z = e^{\frac{2\pi\omega}{\tau}} .$$ (2.8)

In such a way the problem of transforming the original domain bounded by lines onto the exterior of circles - has been reduced to the already known one, i.e. to transformation of a symmetric system of profiles.

2.2 The Complex Velocity

As it is generally known, the flow in the original z-plane can be described by a holomorphic function:

$$w = w(z) = \varphi(x,y) + i\psi(x,y); \qquad (z = x + iy)$$ (2.9)

called complex potential, the real part $\varphi(x,y)$ and the imaginary one $\psi(x,y)$ representing velocity potential and stream function respectively.

Substituting z in (2.9) by the mapping function, which is also holomorphic, one arrives at another complex potential:

$$w(z(\zeta)) = W(\zeta) = \Phi(\xi,\eta) + i\Psi(\xi,\eta); \qquad (\zeta = \xi + i\eta)$$ (2.10)

which describes the flow in the auxiliary ζ-plane. Derivation of (2.10) with respect to ζ gives the known relation between complex velocities in the both planes:

$$\frac{dw}{dz} = \frac{dW}{d\zeta} \cdot \left(\frac{dz}{d\zeta}\right)^{-1} ,$$ (2.11)

the last factor denoting derivative of the mapping function. The problem of determing the velocity field in the original plane reduces therefore to determing the velocity field in the auxiliary plane, the mapping function being already known.

The general expression for the complex velocity in the last-mentioned plane will be assumed again in form of a series of rational fractions:

$$\frac{1}{V_\infty} \frac{dW}{d\zeta} = F(\zeta) + \sum_{k=1}^{K} \sum_{n=1}^{\infty} c_{kn} \left(\frac{a_k}{\zeta - \zeta_k}\right)^n$$ (2.12)

quite similar to (2.3). Here V_∞ denotes an appropriate velocity scale. The symbols a_k and ζ_k represent again the radius and the centre of the k-th profile, their values following from the known mapping function. The function $F(\zeta)$ contains - first of all - the boundary conditions to be satisfied beyond the circumferences of the circles. If e.g. the flow has to be uniform at $\zeta = \infty$, then

$$F(\zeta) = e^{-i\alpha} ,$$ (2.13)

the symbol α denoting the angle of inclination of this flow with respect to the real axis. If the flow is induced by a source and a vortex located at $\zeta = 0$, then

$$F(\zeta) = \frac{Q - i\Gamma}{\zeta} \cdot B , \qquad (2.14)$$

Q and Γ denoting, respectively, known intensities of the source and the vortex, and B representing a length scale. However, if sources, and sinks are distributed along the circumference of the profile in order to simulate the wake, the function $F(\zeta)$ will contain appropriate terms. The same is true as far as other singularities are concerned distributed in known manner in the whole flow domain.

The problem of determination of the complex velocity (2.12) reduces therefore to the determination of the complex coefficients:

$$c_{kn} = \mu_{kn} + i\, \nu_{kn} ; \qquad k = 1,2,\ldots K; \; n = 1,2,\ldots \qquad (2.15)$$

The set of linear equations containing these coefficients as unknowns, follows from appropriate conditions imposed on velocity components at every circle of the system. Taking into account the l-th circle one can introduce an auxiliary variable Θ, such that a point at this circle is determined by the following expression:

$$\zeta = \zeta_1 + a_1\, e^{i\Theta} , \qquad (2.16)$$

the meaning of a_1, ζ_1 being explained previously.

The velocity components: perpendicular $V_n^{[1]}$ and tangent $V_s^{[1]}$ to the circle at the point (2.6) can now be expressed by means of the derivative (2.12) as:

$$V_n^{[1]}(\Theta) - i\, V_s^{[1]}(\Theta) = e^{i\Theta} \cdot \frac{dw}{d\zeta}\Big|_{\zeta_1 + a_1 e^{i\Theta}} \qquad (2.17)$$

the superscript l referring to the number of the circle.

The conditions mentioned above can now be written in the following simple form:

$$V_n^{[1]}(\Theta) = 0; \qquad (2.18)$$

$$V_s^{[1]}(\epsilon_1) = 0, \qquad 1 = 1,2,\ldots K, \qquad (2.19)$$

representing the impermeability condition, and the Kutta-Joukowski condition, resp. The impermeability condition may easily be modified, if the perpendicular velocity component does not have to vanish, but has to be equal to a given function of Θ .

In order to satisfy the impermeability condition (2.18) the perpendicular velocity component $V_n^{[1]}$ (Θ) in (2.17) must now be developed into Fourier series, and the consecutive coefficients at $\cos n\Theta$, $\sin n\Theta$ (where $n = 0, 1,\ldots N$) have to vanish. Satisfying the Kutta-Joukowski condition (2.19) is much more simple a task: it suffices to introduce $\Theta = \epsilon_1$ into the formula for the tangent velocity component $V_s^{[1]}$ (Θ) stemming from (2.17). The meaning of ϵ_1 was already explained - comp. Fig. 5.

If the function $F(\zeta)$ in (2.12) corresponds to the uniform flow - see (2.13) - the system of linear equations obtained in the described manner can be written in the following form:

$$\mu_{11} = 0;$$

$$\nu_{11} + \sum_{n=2}^{N} \left[\mu_{1n} \sin(-n+1)\, \varepsilon_1 + \nu_{1n} \cos(-n+1)\, \varepsilon_1 \right] +$$

$$\text{Im}\left\{ e^{i\varepsilon_1} \sum_{k=1}^{K} \sum_{n=1}^{N} (\mu_{kn} + i\nu_{kn}) \left(\frac{a_k}{\zeta_1 - \zeta_k + a_1 e^{i\varepsilon_1}} \right)^n \right\} =$$

$$= \sin\alpha \cos\varepsilon_1 - \cos\alpha \sin\varepsilon_1 \quad ;$$

$$\mu_{12} + \sum_{k=1}^{K} \sum_{n=1}^{N} (\mu_{kn}\, r_{k1n}^{[0]} - \nu_{kn}\, s_{k1n}^{[0]}) = -\cos\alpha \quad ;$$

$$\nu_{12} + \sum_{k=1}^{K} \sum_{n=1}^{N} (\mu_{kn}\, s_{k1n}^{[0]} + \nu_{kn}\, r_{k1n}^{[0]}) = -\sin\alpha \quad ;$$

$$\mu_{13} + \sum_{k=1}^{K} \sum_{n=1}^{N} (\mu_{kn}\, r_{k1n}^{[1]} - \nu_{kn}\, s_{k1n}^{[1]}) = 0 \;;$$

$$\nu_{13} + \sum_{k=1}^{K} \sum_{n=1}^{N} (\mu_{kn}\, s_{k1n}^{[1]} + \nu_{kn}\, r_{k1n}^{[1]}) = 0;$$

$$\cdots\cdots\cdots\cdots\cdots\cdots\cdots\cdots\cdots\cdots\cdots\cdots\cdots\cdots\cdots$$

$$\mu_{1N} + \sum_{k=1}^{K}{}' \sum_{n=1}^{N} (\mu_{kn}\, r_{k1n}^{[N-2]} - \nu_{kn}\, s_{k1n}^{[N-2]}) = 0;$$

$$\nu_{1N} + \sum_{k=1}^{K}{}' \sum_{n=1}^{N} (\mu_{kn}\, s_{k1n}^{[N-2]} + \nu_{kn}\, r_{k1n}^{[N-2]}) = 0.$$

$$\left. \right\} \quad (2.20)$$

The second equation stems from the Kutta-Joukowski condition. The prime (') denotes omission of the term k = 1 in the respective sum.

Finally, it is to be understood, that $l = 1, 2, \ldots K$, so that the system (2.20) contains in fact 2KN equations with the same number of real unknowns μ_{kn}, ν_{kn}.

The constants $r_{k1n}^{[q]}$, $s_{k1n}^{[q]}$ are real and depend on four indices. They are entirely determined by the geometry of the system of circles, and they are defined by the following recursive formula:

$$r_{k1n}^{[q]} + i\, s_{k1n}^{[q]} = h_{k1n}^{[q]} = h_{k1n}^{[q-1]} \left[-\frac{n+q-1}{q} \frac{a_1}{\zeta_1 - \zeta_k} \right] \qquad (2.21)$$

where $q = 1,2,.....$ and

$$h_{k1n}^{[0]} = \left(\frac{a_k}{\zeta_1 - \zeta_k} \right)^n .$$

(the function $F(\zeta)$ is represented by (2.14) the derivation of the system of equations (2.20) as well as the left-hand sides of the equations do not change. On the right-hand sides, however, certain constants will appear instead of $\cos \alpha$, $\sin \alpha$, representing Fourier coefficients of the development of the function

$$F(\zeta_1 + a_1 e^{i\theta}) = \frac{Q - i\Gamma}{\zeta_1 + a_1 e^{i\theta}} B \qquad (2.22)$$

into Fourier series with respect to θ.

2.3 Numerical Examples

Two examples will be presented corresponding to two different forms (2.13), (2.14) of the function $F(\zeta)$ in (2.13).

The first one of them concerns flow around a system of three profiles (Fig. 8), uniform at infinity and inclined at $\alpha = 14^o$ with respect to the chord of the largest profile. The geometry of the system is borrowed from a paper by K. Jacob [5]. After determination of the mapping function (2.3) the system of circles shown in Fig. 9 is obtained. Solving the system (2.20) of equations one arrives at the velocity field around the circles, represented by the complex velocity (2.12). Integrating (2.12) with respect to ζ one obtains the complex potential $W(\zeta)$ - comp. (2.10). Now, a family of streamlines

$$\Psi(\xi, \eta) = const \qquad (2.23)$$

can easily be computed, the results being shown in Fig. 9. The streamlines around the profiles follow from transformation of those around the circles by the use of the mapping function (2.3). The patterns of streamlines in the Figs. 8. and 9. are taken from [6].

The second numerical example concerns flow around a profile, situated between parallel lines [7], as it is shown in Fig. 10. The upper profile of the corresponding symmetric "biplane" obtained by means of the exponential function (2.8) in the z-plane as well as the upper circle obtained in the ζ- plane by conformal mapping of the "biplane" by the use of the mapping function (2.5) are presented in Fig. 11. In the case considered, the flow in the ω-plane uniform at $\bar{\rho} = -\infty$ transforms into a source in $z = 0$, the intensity of the source being:

$$Q = u_\infty \tau,$$

where u_∞ denotes velocity of the uniform flow. This source is denoted by S in the ζ- plane. Consequently, the flow to be determined in the last mentioned plane is generated indeed by the function $F(\zeta)$ of the type (2.14).

After obtaining the complex velocity in the original ω - plane:

$$\frac{dw}{d\omega} = \frac{dW}{d\zeta} \cdot \left(\frac{dz}{d\zeta}\right)^{-1} \cdot \frac{dz}{d\omega} \quad ; \qquad \frac{dz}{d\omega} = \frac{2\pi}{\tau} \cdot e^{\frac{2\pi\omega}{\tau}} \quad ; \tag{2.24}$$

the factors $dz/d\zeta$ and $dW/d\omega$ being computed in the manner described in the two preceding Sections, the pressure can be determined from the Bernoulli equation. Using the nondimensional pressure coefficient:

$$c_p = 2 \frac{p-p_\infty}{\rho u_\infty^2} = 1 - \left|\frac{1}{u_\infty} \frac{dw}{d\omega}\right|^2 \quad , \tag{2.25}$$

where ρ denotes the fluid density, the pressure distributions on both walls as well as on the profile can be presented as functions of $\bar{p}$. Such distributions are shown in the Fig. 12.

The next Figure illustrates the influence of the walls on the pressure distribution along the profile, the continous line corresponding to flow around the profile situated between the walls, and the dashed one - to flow around the same profile and at the same angle of attack, however, in the unbound domain.

2.4. Force and Moment

Considering now again the flow around a system of K profiles in an unbounded domain, the flow being uniform at infinity, and denoting the incidence, one can easily arrive at a generalization of the known Joukowski formula expressing lift, drag and moment acting on a single profile as functions of incidence.

It can be done in the following manner: The Blasius-Chaplygin formulae derived for the case of a single profile remain valid, in fact, also for the case of a system of profiles. Applying them to the k-th profile of the system one can represent the lift $L^{[k]}$, the drag $D^{[k]}$, and the moment $M^{[k]}$, acting on this profile, as:

$$L^{[k]} + i \, D^{[k]} = - \frac{\rho}{2} \, e^{i\alpha} \oint_k \left(\frac{dw}{d\zeta}\right)^2 \left(\frac{dz}{d\zeta}\right)^{-1} d\zeta \quad ; \tag{2.26}$$

$$M^{[k]} = - \frac{\rho}{2} \, \text{Re}\left\{ \oint_k \left(\frac{dw}{d\zeta}\right)^2 \left(\frac{dz}{d\zeta}\right)^{-1} \cdot z \, (\zeta) \cdot d\zeta \right\} \quad ,$$

the integrals being taken along the circumferences of the k-th circle in the auxiliary ζ-plane. The incidence appears in the formulae (2.26) either directly, i.e. through the exponential factor, or indirectly, i.e. through the complex velocity $(dW/d\zeta)$. The mapping function obviously does not depend on the incidence.

However, the general relationship between the complex velocity and the incidence can be established very simply by consideration of the system of equations (2.20). They contain $\cos\alpha$, $\sin\alpha$ linearly on the right-hand sides, therefore, the solution of the system must also depend linearly on these two values:

$$c_{kn} = c'_{kn} \cos \alpha + c''_{kn} \sin \alpha \; ; \quad k = 1,2,\ldots K; \quad n = 1,2,\ldots N \qquad (2.27)$$

the symbols c'_{kn}, c''_{kn} denoting solutions of the system (2.20) at $\alpha = 0$, and $\alpha = \pi/2$, resp. The complex velocity (2.12) depends linearily on the coefficients c_{kn}, therefore it can be expressed as

$$\frac{dW}{d\zeta} = V_\infty \left[E_c(\zeta) \cos \alpha + E_s(\zeta) \sin \alpha \right] \; , \qquad (2.28)$$

the functions E_c and E_s being independent of the incidence.

Substitution of (2.28) into (2.26) and introduction of nondimensional coefficients of lift, drag and moment in the usual way leads to the following formulae:

$$\left. \begin{aligned}
c_L^{[k]} + i\, c_D^{[k]} &= j_{k1} \cos \alpha + j_{k2} \sin \alpha + j_{k2}(\cos 3\alpha + i \sin 3\alpha) \\
c_M^{[k]} &= j_{k4} + j_{k5} \cos 2\alpha + j_{k6} \sin 2\alpha,
\end{aligned} \right\} \qquad (2.29)$$

representing the above mentioned <u>generalization</u> of the Joukowski formulae.

The symbols j_{km} denote complex (m= 1,2,3), or real (m = 4,5,6) constants.

The formulae have been derived formerly [8] for the case k = 2, and generalized afterwards for an arbitrary number K.

For a single profile the constant j_{k3} disappears, the constants j_{k1} and j_{k2} becoming real.

3. Unsteady Flow in Time-Depending Domains

Extension of the method of conformal mapping to investigation of <u>unsteady</u> flows satisfying, however, all the remaining assumptions introduced in Chapter 1, - is almost obvious.

The additional difficulties which arise in comparison with the steady case stem from the peculiar properties of unsteady plane flows in general, not from the method itself.

One of the most important peculiarities of this kind concerns the non-uniqueness of pressure, caused by the local derivative of the multi-valued velocity potential appearing in the Cauchy-Lagrange integral. As it is well-known, the uniqueness of pressure may be restored in many ways. Confining the considerations to unsteady flows with constant circulation around every profile represents one of such possibilities. Introduction of vorticity shedded behind every profile in form of a continuous "sheet" in form of separate vortices - represents another one. Assumptions concerning the manner in which the circulations around profiles change with time are also conceivable.

It is not intended to discuss and to compare these concepts in the frame of the present paper. Our aim is restricted to showing that all of them can equally easy be realized by means of the method of conformal mapping.

3.1 The Time-Depending Mapping Function

Let us consider a system of K profiles moving in a known manner with respect to a frame of reference adopted in the complex z-plane. The profiles may be rigid or not, however, at any instant of the considered period of time they must satisfy the assumptions introduced in the Section 2.1. It is obvious that at any fixed instant the exterior of such a system of profiles may be transformed onto the exterior of K circles, the transformation being represented by the formerly assumed mapping function (2.3), the parameters (2.4) of this function referring now to the selected instantaneous configuration of the system or - which is equivalent -to the selected instant t:

$$\left. \begin{array}{l} C_o = C_o(t) ; \\[2mm] C_{kn} = C_{kn}(t) ; \\[2mm] a_k = a_k(t) ; \\[2mm] \zeta_k = \zeta_k(t) ; \end{array} \right\} \tag{3.1}$$

where $k = 1,2,...,K; \ n = 1,2,...$

It means that these parameters represent now functions of time which can be evaluated instant by instant by exactly the same algorithms, and programs which were developed for the steady case. Consequentely, the mapping function itself must be regarded as depending not only on the complex variable ζ, but also on the time t:

$$z = z(\zeta;t) = \zeta + C_o + \sum_{k=1}^{K} \sum_{n=1}^{\infty} C_{kn} \left(\frac{a}{\zeta-\zeta_k} \right)^n , \tag{3.2}$$

the symbols $C_o, C_{kn}, a_k, \zeta_k$ representing here the functions (3.1).

It is advisable to describe the instantaneous positions of the considered system of profiles not directly by means of time, but by means of an auxiliary variable X, which can be regarded afterwards as a given function of time:

$$X = X(t) \tag{3.3}$$

An example of such an auxiliary variable is shown in the Fig. 14, which represents a system of 9 profiles modelling a radial-flow hydraulic machine. The profiles numbered 1-5 represent blades of the diffuser-ring, and do not move with respect to the system of axes x, y. The remaining four profiles, numbered 6-9 represent baldes of the impeller, which rotates as a whole around the origin $x = y = 0$. The variable X describing the instantaneous configuration of profiles is assumed as an angle between the y-axis, and the radius connecting the said origin with the middle point of the chord corresponding to the blade No 6.

The system of profiles shown in the Fig. 14 can be transformed onto a system of circles presented in the Fig. 15 true to scale. The circles should be regarded as pulsating and translating with changing X, no matter whether they correspond to steady profiles or to the rotating ones. Some results of computations borrowed from [9] illustrate this fact.

It should be mentioned that it is advisable not to consider them any more as self-sustaining "profiles" but rather as auxiliary lines of no physical meaning, convenient from the point of view of formulating the transformed boundary conditions.

In the Fig. 16 the coordinates of the center of the circle No 1 are shown as functions of X. It can be seen that both coordinates change considerably although the circle in question represents the image of a steady profile.

The variation of radii is illustrated in the Fig. 17, the upper curve corresponding also to the circle No 1, and the lower one - to the circle No 6. The differences between maximal and minimal values of the radius are about the same for both circles.

The centers of circles corresponding to rotating profiles perform by no means circulatory motion. This fact is illustrated in the Fig. 18 wherein the radius of the center corresponding to the circle No 6 is presented as function of X.

Finally, a certain lag between the rotation of a profile and the corresponding circle follows from Fig. 19, referring to the profile No 6 and its image.

All the functions presented in the Figs. 16-20 are periodic, the period concerning the impeller and the diffuse being, however, different. The explanation follows directly from Fig. 14.

Summarising, we should stress the advantage of using the auxiliary variable X instead of time t. By virtue of this variable the computational results represented by graphs in Figs. 16-20 and alike can be applied directly to quite arbitrary functions (3.3), expressing not only rotation of the impeller with constant angular velocity, but also various accelerated or decelerated motions. It enables one to study different forms of flow in the hydraulic machine at no additional cost as far as computation of the mapping function is concerned.

3.2. The Time-Depending Complex Velocity

The complex potential of flow in the physical domain depends now obviously on the complex variable z as well as on time:

$$w = w(z;t) \tag{3.4}$$

Hence, the instantaneous complex velocity in this domain can be represented by the formula similar to (2.11) containing, however, partial derivatives:

$$\frac{\partial w}{\partial z} = \frac{\partial W}{\partial \zeta} \left(\frac{\partial z}{\partial \zeta} \right)^{-1} . \tag{3.5}$$

The last factor denotes such a derivative of the mapping function (3.2), which will be regarded as known in further considerations. The problem of determination of the complex velocity in the physical plane reduces therefore exactly as in the steady case to determination of $\partial W/\partial \zeta$.

This derivative will be assumed again in the form analogue to (2.12):

$$\frac{1}{V_\infty} \frac{\partial W}{\partial \zeta} = F(\zeta;t) + \sum_{k=1}^{K} \sum_{n=1}^{\infty} c_{kn}(t) \left[\frac{a_k(t)}{\zeta - \zeta_k(t)}\right]^n \qquad (3.6)$$

where a_k, ζ_k denote known functions of time, appearing already in the mapping function.

The function $F(\zeta;t)$ contains again the boundary conditions to be satisfied beyond the circumferences of the circles as well as known singularities distributed in the whole domain of flow. In particular, it may contain in the shedded vorticity distributed either continuously along a line trailing behind every profile, or discontinuously. This line - or point vortices move with respect to the frame of reference in an appropriate manner, defined by the investigator.

The problem of determination of the auxiliary complex velocity (3.6) reduces again to determination of the complex coefficients:

$$c_{kn}(t) = \mu_{kn}(t) + i \, \nu_{kn}(t) , \qquad k = 1,2,\ldots K; \quad n = 1,2,\ldots \qquad (3.7)$$

As previously, a set of linear equations containing instantaneous values of $\mu_{kn}(t)$, $\nu_{kn}(t)$ can easily be derived from the condition of impermeability, and the Kutta-Joukowski condition. It can be done quite similarily as in the steady case, however, the instantaneous velocities of the boundaries will appear now in the system.

Let us consider an 1-th profile of the system as well as its circular image (Fig. 20). The super-script 1 indicating the number of the profile and the circle will be omitted for the sake of brevity. The motion of the profile is assumed to be known, which means, that the velocity $\vec{c}$ of any point of the profile as well as its components c_n, c_s are known.

The condition of impermeability requires that the fluid velocity component v_n perpendicular to the profile equals to the normal velocity component c_n:

$$v_n = c_n \qquad (3.8)$$

along the whole circumference of the profile.

On the other hand, the Kutta-Joukowski condition requires similar equality of the tangent velocity components:

$$v_s = c_s, \qquad (3.9)$$

however, at the trailing edge T only.

Transformation of these conditions to the auxiliary plane does not involve any difficulties. Due to general properties of conformal mapping the corresponding velocity components in both domains are connected by means of the following simple relationship:

$$C_n = C_n(\theta) = c_n \left|\frac{\partial z}{\partial \zeta}\right|_{\zeta_1 + a_1 e^{i\theta}} \quad ; \quad C_s = C_s(\theta) = c_s \left|\frac{\partial z}{\partial \zeta}\right|_{\zeta_1 + a_1 e^{i\theta}} \quad . \tag{3.10}$$

The instantaneous fluid velocity components in the auxiliary plane follow from the formula similar to (2.17):

$$V_n(\theta) - i V_s(\theta) = e^{i\theta} \left.\frac{\partial w}{\partial \zeta}\right|_{\zeta_1 + a_1 e^{i\theta}} , \tag{3.11}$$

the superscript 1 being omitted again. Consequently, both conditions (3.8) and (3.9) can be reduced to the following ones:

$$V_n(\theta) = C_n(\theta) \quad ; \tag{3.12}$$

$$V_s(\epsilon_1) = C_s(\epsilon_1) \quad ; \qquad 1 = 1,2,\ldots K. \tag{3.13}$$

If the profile concerned remains still with respect to the system of coordinates, the right-hand sides disappear, and the both conditions assume the form (2.18), (2.19) corresponding to the steady case.

Finally, in order to arrive at the said system of linear equations one should develop again into Fourier series - with respect to θ - the function (3.11) as well as $C_n(\theta)$ in (3.12), equating afterwards coefficients at $\cos n\theta$, $\sin n\theta$ (where n = 1,2,...N). The system obtained in such a way appears to be quite similar to (2.20), the left-hand sides remaining exactly the same. The right-hand sides, however, with contain the appropriate Fourier coefficients of $C_n(\theta)$ as well as of $F(\zeta_1 + a_1 e^{i\theta}; t)$.

The method of investigation of unsteady flows just described is sometimes understood falsely as being "quasi-steady", which implicates that a "really unsteady" approach is conceivable. This is, however, not quite so, because the apparent "quasi-steadiness" follows directly from a fundamental property of inviscid and incompressible flow. Namely, the Laplace equation governing such a flow does not contain derivatives with respect to time. In consequence, the physical process described by this equation, must necessarily be regarded as a process "without history". It consists in fact of a sequence of instantaneous states, isolated in such a sense, that no transfer in time between the neighbouring ones is described by the equation itself. The presented method which consists really in seeking instantaneous solutions of the Laplace equation satisfying instantaneous boundary conditions corresponds, therefore, closely to the nature of the problem under investigation, and does not follow from any simplifying assumptions of "quasi-steadiness".

3.3. Instantaneous Pressure, Force and Moment

Acting on a selected profile of the system can be determined by means of the Cauchy-Lagrange integral after determination of the velocity field in the physical z-plane. General formulae for the aerodynamic coefficients of lift, drag and moment can easily be derived in analogy to (2.29), involving, however, suitable functions of time instead of constants, [10]. They depend obviously on the adopted particular form of the function $F(\zeta, t)$ in (3.6).

4. The Design Problem for Systems of Profiles

The problem of designing a system of profiles, which possesses some aerodynamic properties given in advance, may be formulated in a variety of ways, more or less complicated and sophisticated. The simplest possible formulation will be given in the present Chapter, the interesting but rather separate question concerning the choice of the desirable aerodynamic properties being left wholly aside. The theory presented may be regarded as generalization of the one given by Mangler [11] and Lighthill [12] in the case of a single profile. Some concepts of the theory have already been published in [13] and [14]. The presentation given in this paper will be restricted to a system of profiles in a stream uniform at infinity.

4.1. Formulation

In the considered design problem two elements are supposed to be given.

The first one consists of a system of K circular profiles in a steady unbounded stream of inviscid and incompressible fluid, uniform at infinity, and inclined at an angle α with respect to the real axis in an assumed frame of reference. Two stagnation points S_k, T_k are given on every circle, as it is shown in the Fig. 21, the subscript k denoting the number of the circle. Velocity distribution $V_c^{[k]}(\theta)$ on every _circle_ can easily be computed for these data by the use of slightly modified method, described in the Section 2.2.

The second element consists of velocity distributions $V_P^{[k]}(\theta)$ given along every sought profile as a function of the auxiliary variable θ. The stagnation points of such distributions shown schematically in the Fig. 22 must coincide with those given formerly on the circles. Moreover, the velocity distribution in the immediate vicinity of the stagnation point T_k must be consistent with the given angle δ_k of the trailing edge of the profile:

$$V_P^{[k]}(\theta) = - \text{ const } \left(\theta - \epsilon_k\right)^{\delta_k/\pi} ;$$

$$V_P^{[k]}(\theta) = + \text{ const } \left(2\pi + \epsilon_k - \theta\right)^{\delta_k/\pi} \qquad k = 1,2,\ldots K.$$

In other words, the two given elements of the problem are identical with functions:

$$V_c^{[k]}(\theta) \quad , \quad V_P^{[k]}(\theta) ; \qquad 0 \le \theta \le 2\pi ; \quad k = 1,2,\ldots K. \qquad (4.1)$$

representing velocity distributions on the circles and on the corresponding profiles. Sought is the mapping function

$$z = z(\zeta)$$

which would transform the given system of circles into a system of profiles.

4.2. Method of Solution

The mapping function representing the solution to the design problem will be sought in a form equivalent to (2.3), and modified only slightly by exclusion of terms representing the singularities due to trailing edges. After such a modification the derivative of the mapping function can be written as:

$$\frac{dz}{d\zeta} = \left[1 + \sum_{k=1}^{K} \sum_{n=1}^{N} c'_{kn} \left(\frac{a_k}{\zeta-\zeta_k}\right)^n\right] \prod_{k=1}^{K} \left[1 - \frac{a_k \exp(i\varepsilon_k)}{\zeta-\zeta_k}\right]^{1-\frac{\delta_k}{\pi}} \tag{4.2}$$

where the angles δ_k of trailing edges as well as angles ε_k denoting images of the trailing edges - are known. In case of a rounded off trailing edges, when $\delta_k = \pi$ for all $k = 1,2,...K$, the right-hand side of (4.2) reduces to the derivative of (2.3), with the inner series truncated to N terms only.

It can easily be seen that the derivative (4.2) satisfies automatically the condition (2.2). There is, however, another important condition which must be satisfied by this derivative, if it has to represent a one-valued transformation. In order to arrive at this condition one may expand every factor in the product

$$\prod_{k=1}^{K} \left[1 - \frac{a_k \exp(i\varepsilon_k)}{\zeta - \zeta_k}\right]^{1-\delta_k/\pi}$$

into series of rational fractions and perform afterwards multiplications indicated in (4.2). The following expression will result:

$$\frac{dz}{d\zeta} = 1 + \sum_{k=1}^{K} \sum_{n=1}^{\infty} c''_{kn} \left(\frac{a_k}{\zeta-\zeta_k}\right)^n \tag{4.3}$$

which represents a one-valued mapping function (2.3) only if

$$c''_{k1} = 0; \qquad k = 1,2,...K. \tag{4.4}$$

The logarithm of (4.2) will also be assumed as series of rational fractions:

$$\ln \frac{dz}{d\zeta} = \sum_{k=1}^{K} \left(1 - \frac{\delta_k}{\pi}\right) \ln \left[1 - \frac{a_k \exp(i\varepsilon_k)}{\zeta - \zeta_k}\right] + g_0 + \sum_{k=1}^{K} \sum_{n=1}^{\infty} g_{kn} \left(\frac{a_k}{\zeta-\zeta_k}\right)^n$$

in view of further considerations. Obviously, there must be:

$$g_0 = 0, \tag{4.6}$$

if the mapping function represented by the expression (4.5) has to retain the property (2.2).

The constants C_{k1}'' in (4.3) depend obviously on the constants appearing in (4.5)

$$C_{k1}'' = C_{k1}'' (g_{\kappa n}, a_\kappa, \zeta_\kappa, \epsilon_\kappa, \delta_\kappa); \quad k,\kappa = 1,2,\ldots K; \quad n = 1,2,\ldots$$

The relationship is rather complicated, nevertheless, it can always be established either analytically or numerically.

The condition (4.4) of one-valuedness of the mapping imposed now on the coefficients of the logarithm (4.5) may therefore be rewritten as:

$$C_{k1}'' (g_{\kappa n}, a_\kappa, \zeta_\kappa, \epsilon_\kappa, \delta_\kappa) = 0; \quad k,\kappa = 1,2,\ldots K; \quad n = 1,2,\ldots \qquad (4.7)$$

Solution to the design problem formulated in the preceding Section must be founded obviously on an appropriate relationship between the derivative (4.2) illustrated by means of Figs. 21 and 22. These relationships follow easily from (2.11), where complex velocities as well as the derivative of the mapping function appear.

Application of (2.11) to a k-th circle and the corresponding profile yields:

$$\left. \frac{dz}{d\zeta} \right|_k = \frac{V_c^{[k]}(\theta) \cdot \exp\left(- i\beta_c^{[k]}(\theta)\right)}{V_p^{[k]}(\theta) \exp\left(- i\beta_p^{[k]}(\theta)\right)}, \qquad k = 1,2,\ldots K$$

the symbol $\beta_c^{[k]}(\theta)$ denoting the (known) angle between the velocity at the circle and the real axis; the symbol $\beta_p^{[k]}(\theta)$ - the analogous (unknown) angle between the velocity $\vec{V}_p^{[k]}(\theta)$ and the same axis.

By virtue of the last formula the logarithm of the mapping function can be represented as follows:

$$\left. \ln \frac{dz}{d\zeta} \right|_k = \ln \frac{V_c^{[k]}(\theta)}{V_p^{[k]}(\theta)} + i\sigma^{[k]}(\theta) \qquad (4.8)$$

The real part containing given elements (4.1) only. The new symbol referring to the imaginary part denotes an unknown function:

$$\sigma^{[k]}(\theta) = \beta_p^{[k]}(\theta) - \beta_c^{[k]}(\theta),$$

representing the difference of the two angles formerly mentioned.

Comparing now (4.1) and (4.8) for k = 1 results in the following equation:

$$\sum_{k=1}^{K} \left(1 - \frac{\delta_k}{\pi}\right) \ln\left[1 - \frac{a_k \exp(i\epsilon_k)}{\zeta_1 - \zeta_k + a_1 \exp(i\theta)}\right] + g_0 +$$

$$+ \sum_{k=1}^{K} \sum_{n=1}^{\infty} g_{kn} \left[\frac{a_k}{\zeta_1 - \zeta_k + a_1 \exp(i\theta)} \right]^n =$$

$$\ln \frac{V_c^{[1]}(\theta)}{V_p^{[1]}(\theta)} + i\sigma^{[1]}(\theta) . \tag{4.9}$$

It is easy to get rid of the unknown function $\sigma^{[1]}(\theta)$ just by adding to the equation (4.9) its complex conjugate. In such a way the fundamental relationship sought follows as:

$$G^{[1]}(\theta) = g_0 + \bar{g}_0 + \sum_{k=1}^{K} \sum_{n=1}^{\infty} \left\{ g_{kn} \left[\frac{a_k}{\zeta_1 - \zeta_k + a_1 \exp(i\theta)} \right]^n \right.$$

$$\left. + \bar{g}_{kn} \left[\frac{a_k}{\bar{\zeta}_1 - \bar{\zeta}_k + a_1 \exp(-i\theta)} \right]^n \right\} , \tag{4.10}$$

where the dashes refer to complex conjugates, and

$$G^{[1]}(\theta) = 2 \ln \frac{V_c^{[1]}(\theta)}{V_p^{[1]}(\theta)} + \sum_{k=1}^{K} \left(1 - \frac{\delta_k}{\pi} \right) \ln \left\{ \left[1 - \frac{a_k \exp(i\epsilon_k)}{\zeta_1 - \zeta_k + a_1 \exp(i\theta)} \right] \right.$$

$$\left. \cdot \left[1 - \frac{a_k \exp(-i\epsilon_k)}{\bar{\zeta}_1 - \bar{\zeta}_k + a_1 \exp(-i\theta)} \right] \right\}$$

denotes a known function.

Using our usual technique of developing (4.10) into power series in $\exp(\mp i\theta)$ one derives easily a linear system of equations for the unknown complex coefficients g_{kn}, which can be written in the following form:

$$g_{10} + \bar{g}_{10} + \sum_{k=1}^{K}{}' \sum_{n=1}^{N} (g_{kn} h_{k1n}^{[0]} + \bar{g}_{kn} \bar{h}_{k1n}^{[0]}) = \int_{0}^{2\pi} G^{[1]}(\theta) \, d\theta ; \tag{4.11}$$

$$\left. \begin{array}{l} g_{11} + \sum_{k=1}^{K}{}' \sum_{n=1}^{N} \bar{g}_{1n} \bar{h}_{k1n}^{[1]} = \int_{0}^{2\pi} G^{[1]}(\theta) e^{i\theta} \, d\theta ; \\ \\ g_{12} + \sum_{k=1}^{K}{}' \sum_{n=1}^{N} \bar{g}_{1n} \bar{h}_{k1n}^{[2]} = \int_{0}^{2\pi} G^{[1]}(\theta) e^{2i\theta} \, d\theta ; \end{array} \right\} \tag{4.12}$$

. . . .

$$g_{1N} + \sum_{k=1}^{K}{}' \sum_{n=1}^{N} \bar{g}_{1n} \bar{h}_{k1n}^{[N]} = \int_{0}^{2\pi} G^{[1]}(\theta) e^{Ni\theta} d\theta; \quad 1=1,2,\ldots K \qquad (4.12)$$

The constants $h_{k1n}^{[q]}$ depend exclusively on the geometry of the system of circles, they are defined by (2.21).

The linear subsystem (4.12) contains $K N$ complex unknowns g_{kn}, and can be solved separately. Afterwards, the real part of g_{10} can be evaluated from every one of the K equations (4.11); this is the reason why the new symbol g_{10} has been introduced instead of the old one g_{0} which appears in the original expression (4.5).

Splitting the equations of the system (4.11), (4.12) into real and imaginary part one arrives at a system of real equations resembling (2.20).

4.3 Conditions for Solvability of the Problem

The solution to the system of equations (4.11), (4.12) obtained for quite arbitrarily assumed two elements (4.1) may not necessarily represent an acceptable solution to the design problem. It is clear that such arbitrary elements (4.1) will not automatically supply values of g_{0}, g_{kn}, satisfying the conditions (4.6) and (4.7). In consequence, the mapping function corresponding to (4.5) with such coefficients g_{0}, g_{kn} stemmig from (4.11), (4.12) may be not unique, the nonuniqueness manifesting itself by "profiles" not closed or intersecting themselves.

In order to illustrate this point a numerical example will be presented, borrowed from [15].

Just two profiles are considered for the sake of simplicity, the values of parameters defining the system of corresponding two circles as well as velocity distributions on the circles being contained in Table 2.

Table 2

k	a_k	$\xi_k = Re(\zeta_k)$	$\eta_k = Im(\zeta_k)$	ε_k	θ_{s_k}	δ_k
1	0.25	0	0	-1.700°	201.732°	15°
2	0.25	-0.1	1.1	-2.511°	200.921°	15°

The assumed value of the incidence $\alpha = 10^{\circ}$, the assumed velocity distributions along the sought profiles being shown in Fig. 23. It should be stressed that $|V_P|$ denotes velocity modulus in this Figure which explains the qualitative difference between distributions sketched in Fig. 22, and the one just discussed.

Solution of the system (4.12) together with the "artificial" introduction of

$$g_{10} = g_{20} = 0 ,$$

being justified by (4.6) gives a system of two "profiles" shown in the Fig. 24. They certainly cannot be accepted as solution to the design problem, being not closed.

However, enforcing "artificially" also the condition (4.7) of one-valuedness of the mapping results at once in a pair of closed profiles, shown in the Fig. 25.

The penalty for the artificial closing of the profiles is the only to be expected. It consists in difference between the required velocity distributions $V_p^{[k]}(\theta)$ along the sought profiles, and the really obtained distributions corresponding to the closed profiles. The last-mentioned distributions are shown in Fig. 26., the discrepancy between the assumed ones (Fig. 23) being quite pronounced.

The artificial fulfillment of the conditions (4.6) and (4.7) can hardly be regarded as a satisfactory means of arriving at the solution to a real design problem. A more effective way consists in introduction of certain parameters into the given functions (4.1), the parameters being assumed in such a manner that they allow:

1. to retain the desired and formulated sufficiently "elastic" aerodynamical and geometrical properties of the sought system of profiles;
2. to satisfy the conditions (4.6), (4.7) ensuring the identity of velocity fields at $z=\infty$ and $\zeta=\infty$ as well as the one-valuedness of the mapping.

Certain optimisation criteria may also be included into formulation of the design problem so generalized.

It should be mentioned, too, that the velocity distributions along the sought profiles may be given not necessarily as functions of the auxiliary variable θ, but as functions of the length measured along the circumference of the respective profile. The correspondence between both said variables can easily be established analogically as in the case of the single profile [11]. Application of the variable s may sometimes be more convenient.

5. Concluding Remarks

The method of conformal mapping can be applied to axissymmetric systems of profiled rings. Fig. 6 can be thought of as a cross-section of such a system. Transformation of the infinite exterior of profiled rings by means of the mapping function (2.5) will yield an exterior of a system of tori. The original system may include central bodies beside the profiled rings. Such bodies will be transformed into spheres by the said function.

The transformation represents therefore no difficulty at all. However, even the simplest problem of axisymmetric, steady, irrotational, inviscid and incompressible flow around such a system, the flow being uniform at infinity cannot be solved in such a general and easy way as in the plane case. First of all, the transformed Laplace equation will contain the mapping function (2.5) and secondly construction of the semi-analytic solution in the transformed space is not so obvious. Still, even using purely numerical methods for construction of solution in the trans-formed space, such as the method of boundary elements, some gain is to be expected following from relative simplicity of formulating boundary conditions in this space in comparison with the original one.

6. References

[1] G a i e r, D.: Konstruktive Methoden der konformen Abbildung. Springer-Verlag (1964).

[2] P r o s n a k, W.J.: Conformal representation of arbitrary multiply-connected airfoil sections. Bulletin de l'Academie Polonaise des Sciences, série des sciences techniques, 25 (1977).

[3] P r o s n a k, W.J.: Theory of airfoil systems (in Polish), Ossolineum, Warszawa- Wroclaw (in print).

[4] P r o s n a k, W.J.: Some special cases of potential flow around multiply-connected airfoil sections. Bulletin de l'Academie Polonaise des Sciences, série des sciences techniques, 25 (1977).

[5] J a c o b, K.: Programm zur Berechnung der inkompressiblen Strömung mit Ablösung für Profilsysteme. DFVLR Göttingen, Bericht Nr. IB23073 R 01 (1973).

[6] P o l c h, E.Z.: Doctoral Thesis. Warszawa Technical University (in preparation).

[7] P r o s n a k, W.J.: On a method of computing the plane steady flow around a profile situated between straight parallel lines. Bulletin de l'Academie Polonaise des Sciences, série des sciences techniques, Vol. XX, No. 4, 1972.

[8] P r o s n a k, W.J., K l o n o w s k a, M.E.: Closed formulae for aerodynamic coefficients of arbitrary biplane wing sections. ZAMM 57 (1977), pp. 25-31.

[9] K l a m m e r, J.: Doctoral Thesis, Warszawa Technical University (in preparation).

[10] P r o s n a k, W.J., K l o n o w s k a, M.E.: PLane unsteady flow of inviscid and in-compressible fluid around a system of profiles. In: Notes on Numerical Fluid Mechanics, Vieweg & Sohn, Braunschweig-Wiesbaden, 2F (1980), pp. 231-239.

[11] M a n g l e r, W.: Die Berechnung eines Tragflügelprofiles mit vorgeschriebener Druck-verteilung. Jahrbuch der deutschen Luftfahrtforschung, Bd.I (1938), pp.46-53.

[12] L i g h t h i l l, M.J.: A new method of two-dimensional aerodynamic design. Reports and Memoranda No. 2112, (April 1945).

[13] P r o s n a k, W.J.: The theory of the design problem for multiplanes. Bulletin de l'Academie Polonaise des Sciences, série des sciences techniques, 25 (1978).

[14] P r o s n a k, W.J.: The inverse problem for multiply-connected airfoil systems. In: Recent Developments in Theoretical and Experimental Fluid Mechanics. Springer-Verlag (1979), pp. 517-528.

[15] J e l é n, A.: Doctoral Thesis, Warszawa Technical University (in preparation).

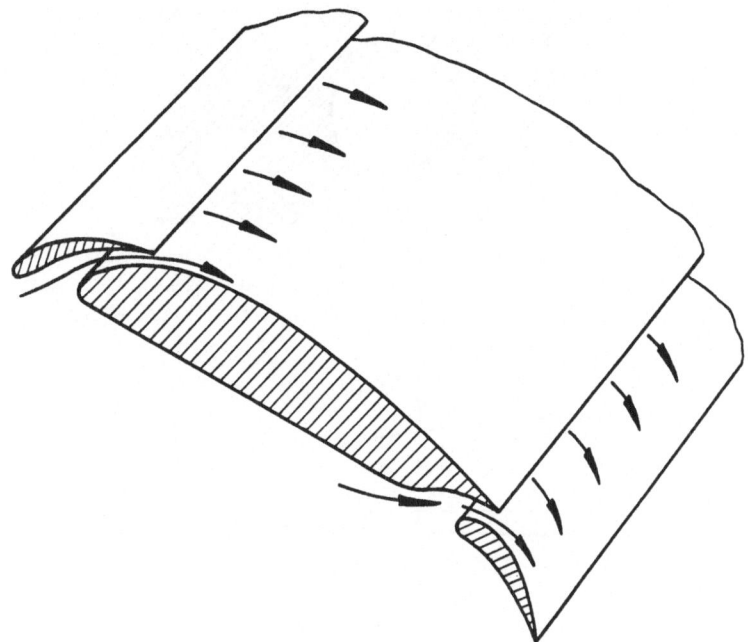

Figure 1

Figure 2

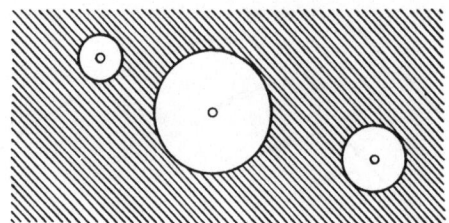

Figure 3

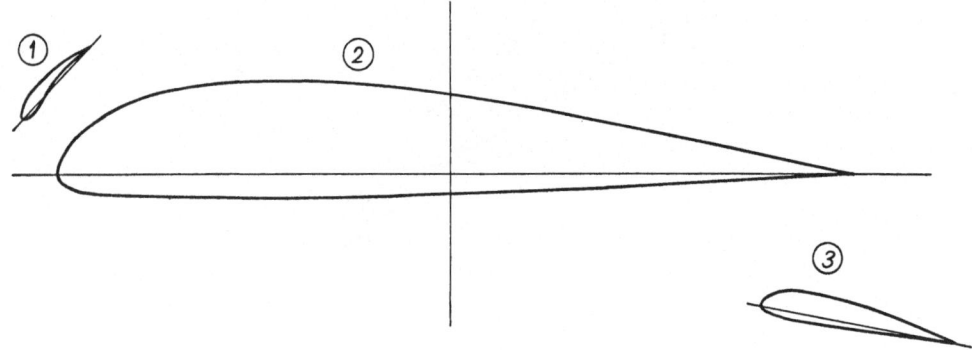

Figure 4

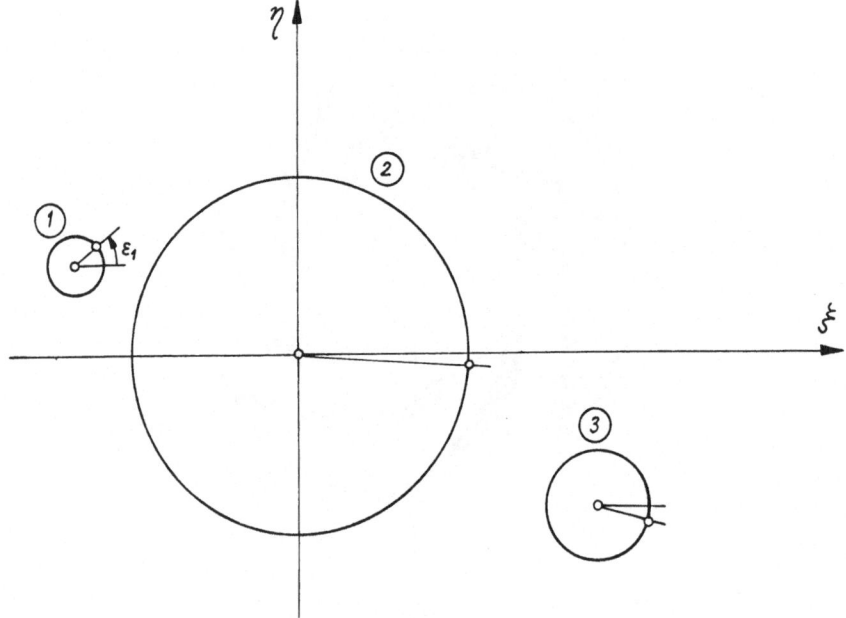

Figure 5

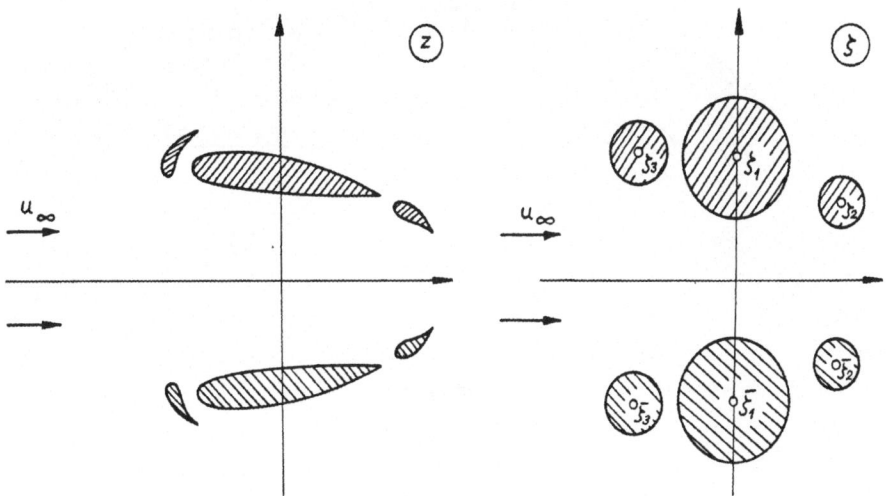

Figure 6

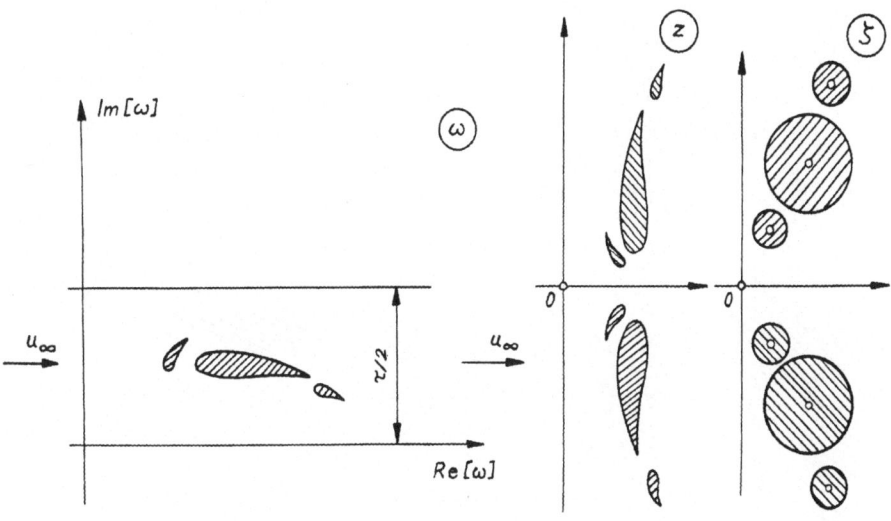

Figure 7

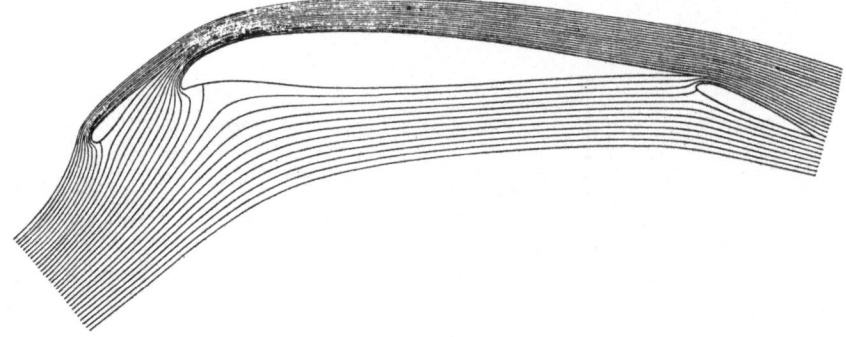

Figure 8

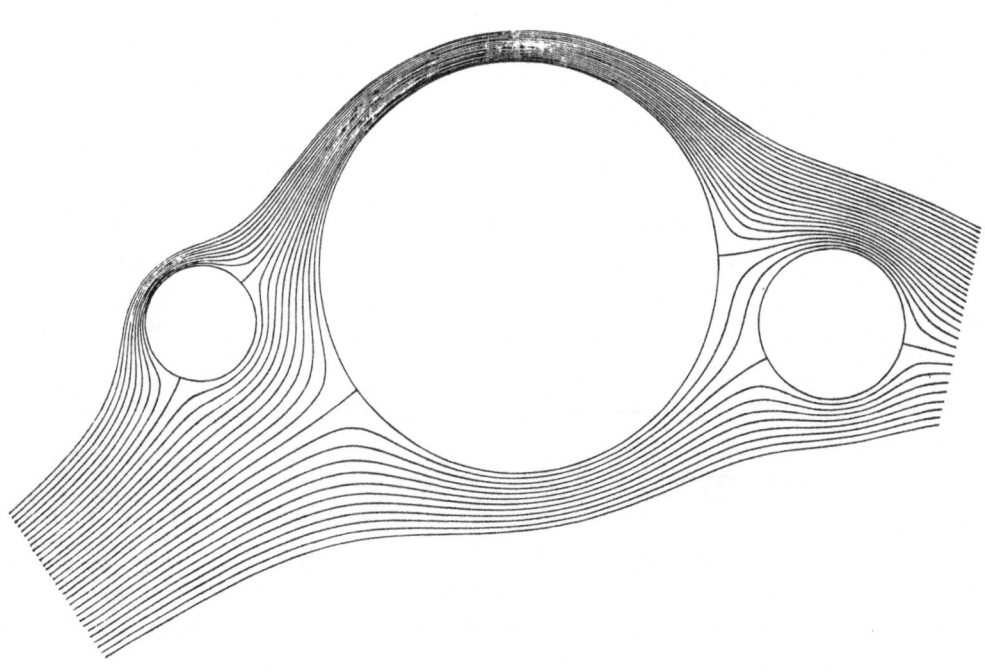

Figure 9

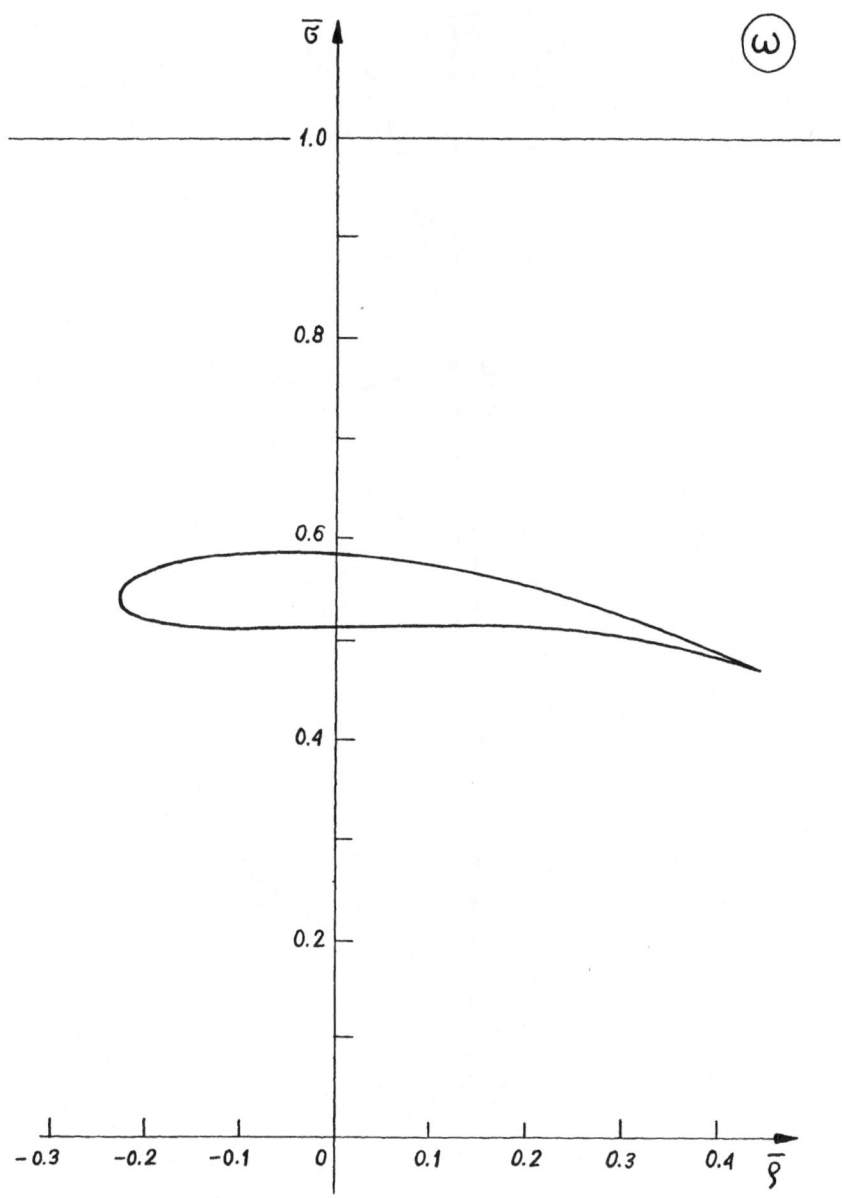

Figure 10

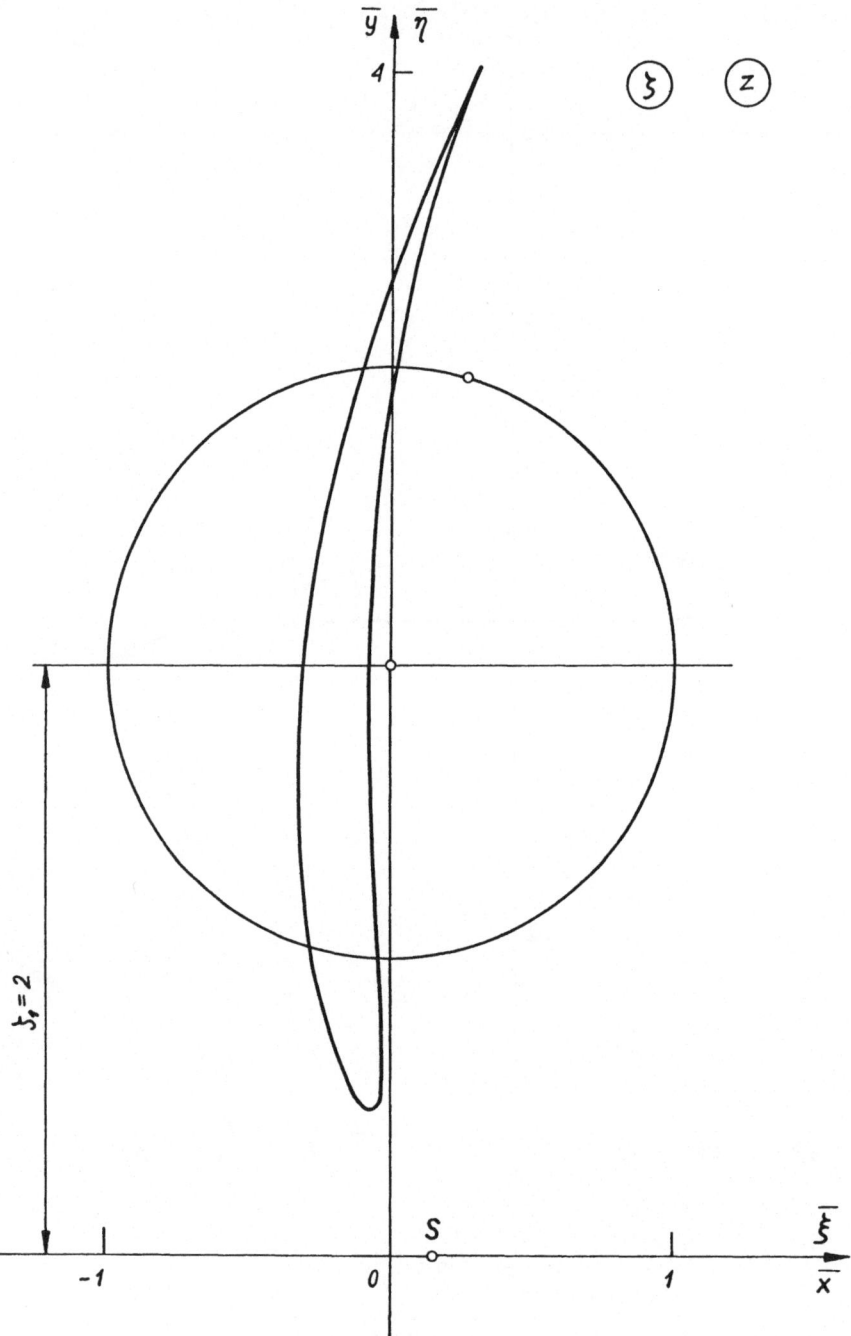

Figure 11

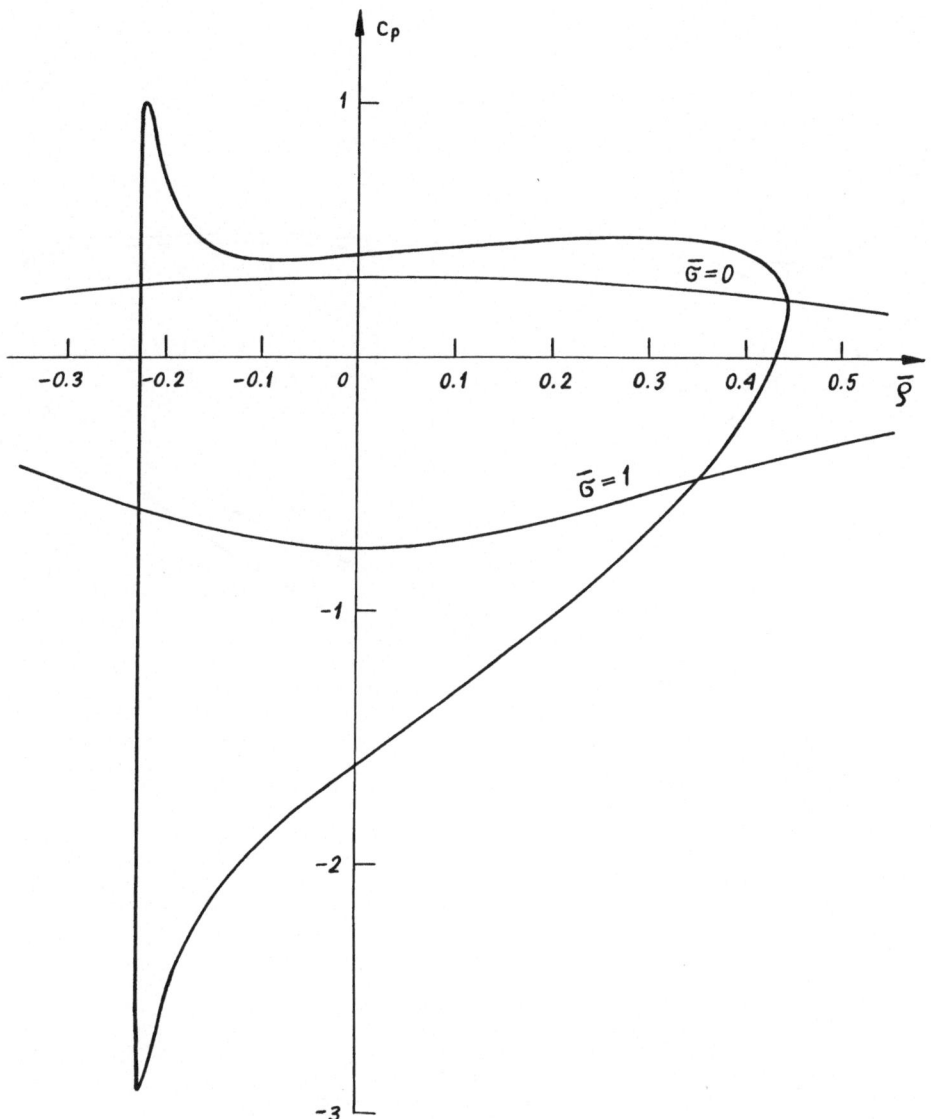

Figure 12

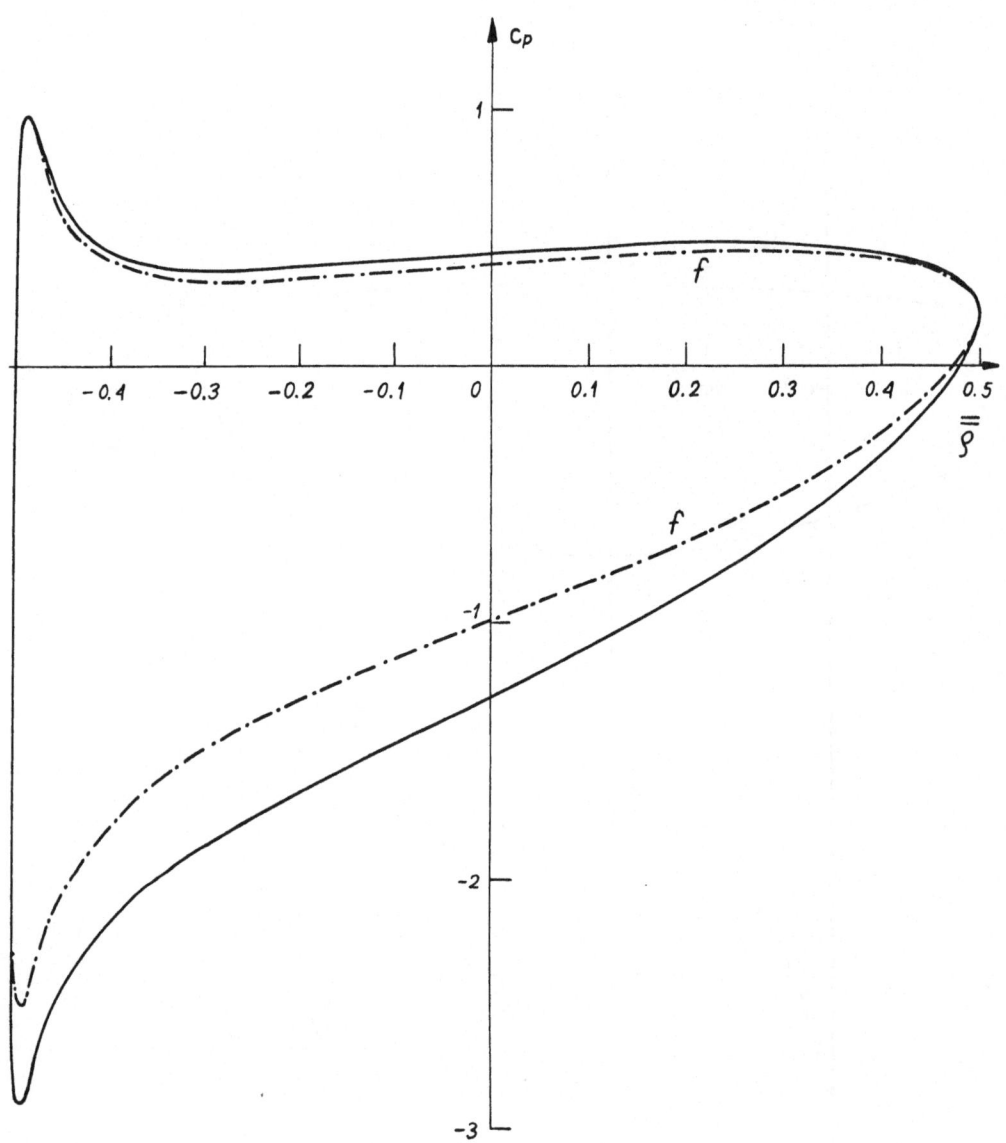

Figure 13

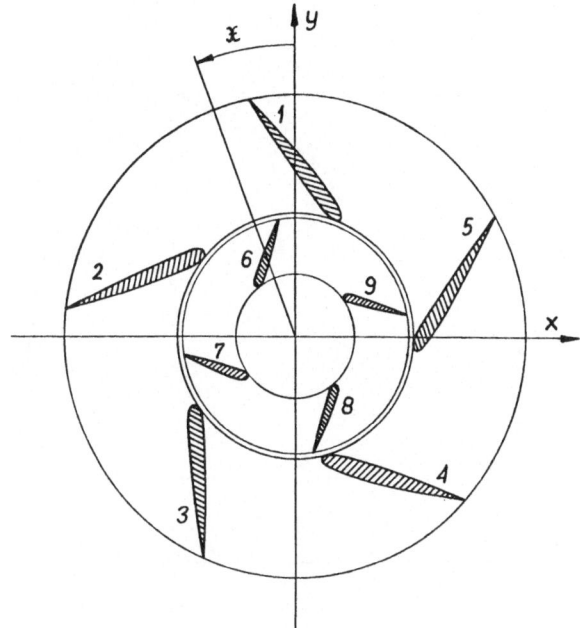

Figure 14

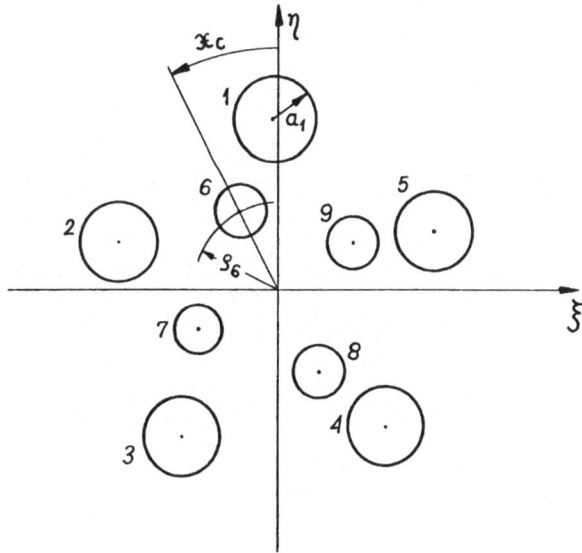

Figure 15

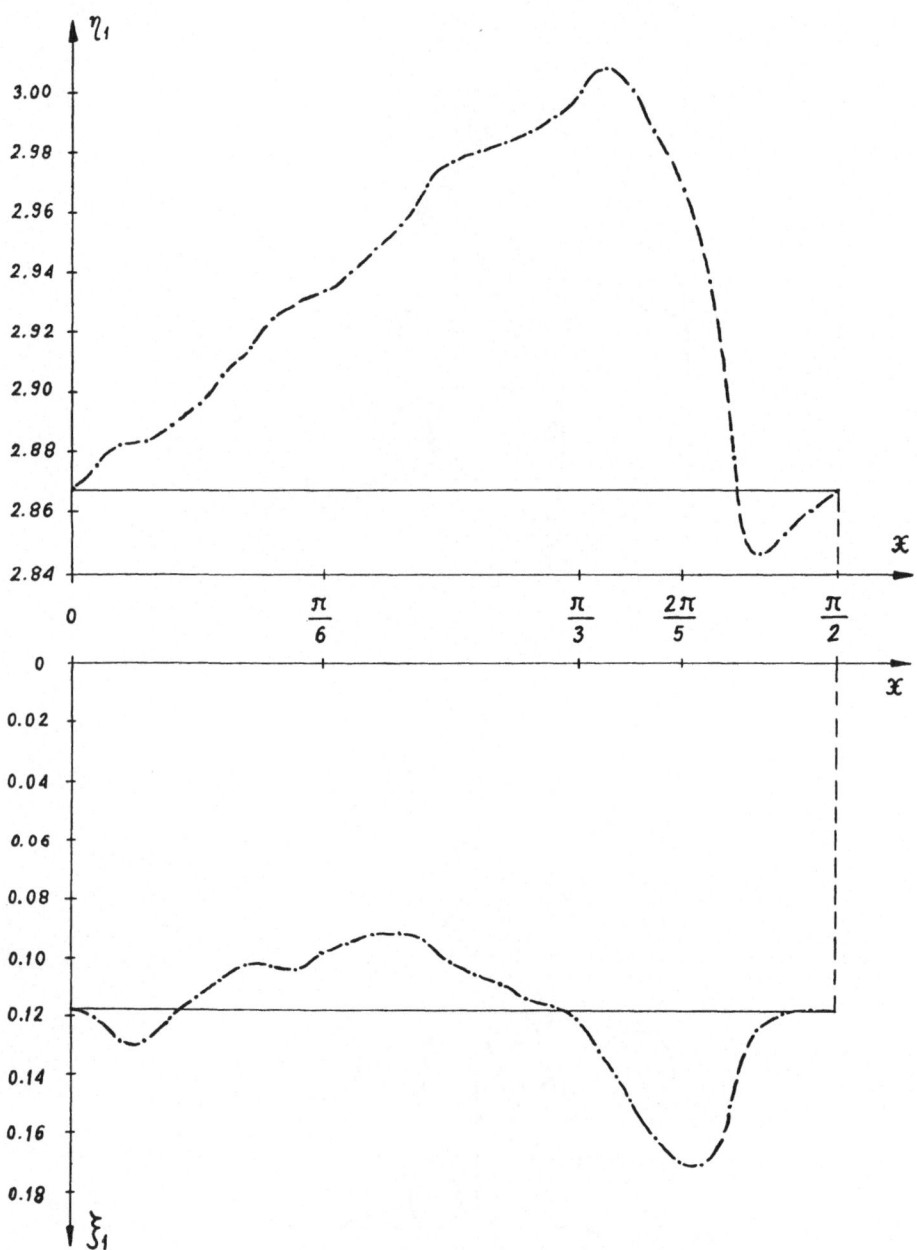

Figure 16

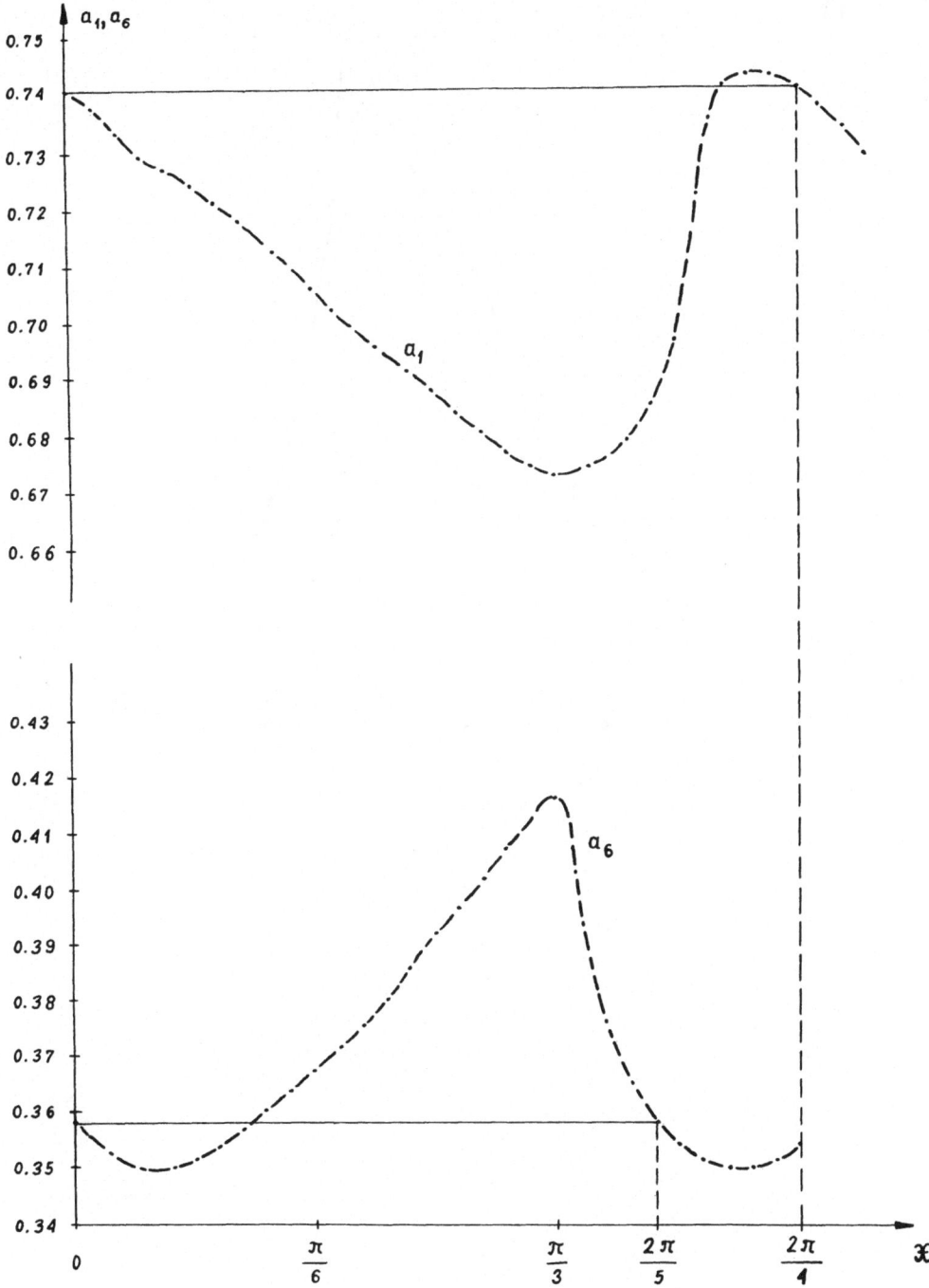

Figure 17

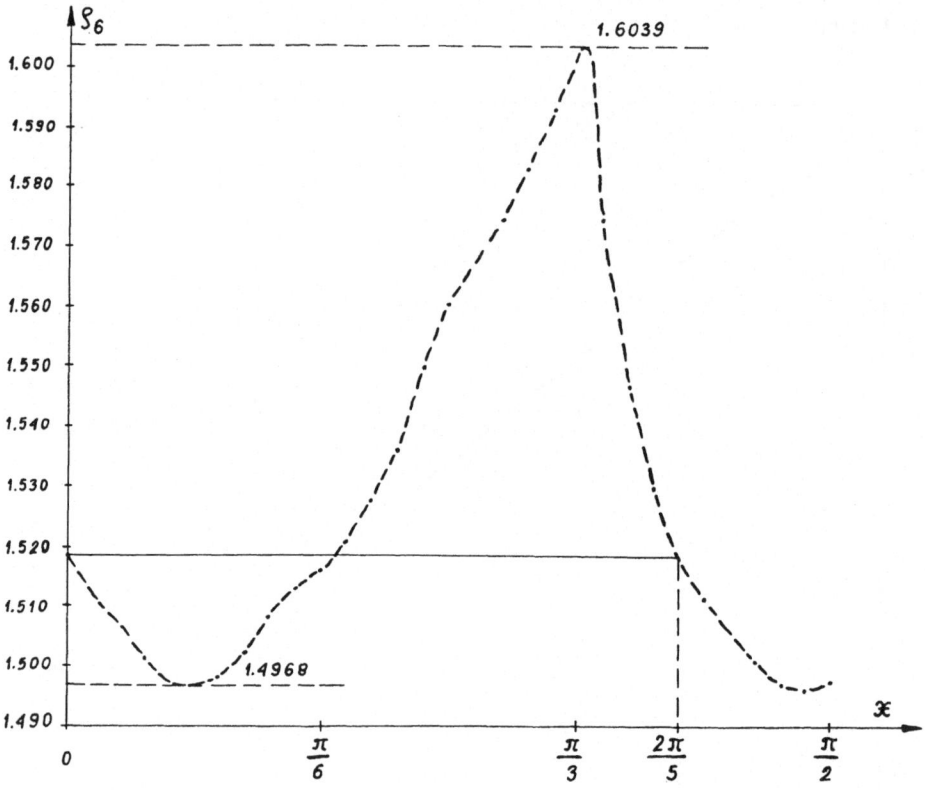

Figure 18

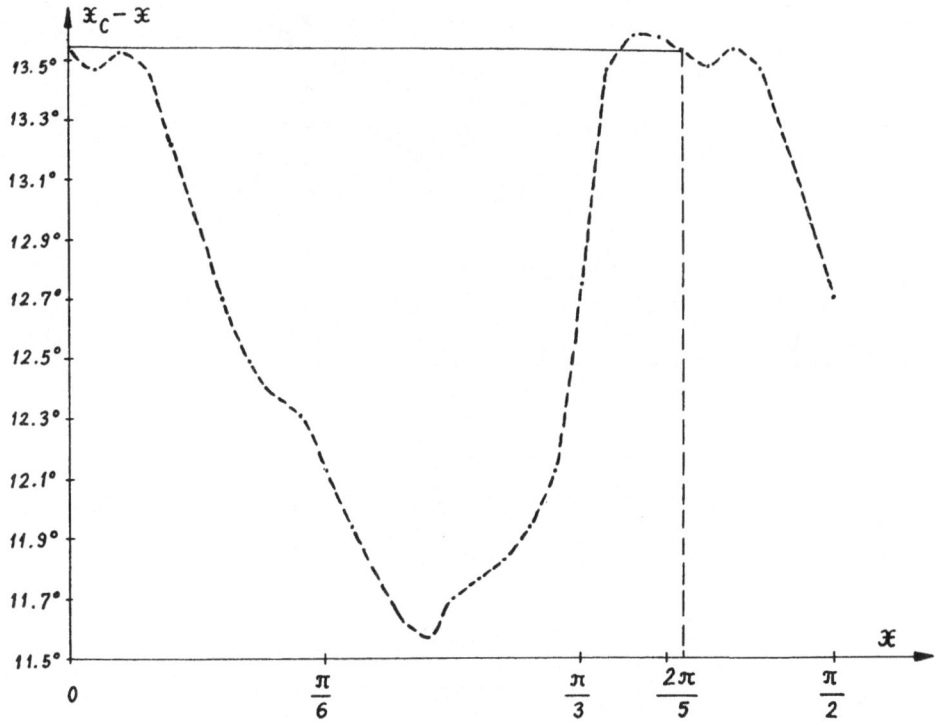

Figure 19

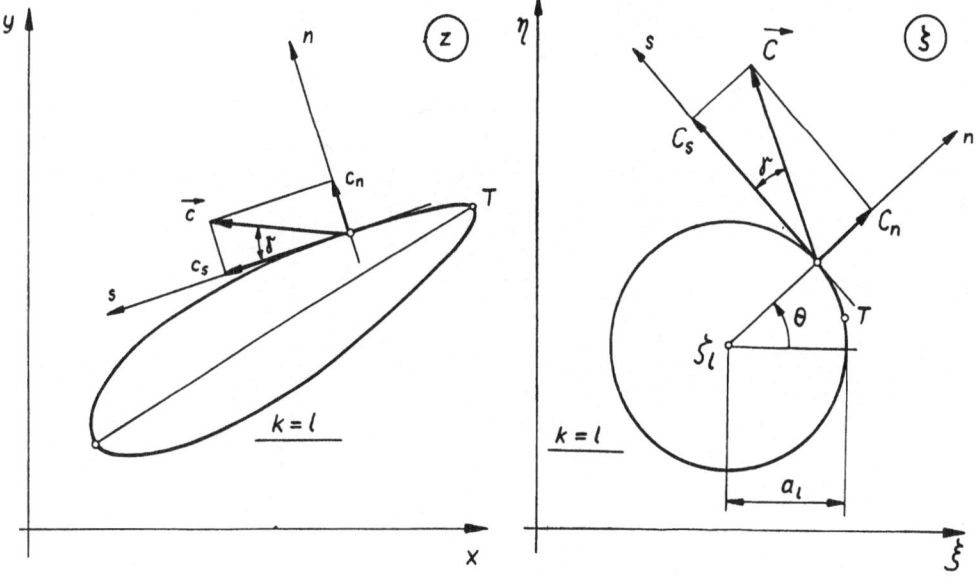

Figure 20

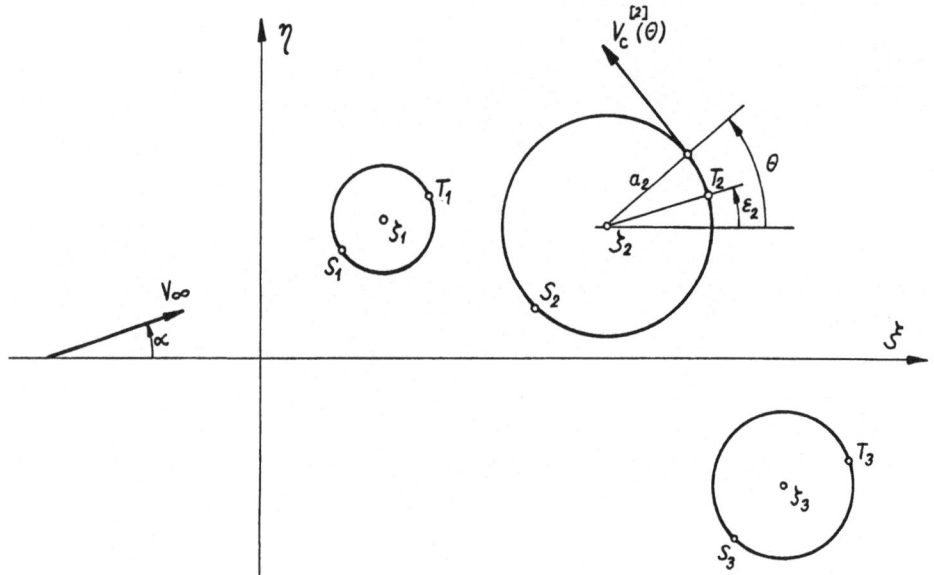

Figure 21

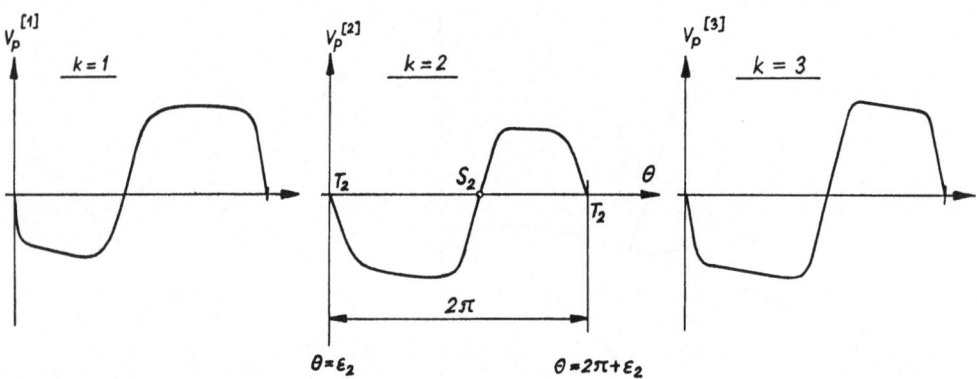

Figure 22

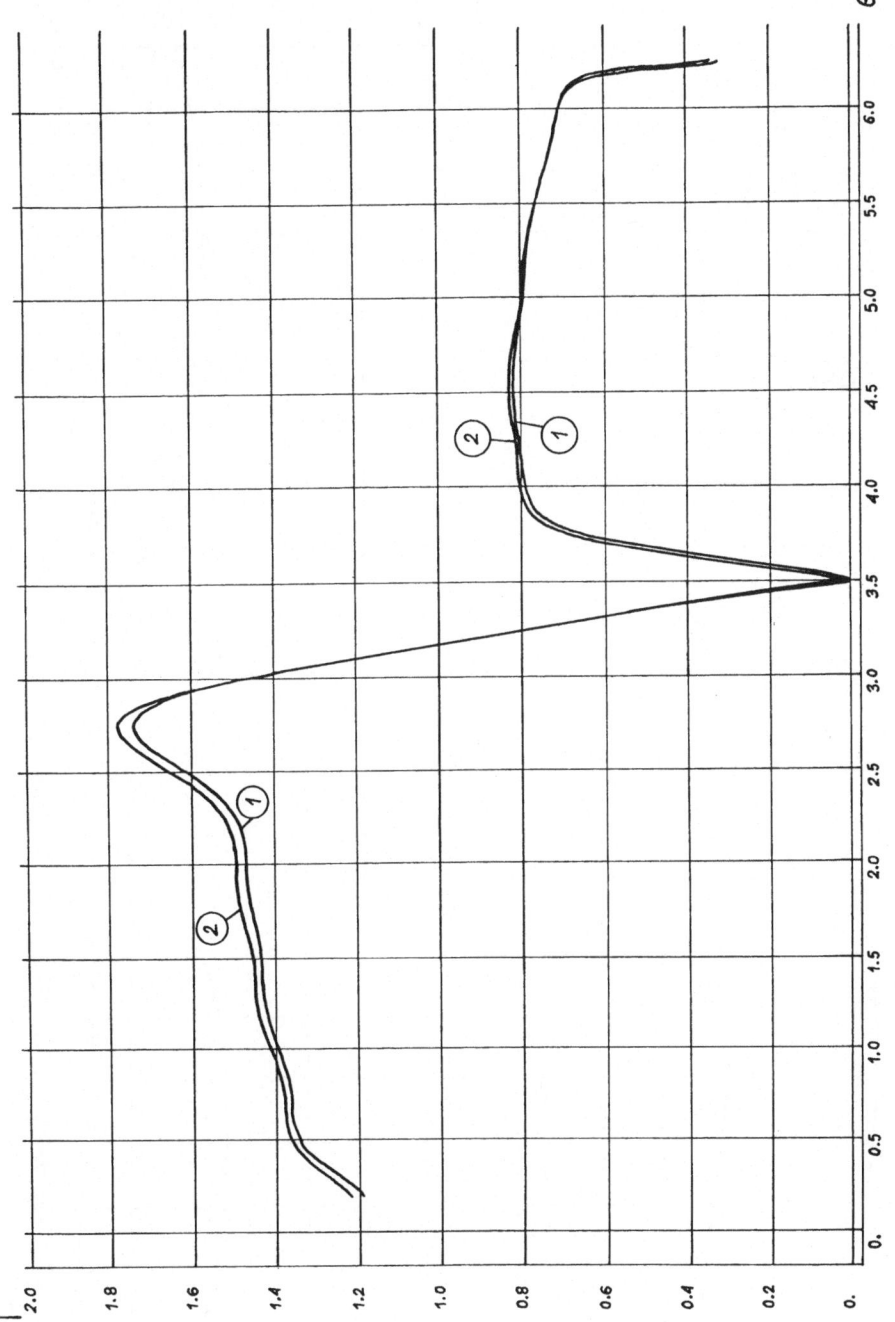

Figure 23

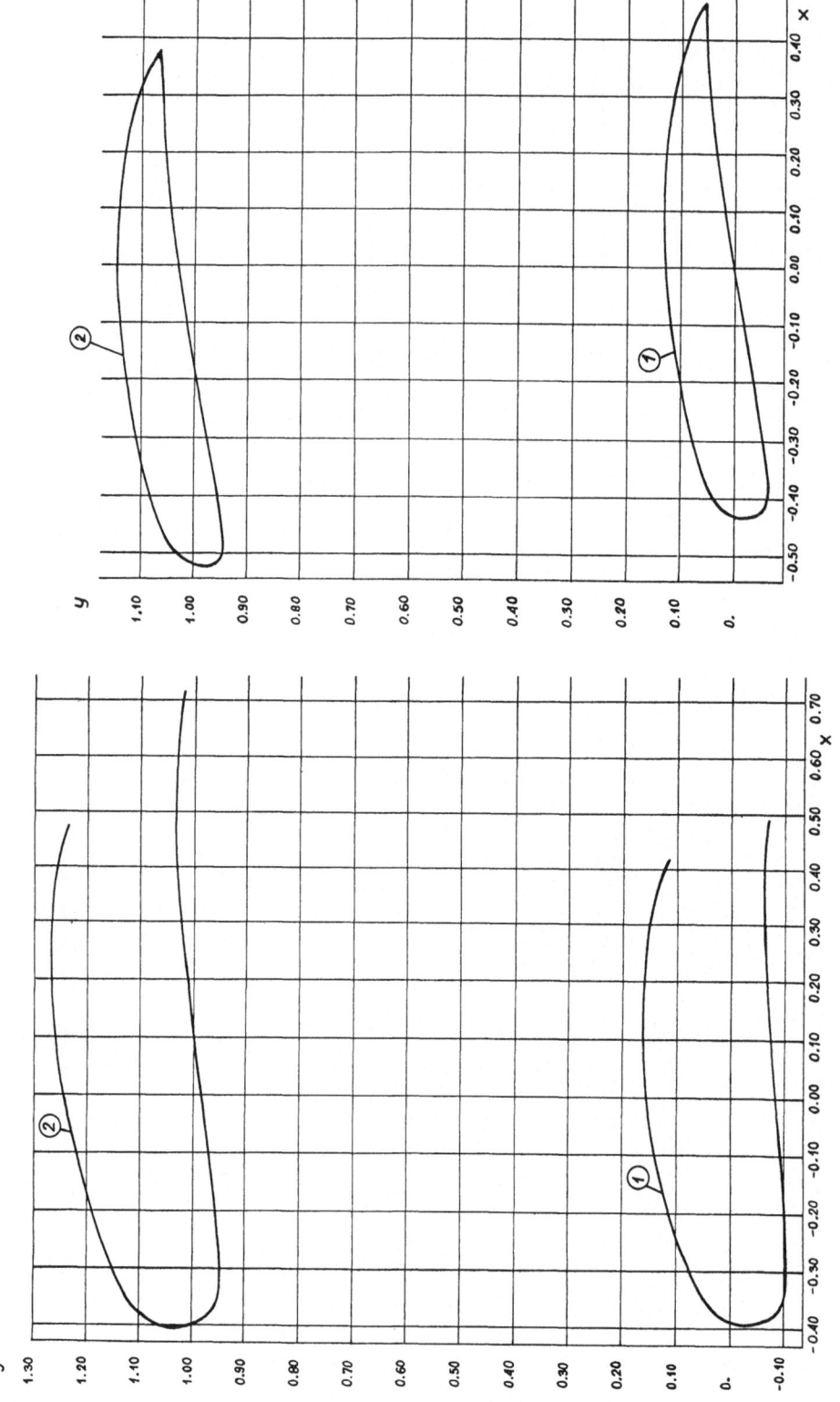

Figure 24

Figure 25

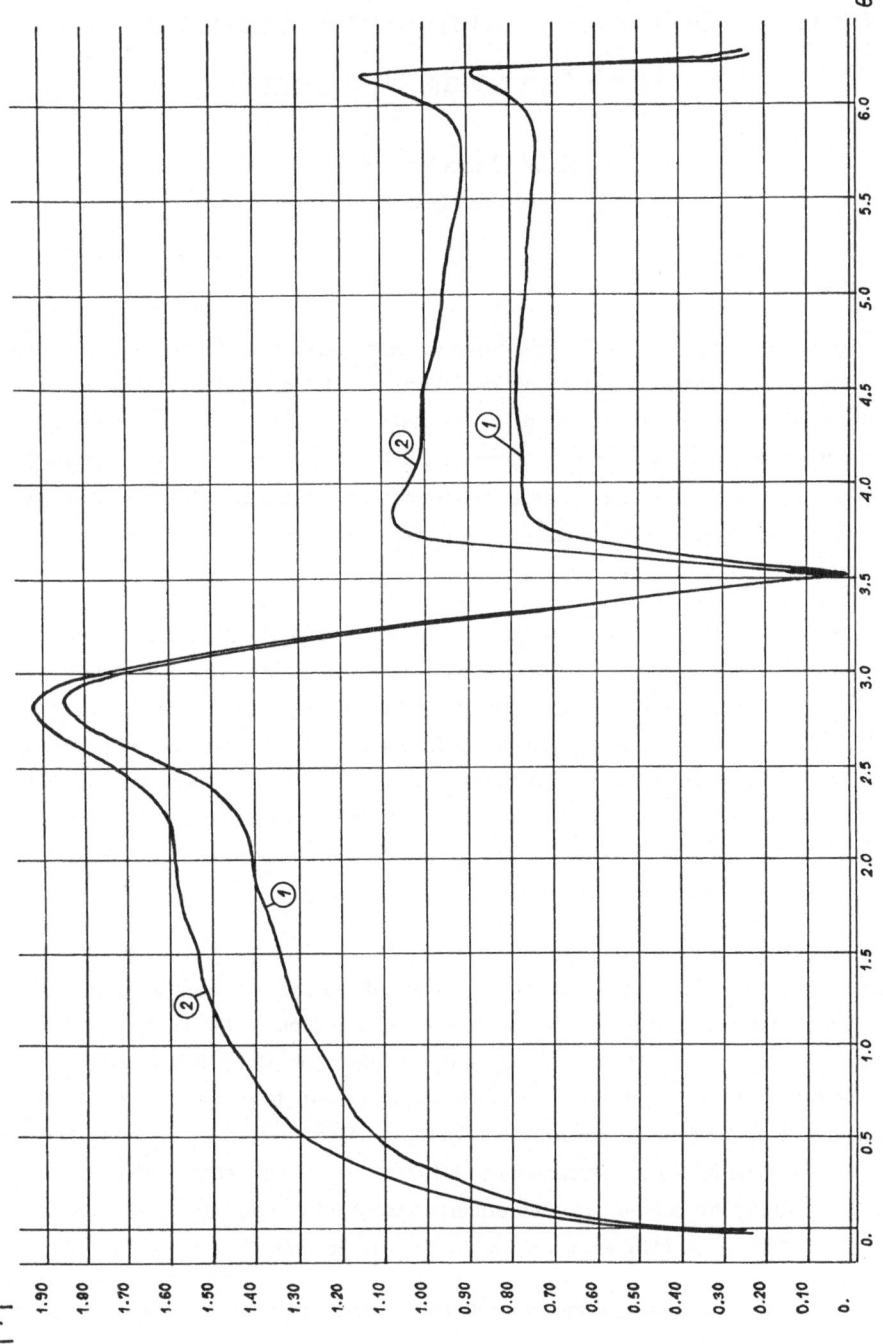

Figure 26

Numerical Solution of Compressible Viscous Flows at High Reynolds Numbers

R. W. MacCormack
Moffett Field

1. Introduction

The equations governing compressible viscous flow have been known for more than a century. Although many special-case solutions have been determined analytically over the years, many others of interest have continued to defy mathematical analysis. Wind tunnels and other experimental facilities have served as invaluable tools in the integration, by physical simulation, of these equations of fluid motion. During the last decade the computer has come to share - through its use of numerical simulations - the work of the earlier analytical and experimental tools in determining new flow solutions.

Like the limits on the range of problems that can be solved analytically, there are limits on the range of flow cases that can be accurately simulated in experimental facilities. The experimental limits are imposed by such factors as tunnel size, wall interference, and stream uniformity[1]. Similarly, the range of computer flow simulations is also limited, principally by computer speed and memory storage. Fortunately, the limits of the theoretical, experimental, and computational techniques are different; as a result, the range of applicability afforded by the three is greater than that attainable with any one of them. Moreover, in regions where they overlap, one approach can be used to verify the results of another.

Nevertheless, we still cannot solve all fluid flow problems of interest, nor can we anticipate that capability in the near future. However, because of the present rapid and potential large growth of computer capabilities, much emphasis is being placed on the development of computational fluid dynamics[2,3]. Before we can calculate the flow field about a complete aircraft configuration at flight Reynolds numbers, there will have to be great progress in developing powerful and reliable computer hardware, in understanding and modeling the physics of turbulent flow, and in devising accurate and efficient numerical methods. That progress will depend, to a significant degree, on theoretical, experimental, and numerical research. One element, the devising of an efficient numerical method, is discussed in this paper.

During the last 10 years many significant contributions have been made in the development of computational methods for solving the equations of compressible viscous flow. Chief among these has been the development of noniterative, block-tridiagonal implicit methods. These methods, which are not subject to restrictive stability conditions, are much more efficient than the earlier explicit methods. The newer methods are, however, much more complicated than the earlier ones and frequently still require long computation times. The goal of the present research is to develop a method for solving the compressible form of the Navier-Stokes equations at high Reynolds number that is unconditionally stable, computationally more efficient than existing methods, and simple and straightforward to program. The method to be described is the implicit analogue of the explicit finite-difference method the author presented in 1969 (see [4]). The new method uses the 1969 method as its first stage. The second stage removes the restrictive stability condition of the 1969 method by recasting the difference equations in an implicit form. The resulting matrix equations to be solved are either upper or lower block-bidiagonal equations and are solved more easily than the block-tridiagonal matrix equations of existing methods. The method is second-order accurate in space and time and is presented in conservation form in two dimensions. Its extension to three dimensions is straightforward.

2. The Navier-Stokes Equations

In two dimensions and by neglecting body force and heat sources, the unsteady compresible form of the Navier-Stokes equations may be written in conservative form as

$$\frac{\partial U}{\partial t} + \frac{\partial F}{\partial x} + \frac{\partial G}{\partial y} = 0$$

where

$$U = \begin{bmatrix} \rho \\ \rho u \\ \rho v \\ e \end{bmatrix}$$

$$F = \begin{bmatrix} \rho u \\ \rho u^2 + \sigma_x \\ \rho uv + \tau_{xy} \\ (e + \sigma_x)u + \tau_{yx}v - k\frac{\partial T}{\partial x} \end{bmatrix} \qquad G = \begin{bmatrix} \rho v \\ \rho uv + \tau_{yx} \\ \rho v^2 + \sigma_y \\ (e + \sigma_y)v + \tau_{xy}u - k\frac{\partial T}{\partial y} \end{bmatrix}$$

and where

$$\sigma_x = p - \lambda\left(\frac{\partial u}{\partial x} + \frac{\partial v}{\partial y}\right) - 2\mu\frac{\partial u}{\partial x}$$

$$\tau_{xy} = \tau_{yx} = -\mu\left(\frac{\partial u}{\partial y} + \frac{\partial v}{\partial x}\right)$$

$$\sigma_y = p - \lambda\left(\frac{\partial u}{\partial x} + \frac{\partial v}{\partial y}\right) - 2\mu\frac{\partial v}{\partial y}$$

in terms of density ρ; x and y velocity components u and v; viscosity coefficients λ and μ; total energy per unit volume e; coefficient of heat conductivity k; and temperature T. Finally, the pressure p is related to the specific internal energy ε and ρ by an equation of state, $p(\varepsilon, \rho)$, where $\varepsilon = e/\rho - (u^2 + v^2)/2$.

The Navier-Stokes equations adequately describe aerodynamic flow at standard temperatures and pressures. If we could efficiently solve these equations there would be no need for experimental tests when designing flight vehicles or other aerodynamic devices. As John Von Neumann said in 1946[5], " Indeed to a great extent, experimentation in fluid dynamics is carried out under conditions where the underlying physical principles are not in doubt, where the quantities to be observed are completely determined by known equations. The purpose of experiment is not to verify a theory but to replace a computation from an unquestioned theory by direct measurements. Thus wind tunnels, for example, are used at present, at least in part, as computing devices to integrate the partial differential equations of fluid dynamics."

Unfortunately the solution of these equations for flows at high Reynolds numbers with srong viscous-inviscid interactions has defied mathematical analysis. Of the two key features of such flow - separation and turbulence - we have been able to make substantial progress during the last decade in the calculation of laminar separation using numerical methods. The calculation of turbulence largely remains an unsolved problem. Although the Navier-Stokes equations adequately describe such flows, computer speed and memory limitation make it impossible for the computational mesh to be fine enough in all spatial directions to resolve all significant eddy length scales of a high-Reynolds-number turbulent flow. As Bradshaw said in 1972[6], " In turbulence studies we are fortunate in having a complete set of equations, the Navier-Stokes equations, whose ability to describe the motion of air at temperatures and pressures near atmospheric is not seriously in doubt (it is easy to show that the smallest significant eddies are many times larger than a molecular mean free path). We are unfortunate because numerical sollution of the full time-dependent equations for turbulent flow is not practical with present computers."

In an approach that circumvents the turbulence problem, the Reynolds or "time-averaged" Navier-Stokes equations are solved. Thus, instead of seeking a time and spacially resolved solution of a rapidly fluctuating turbulent flow, only the time-averaged or mean flow solution is sought. This solution is sufficient to determine the principal quantities of interest, such as lift, drag, and heat transfer. The time-averaged equations look very similar to the original Navier-Stokes equations except that some new terms, called Reynolds stress and turbulent heat transfer terms, appear. These new terms represent the additional mixing caused by turbulence and are determined by models. The models vary from simple algebraic expressions to sets of additional differential equations that need to be solved. Although much progress has already been made in the understanding and modeling of the physics of turbulence, much more is needed before we will have the capability to numericially predict turbulent flow separation with confidence.

3. Numerical Method Applied to a Model Equation

Before discussing the numerical solution of the Navier-Stokes equations it is worthwhile to consider the solution of the following simpler model equation

$$\frac{\partial U}{\partial t} = -c \frac{\partial U}{\partial x} + \nu \frac{\partial^2 U}{\partial x^2}$$

The flow variable U governed by this equation with speed c and diffuses with kinematic viscosity ν. The implicit analogue of the author's 1969 method applied to solve this equation yields the following set of finite-difference equations [7]:

$$p: \begin{cases} \Delta U_i^n = -\frac{\Delta t c}{\Delta x} (U_{i+1}^n - U_i^n) + \frac{\Delta t \nu}{\Delta x^2} (U_{i+1}^n - 2U_i^n + U_{i-1}^n) \\[2mm] \left(1 + \frac{\lambda \Delta t}{\Delta x}\right) \overline{\delta U_i^{n+1}} = \Delta U_i^n + \frac{\lambda \Delta t}{\Delta x} \overline{\delta U_{i+1}^{n+1}} \\[2mm] \overline{U_i^{n+1}} = U_i^n + \overline{\delta U_i^{n+1}} \end{cases}$$

$$c: \begin{cases} \Delta \overline{U_i^{n+1}} = -\frac{\Delta t c}{\Delta x} \left(\overline{U_i^{n+1}} - \overline{U_{i-1}^{n+1}}\right) + \frac{\Delta t \nu}{\Delta x^2} \left(\overline{U_{i+1}^{n+1}} - 2\overline{U_i^{n+1}} + \overline{U_{i-1}^{n+1}}\right) \\[2mm] \left(1 + \frac{\lambda \Delta t}{\Delta x}\right) \delta U_i^{n+1} = \Delta \overline{U_i^{n+1}} + \frac{\lambda \Delta t}{\Delta x} \delta U_{i-1}^{n+1} \\[2mm] U_i^{n+1} = \frac{1}{2} \left(U_i^n + \overline{U_i^{n+1}} + \delta U_i^{n+1}\right) \end{cases}$$

where λ is chosen so that $\lambda \geq \max \{ |c| + (2\nu/\Delta x) - (\Delta x/\Delta t), 0.0 \}$.

The above procedure contains two steps. The first step predicts a new solution at time $t = (n+1)\Delta t$ at each mesh point i from the known solution at time $t = n \Delta t$, using a one-sided difference to approximate the first derivative term and a centered difference for the second derivative. The second stage of the predictor step enables the locally calculated solution changes ΔU_i^n to travel and diffuse throughout the flow field and then calculates implicitly the solution change δU_i^{n+1} to be used in the third and final stage to determine the predicted solution $\overline{U_i^{n+1}}$. The second step, or the corrector step, of the procedure is similar except that it uses opposite one-sided differences to approximate first derivatives.

The second stage of each step represents an implicit approximation to the following equation

$$\frac{\partial \Delta t \frac{\partial U}{\partial t}}{\partial t} = \pm \lambda \frac{\partial \Delta t \frac{\partial U}{\partial t}}{\partial x}$$

with

$$\Delta U_i^n = \Delta t \frac{\partial U^n}{\partial t_i}$$

and

$$\overline{\delta U_i^{n+1}} = \Delta t \, \overline{\frac{\partial U^{n+1}}{\partial t_i}} \quad , \quad \text{etc.}$$

This equation describes the spreading of the solution change $\Delta t(\partial U/\partial t)$, a term of order Δt, with speed $-\lambda$ in the predictor and $+\lambda$ in the corrector. The net effect, if λ is of the order of unity, is the addition to the equations of motion of a term of second order (the difference of two first-order terms). The spreading equation is also related to that obtained by differentiating the model equation by t.

The philosophy behind the procedure is as follows. First the rate of solution change is calculated locally at each mesh point, using an explicit approximation to the governing physical equations. This local rate of change is only valid for a short time, equal approximately to the time required for a flow disturbance to travel from one mesh point to its neighbor. Explicit procedures are restricted, usually for stability reasons, to time steps Δt less than or equal to this characteristic disturbance transit time. Second, this time-step restriction is removed in the second stage by allowing the locally determined rates of solution change to convect and diffuse globally throughout the flow field, governed by an equation related to the physics of the flow. This latter equation is solved implicitly to determine the rate of solution change at each point that approximates the actual during the entire interval Δt.

The method is, according to linear theory, unconditional y stable (unbounded Δt), requires the solution of bidiagonal equations only, and is second-order accurate under the constraint that $v\Delta t/\Delta x^2$ remains bounded as Δt and Δx approach zero (i.e., λ remains of the order of unity).

Note that if the quantity λ is zero, the second stages reduce to

$$\overline{\delta U_i^{n+1}} = \Delta U_i^n$$

and

$$\delta U_i^{n+1} = \Delta U_i^{n+1}$$

or no implicit procedure at all, and the method is identical to the 1969 method. Such a choice for λ results if the chosen time step Δt already statisfies the stability condition of the 1969 method,

$$\Delta t \leq \frac{\Delta x}{|c| + \frac{2v}{\Delta x}}$$

Because of this feature, the method has an advantage in numerical efficiency over existing implicit methods. Not only are the numerical procedures simpler - bidiagonal versus tri-diagonal - but in flow regions for which Δt satisfies the above stability condition, the method reduces from an implicit to a simpler explicit one.

4. Numerical Method Applied to the Navier-Stokes Equations

Applying the method to solve the Navier-Stokes equations we obtain the following implicit predictor-corrector set of finite-difference equations.

$$
p: \begin{cases}
\Delta U^n_{i,j} = -\Delta t \left(\dfrac{\Delta_+ F^n_{i,j}}{\Delta x} + \dfrac{\Delta_+ G^n_{i,j}}{\Delta y} \right) \\[2ex]
I - \Delta t \dfrac{\Delta_+ |A| \cdot}{\Delta x} \left(I - \Delta t \dfrac{\Delta_+ |B| \cdot}{\Delta y} \right) \delta U^{\overline{n+1}}_{i,j} = \Delta U^n_{i,j} \\[2ex]
U^{\overline{n+1}}_{i,j} = U^n_{i,j} + \delta U^{\overline{n+1}}_{i,j}
\end{cases}
$$

$$
c: \begin{cases}
\Delta U^{\overline{n+1}}_{i,j} = -\Delta t \left(\dfrac{\Delta_- F^{\overline{n+1}}_{i,j}}{\Delta x} + \dfrac{\Delta_- G^{\overline{n+1}}_{i,j}}{\Delta y} \right) \\[2ex]
\left(I + \Delta t \dfrac{\Delta_- |A| \cdot}{\Delta x} \right) \left(I + \Delta t \dfrac{\Delta_- |B| \cdot}{\Delta y} \right) \delta U^{n+1}_{i,j} = \Delta U^{\overline{n+1}}_{i,j} \\[2ex]
U^{n+1}_{i,j} = \dfrac{1}{2} \left(U^n_{i,j} + U^{\overline{n+1}}_{i,j} + \delta U^{n+1}_{i,j} \right)
\end{cases}
$$

where $\Delta_+ / \Delta x$, $\Delta_- / \Delta x$, $\Delta_+ / \Delta y$ and $\Delta_- / \Delta y$ are difference operators defined by

$$
\frac{\Delta_+ Z_{i,j}}{\Delta x} = \frac{Z_{i+1,j} - Z_{i,j}}{\Delta x}
$$

$$
\frac{\Delta_- Z_{i,j}}{\Delta x} = \frac{Z_{i,j} - Z_{i-1,j}}{\Delta x}
$$

$$
\frac{\Delta_+ Z_{i,j}}{\Delta y} = \frac{Z_{i,j+1} - Z_{i,j}}{\Delta y}
$$

and

$$
\frac{\Delta_- Z_{i,j}}{\Delta y} = \frac{Z_{i,j} - Z_{i,j-1}}{\Delta y}
$$

As for the model equation, the first derivative terms are one-sided differenced (as shown above) and the second derivative terms are centrally differenced. The matrices $|A|$ and $|B|$ are matrices with positive eigenvalues and are related to the Jacobians of F and G. Let S_x, S_y, and their inverses denote the matrices that diagonalize A and B with $\mu = \lambda = k = 0$ (viscous terms neglected). If the gas equation of state is perfect, $p = (\gamma - 1)\rho \varepsilon$, $A = S_x^{-1} \Lambda_A S_x$, and $B = S_y^{-1} \Lambda_B S_y$, where

$$S_x = \begin{bmatrix} 1 & 0 & 0 & -1/c^2 \\ 0 & \rho c & 0 & 1 \\ 0 & 0 & 1 & 0 \\ 0 & -\rho c & 0 & 1 \end{bmatrix} \begin{bmatrix} 1 & 0 & 0 & 0 \\ -u/\rho & 1/\rho & 0 & 0 \\ -v/\rho & 0 & 1/\rho & 0 \\ \alpha\beta & -u\beta & -v\beta & \beta \end{bmatrix}$$

$$S_y = \begin{bmatrix} 1 & 0 & 0 & -1/c^2 \\ 0 & 1 & 0 & 0 \\ 0 & 0 & \rho c & 1 \\ 0 & 0 & -\rho c & 1 \end{bmatrix} \begin{bmatrix} 1 & 0 & 0 & 0 \\ -u/\rho & 1/\rho & 0 & 0 \\ -v/\rho & 0 & 1/\rho & 0 \\ \alpha\beta & -u\beta & -v\beta & \beta \end{bmatrix}$$

$$\Lambda_A = \begin{bmatrix} u & 0 & 0 & 0 \\ 0 & u+c & 0 & 0 \\ 0 & 0 & u & 0 \\ 0 & 0 & 0 & u-c \end{bmatrix} \qquad \Lambda_B = \begin{bmatrix} v & 0 & 0 & 0 \\ 0 & v & 0 & 0 \\ 0 & 0 & v+c & 0 \\ 0 & 0 & 0 & v-c \end{bmatrix}$$

and where $c = \sqrt{\gamma p/\rho}$ is the speed of sound, $\alpha = (1/2)(u^2 + v^2)$ and $\beta = \gamma - 1$.

The matrices S_x and S_y are each given above as the product of two matrices. For each, the right matrix represents a transformation from conservative to nonscervative variables, for example, from $(\delta\rho, \delta\rho u, \delta\rho v, \delta e)$ to $(\delta\rho, \delta u, \delta v, \delta p)$. The left matrix transforms from nonconservative to characteristic form $(\delta\rho - \delta p/c^2, \rho c \delta u + \delta p, \delta v, -\rho\delta u + \delta p)$ and $(\delta\rho - \delta p/c^2, \delta u, \rho c \delta v + \delta p, -\rho c \delta v + \delta p)$ for the S_x and S_y matrices, respectively. The inverses S_x^{-1} and S_y^{-1} are simple to derive. The matrices $|A|$ and $|B|$ are defined by

$$|A| = S_x^{-1} D_A S_x \qquad \text{and} \qquad |B| = S_y^{-1} D_B S_y$$

where

$$D_A = \begin{bmatrix} \lambda_{A1} & 0 & 0 & 0 \\ 0 & \lambda_{A2} & 0 & 0 \\ 0 & 0 & \lambda_{A3} & 0 \\ 0 & 0 & 0 & \lambda_{A4} \end{bmatrix}, \qquad D_B = \begin{bmatrix} \lambda_{B1} & 0 & 0 & 0 \\ 0 & \lambda_{B2} & 0 & 0 \\ 0 & 0 & \lambda_{B3} & 0 \\ 0 & 0 & 0 & \lambda_{B4} \end{bmatrix}$$

$$\lambda_{A1} = \max \left\{ |u| + \frac{2\nu}{\rho\Delta x} - \frac{1}{2}\frac{\Delta x}{\Delta t} , \quad 0.0 \right\}$$

$$\lambda_{A2} = \max \left\{ |u + c| + \frac{2\nu}{\rho\Delta x} - \frac{1}{2}\frac{\Delta x}{\Delta t} , \quad 0.0 \right\}$$

$$\lambda_{A3} = \max \left\{ |u| + \frac{2\nu}{\rho\Delta x} - \frac{1}{2}\frac{\Delta x}{\Delta t} , \quad 0.0 \right\}$$

$$\lambda_{A4} = \max \left\{ |u - c| + \frac{2\nu}{\rho\Delta x} - \frac{1}{2}\frac{\Delta x}{\Delta t} , \quad 0.0 \right\}$$

$$\lambda_{B1} = \max \left\{ |v| + \frac{2\nu}{\rho\Delta y} - \frac{1}{2}\frac{\Delta y}{\Delta t} , \quad 0.0 \right\}$$

$$\lambda_{B2} = \max \left\{ |v| + \frac{2\nu}{\rho\Delta y} - \frac{1}{2}\frac{\Delta y}{\Delta t} , \quad 0.0 \right\}$$

$$\lambda_{B3} = \max \left\{ |v + c| + \frac{2\nu}{\rho\Delta y} - \frac{1}{2}\frac{\Delta y}{\Delta t} , \quad 0.0 \right\}$$

$$\lambda_{B4} = \max \left\{ |v - c| + \frac{2\nu}{\rho\Delta y} - \frac{1}{2}\frac{\Delta y}{\Delta t} , \quad 0.0 \right\}$$

and

$$\nu = \max \{\mu, \lambda + 2\mu, k\}$$

Viscous effects are included through the use of the viscous coefficient ν. For some test problems this coefficient had to be increased during the initial part of the calculation when large transients in the solution occured.

For regions of the flow in which Δt satisfies the following explicit stability conditions,

$$\Delta t \leq \frac{1}{2} \frac{1}{(|u| + c)/\Delta x + (2\nu/\rho\Delta x^2)}$$

and

$$\Delta t \leq \frac{1}{2} \frac{1}{(|v| + c)/\Delta y + (2\nu/\rho\Delta y^2)}$$

all λ_A and all λ_B vanish and the set of difference equations reduces to the 1969 explicit equations. For other regions in which neither relation is satisfied, the resulting difference equations are either upper or lower block bidiagonal equations with fairly straightforward solutions.

5. Numerical Results

The method was applied to solve for the interaction of a shock wave incident upon a boundary layer. The flow field is sketched in Fig. 1. As shown in Fig. 2, the initial condition was that of uniform flow, and the condition at the top mesh boundary was such that a shock wave given strength would be generated and impinge upon a flat plate at a given point. The conditions at the upstream boundary were held fixed at their initial supersonic free-stream values; the downstream boundary conditions were obtained by zero-order interpolation; the lower boundary conditions were obtained by reflection.

The mesh contained 32 x 32 points, with 16 spanning the boundary layer. The time step was chosen so that the free stream moved approximately 1% during each time step. With this choice the time step satisfied the explicit stability conditions everywhere except in the fine mesh spanning the boundary layer.

In Fig. 3., the computed results are compared with experiment [8] and with boundary- layer theory in the absence of a shock wave [9]. The results are for Mach 2 laminar flow at a Reynolds number of 2.95×10^5. The calculation (1) used Sutherland's formula to calculate molecular viscosity; (2) was run for 256 time steps, at which time the mesh was rezoned to cover just the interaction region; and (3) was run for an additional 256 time steps. It required about 1.5 min of computer time on a CDC 7600 computer. The results for skin friction and surface pressure compare favorably with those of theory and experiment.

In Fig. 4 the calculated velocity profiles ahead, within, and aft of the separation region are compared with the computed results, using the 1969 method alone. The two sets of results agree closely; however, the computer time required by the newer method was more than an order of magnitude less than that required by the 1969 method.

The computation times for a series of laminar and turbulent boundary-layer interactions with shock waves are given in Table 1. For each problem the flow was computed to the same physical time, which for the new method required 256 time steps. For the turbulent flow cases, a simple algebraic eddy viscosity model [10] was used to account for the effects of turbulence. For each case, the table shows the Reynolds number, the ratio of the time step used to the maximum allowed by explicit stability conditions (Courant-Friedrich-Lewy number, or CFL), the computer time required per time step per grid point on a CDC 7600 computer, and the total computer time required. The tabulated results show that the new method is one to three orders of magnitude more efficient than the 1969 method.

For the test cases considered, the new method is more than twice as fast per time step per grid point as the block-tridiagonal implicit methods in use today. Part of the reason for this is the mesh-point spacing and time step chosen; of the total number of mesh points more than half required only the use of the 1969 explicit method. At these mesh points the chosen time step already satisfied the local explicit stability condition; therefore, the implicit procedure, the second stage, was omitted. The implicit procedures were required only in the fine mesh spanning the boundary layer, where explicit stability conditions would have imposed a severe time-step restriction. It is estimated that if the implicit procedures were used at each grid point, the time step per grid point for a two-dimensional calculation would be 2.45×10^{-4} sec for laminar flow and 2.75×10^{-4} sec for turbulent flow. The difference between these last two values represents the additional computation needed to evaluate the turbulence model relations.

6. Conclusion

A new numerical method has been devised for solving the equations of compressible viscous flow. The method represents the implicit analogue of the explicit method presented by the author in 1969. It is unconditionally stable, second-order accurate, and, for many applications, is more efficient than other methods in use today. Because the new method uses the 1969 method as its first stage, many existing computer programs in which the 1969 method is used can be updated by adding the described implicit second stage.

7. References

[1]   Chapman, D.R., Mark, H., and Pirtle, M.W.: Computer vs. wind tunnel. Astronautics and Aeronautics (Apr. 1975), pp. 22-35.

[2]   Chapman, D.R.: Computational aerodynamics development and outlook. AIAA Paper 79-0129 (Jan. 1979).

[3]   Chapman, D.R.: Trends and pacing items in computational aerodynamics. Proceedings of the Seventh International Conference on Numerical Methods in Fluid Dynamics. Lecture Notes in Physics, Springer-Verlag, (1981).

[4]   MacCormack, R.W.: The effect of viscosity in hypervelocity impact cratering. AIAA Paper 69-354, Cincinnati, Ohio (Apr. 1969).

[5]   Von Neumann, John: On the Principles of Large Scale Computing Machines. Collected Works, Vol. V, Macmillan Co. (1963).

[6]   Bradshaw, P. : The understanding and prediction of turbulent flow. (Sixth Reynolds-Prandtl-Lecture). Aeronaut. J., 76 (1972), pp. 403-418.

[7]   MacCormack, R. W.: A numerical method for solving the equations of compressible viscous flow. AIAA Paper 81-0110, St. Louis, Missouri (Jan. 1981).

[8]   Hakkinen, R.J., Greber, I., Trilling, L., and Abarbanel, S. S.: The interaction of an oblique shock wave with a laminar boundary layer. NASA Memorandom 2-18-59w (1959).

[9]   Van Driest, F. R.: Investigation of laminar boundary layers in compressible fluids using the Crocco method. NACA TN-2597 (Jan. 1952).

[10]   Cebeci, T. and Smith, A.M.O.: Analysis of Turbulent Boundary Layers. Academic Press, New York (1974).

Table 1 Computation time

Case	Method	CFL	CDC 7600 time / step grid point	Total time
Laminar $R = 3 \times 10^5$	1969	0.9	1.25×10^{-4} sec	12 min
	New	20	1.55×10^{-4} sec	41 sec
Turbulent $R = 3 \times 10^6$	1969	.9	1.55×10^{-4} sec	2 hr*
	New	160	1.85×10^{-4} sec	48 sec
Turbulent $R = 3 \times 10^7$	1969	.9	1.55×10^{-4} sec	15 hr*
	New	1200	1.85×10^{-4} sec	48 sec

* Estimated

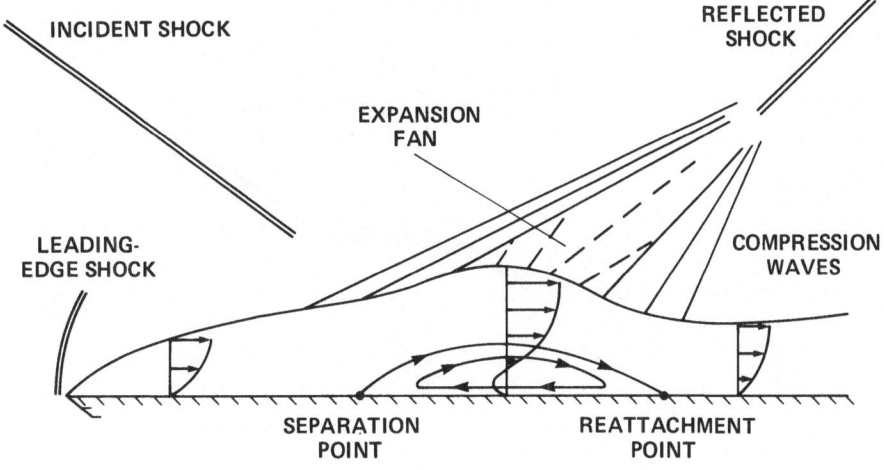

Fig. 1 Sketch of shock, boundary-layer interaction

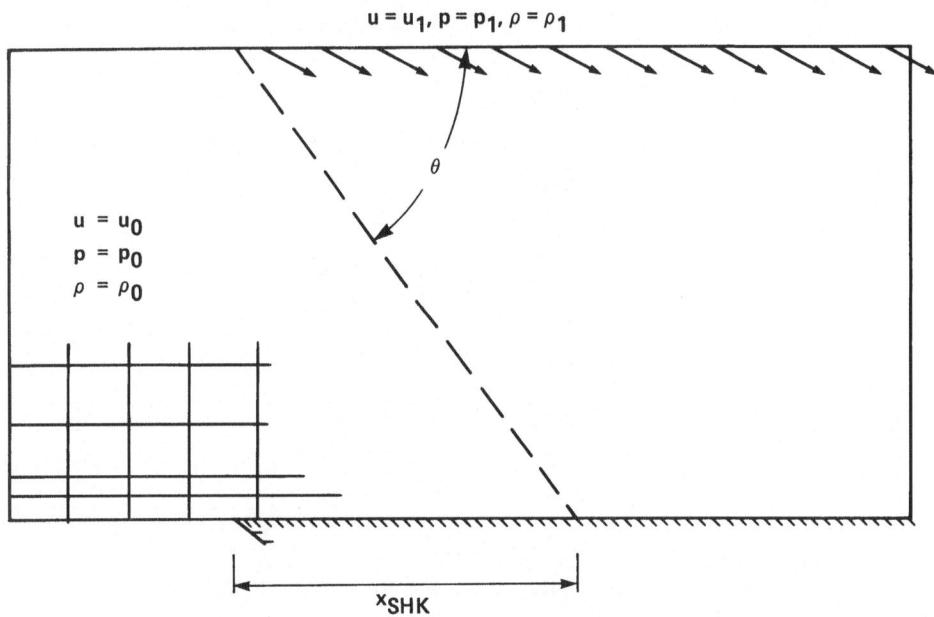

Fig. 2 Initial flow field for shock, boundary-layer interaction

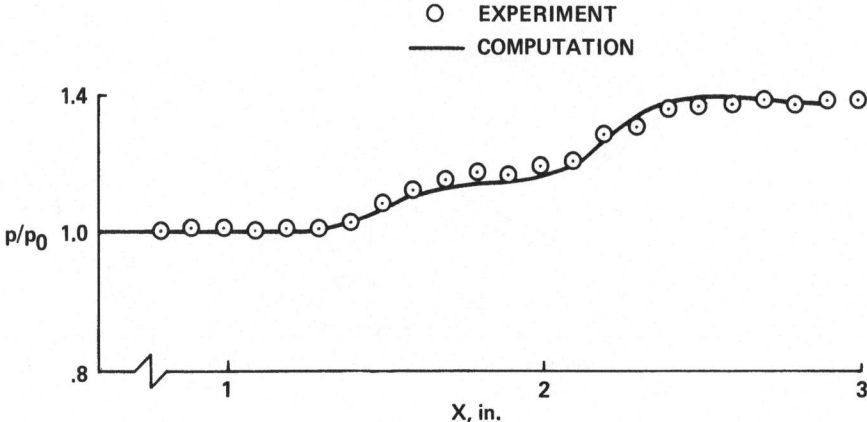

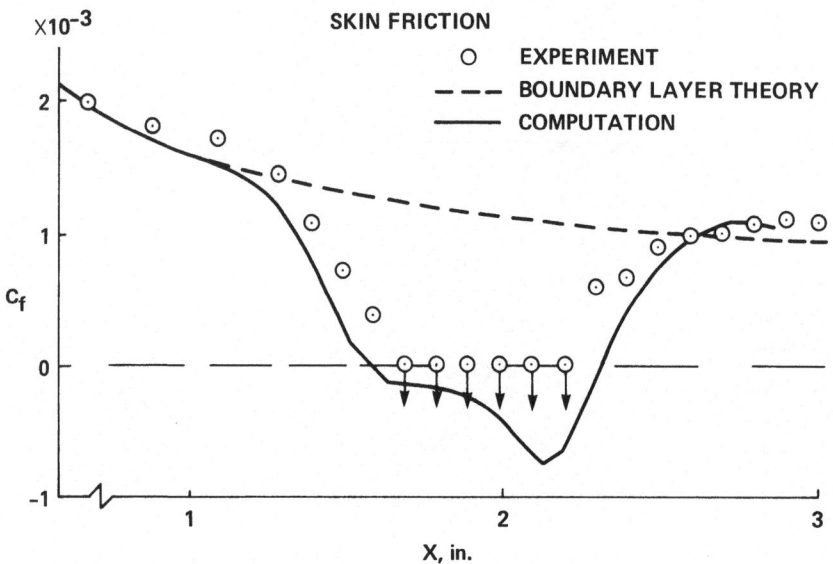

Fig. 3   Comparison of results. a) Surface pressure.   b) Skin friction

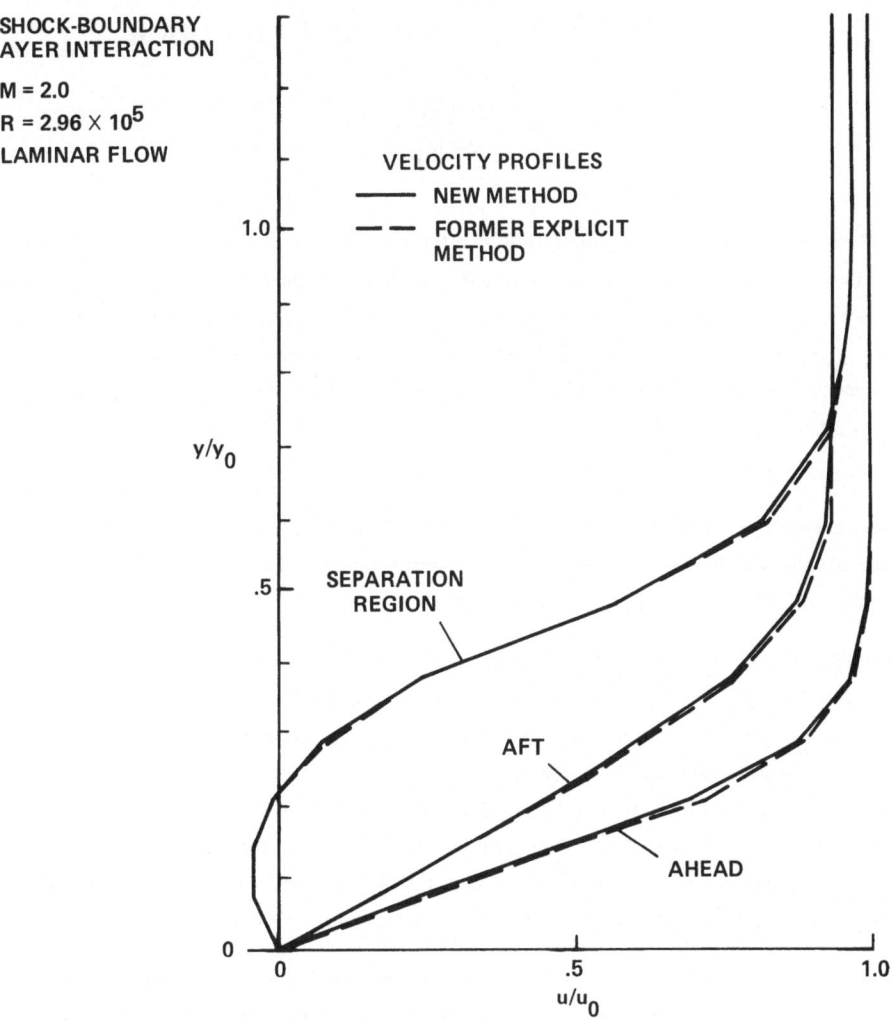

Fig. 4 Comparison of velocity profiles

Boundary-Layer Waves and Transition

F. X. Wortmann
Stuttgart

Transition from laminar to turbulent flow is a fundamental flow phenomenon of fluid mechanics which is still only partly understood. It was Reynolds who nearly hundred years ago put forward the idea that turbulence is the consequence of a flow instability which terminates the laminar state. Today we know of many different types of instability in laminar flow, may it be compressible or incompressible, a free shear layer or a plane or circular Couette flow or the flat plate boundary layer flow.

In this lecture I would like to restrict myself to the case of the flat plate boundary layer. Under two-dimensional conditions, in 1929 Tollmien [1] was able to show theoretically that the Blasius boundary layer can oscillate in certain eigenmodes which may be damped or amplified depending on their frequency and the local Reynolds number of the boundary layer.

This theory which was further developed in Germany by Schlichting, Pretsch and others before and in WW II is today, after twenty years of doubts and misbelief, one of the well established facts in fluid mechanics. Meanwhile many different theoretical approaches have investigated the case of temporal and spatial amplification, the parallel and nonparallel flow and finally the expansion to the nonlinear theory for two-dimensional waves with large velocity amplitudes. Several careful "microscopic" experiments have proved many of these theoretical results except one: the nonlinear theory predicts at large amplitudes of the order of 10% a saturation effect of the two-dimensional waves which has never been observed in any experiment.

Nature seems not to follow the two-dimensional theory. Instead it develops another way to produce transition. The facts we know about these phenomena can shortly be described in the following way: in any experiment there exist small spanwise variations in the velocity field which may also be steady. Regions with an excess of velocity are usually called "valley" and regions with a defect of velocity and a larger displacement thickness are called "peaks". Inside a boundary layer the velocity defect or peak can and must be interpreted as a consequence of a momentum transport due to a longitudinal vorticity component, see Fig. 1.

The spanwise variation of the boundary layer may be modified by a Tollmien wave, depending on the local Reynolds number and the rate of amplification in flow direction. In any case it can be observed that beyond a certain threshold, where the wave velocity amplitude has grown to about 1% - 2%, the wave amplification in the peak and valley region diverge distinctly. From this splitting point it takes only one to three wave lengths downstream to produce in the peak region the transition into turbulence which is characterized by the appearance of a single and then double and multi spike signals on a hot wire probe and local velocity amplitudes as high as 40% of the outer velocity.

These features have been primarily reveiled in the sixties by the work of Klebanoff and his coworkers at the National Bureau of Standards. He summarizes his work in the 1962 publication [2] as follows: "The quantitative physical picture which emerges is that there is in effect a bulge of displacement thickness which increases as it travels downstream with the wave velocity. It moves slower than the outer flow and at breakdown a critical flow condition is reached".

A cautious statement which leaves open many questions and has not been improved so far.

It is my intention to give you now a more detailed description of the flow phenomenon between splitting point and the appearance of the single spike signal.

Fig. 2 illustrates this region with a perspective view over four wave lengths. Some streaklines indicate at the left side a more or less two-dimensional Tollmien wave which deteriorates at the right side into turbulence. What are the details of what happens in this region?

In order to explain the sequence of events I refer to some numerical results of two-dimensional Tollmien waves which are illustrated in a recent IUTAM paper of Hama, Williams and Fasel [3] .

Fig. 3 gives the streamline pattern of a Tollmien wave at $Re_{\delta*} = 700$ and a velocity amplitude of 3%. This so-called cat's eye pattern will be seen by an observer moving with wave speed. At a critical wall distance $\eta_c \approx .8$ the center of the cat's-eye at $x/\lambda \approx .3$ and the cross-point at $x/\lambda \approx .8$ are at rest.

The velocity pattern of this figure can be translated into a pattern of total vorticity ω_z which is shown in Fig. 4. In [3] Hama points out that the total vorticity ($\partial u/\partial y + \omega'$) concentrates mainly to the wall region at $x/\lambda \approx .8$ and that the region at the critical wall distance is mostly void of vorticity. However, as Fig. 4 reveals, a weak secondary vorticity peak develops around $x/\lambda \approx .3$ and $\eta = .9$ for the indicated Reynolds number and wave amplitude.

This weak vorticity peak near the center of the cat's eye is distinguished from the rest of the total vorticity distribution by two facts: it emerges only beyond a certain threshold of Reynolds number and wave amplitude, and it moves along with the local speed of the particles. No other region in the wave is able to develop the combination of a local vorticity peak which is bound not only to the wave but also to particles. Fig. 5 taken also from [3] illustrates the situation from another point of view: it shows the total vorticity distribution in the two-dimensional flat plate boundary layer for two Reynolds numbers and three wave amplitudes at $x/\lambda \approx .3$. It can be seen that for the lower Reynolds number $Re_{\delta*} = 350$ even a large amplitude is not able to establish a pronounced vorticity peak near the critical wall distance $\eta_c \approx .8$. On the other side it is felt that for higher Reynolds numbers, say $Re_{\delta*} > 1000$, an amplitude of about 1% will produce a similar distribution as the 3% amplitude at $Re_{\delta*} = 700$. With respect to the behaviour of a local vorticity peak moving along with particle speed it will behave probably like a bundle of ordinary vortex tubes with different circulation: the vortex tube with the strongest vorticity will always dominate the environment.

Now we perceive the first step. The amplified two-dimensional Tollmien wave modulates the total vorticity distribution, and beyond a certain threshold, it is able to produce a situation which locally resembles an ordinary vortex tube embedded in a shear layer.

In a purely two-dimensional flow this is not a salient feature. With respect to a three-dimensional perturbation, however, the situation becomes now extremely unstable. Any deformation of the weak vorticity peak near the center of the cat's-eye will, due to the surrounding shear field, lead to transport processes which in turn cause an increased deformation. Therefore, we have to expect a strong exponential growth of any three-dimensional perturbation, once the Reynolds number and wave amplitude of the Tollmien wave have established a local vorticity peak of sufficient strength near the critical layer.

When the local ω_z concentration at $x/\lambda \approx .3$ gains small ω_y-components or, in other words, when the vortex tube is slightly bent vertically in a y-z plane, obviously the velocity gradient of the basic flow will transport upper parts of the tube faster than lower parts yielding a stretched vortex tube with ω_x-components. When on the other side the vortex tube is at first only bent horizontally (in a plane parallel to the wall) it cannot stay in this plane because the ω_x-components will transport parts of the vortex tube away from the initial plane. This is the well-known retrograde rotation of a warped vortex tube, see for example [4]. Inside a boundary layer this leads again to a differential transport velocity in the x-direction.

In any case the net result of a slightly bent and conveyed vortex tube in a boundary layer is a stretched vortex tube which will increase the ω_x-components which again increases the convection and so on. This process is most pronounced in the apex positions of a warped vortex tube. As a consequence we should finally expect two types of strongly curved and highly concentrated vortices: one raising through the boundary layer with increasing transport velocity and wall distances and another one which slows down and moves to the wall. The first is the well-known hairpin vortex associated with the single spike signal of a hot wire probe. The second is shown in Fig. 13.

The total development of the flat plate boundary layer below a smooth potential flow can now be seen in three stages: in the first one viscosity builds up a stock of ω_z-vorticity. Then a Tollmien wave with relatively weak instability which spreads over twenty and more wave lengths will modulate the two-dimensional vorticity distribution up to a point where near the center of the cat's eye a local vorticity peak is established which moves along with wave and particle speed. The local vorticity peak may then behave like a ordinary vortex tube and, being embedded in a boundary layer, becomes extremely unstable against any bending or warping. this situation is here identified with the splitting point.

The associated stretching process not only concentrates the vorticity into top and tail vortices but also establishes locally new flow fields with a scale of nearly an order of magnitude smaller than the wave length of the Tollmien wave. The phenomenon originally bound to the wave decouples now from the wave and starts its own independent development which finally will destroy the wave.

In order to support the described development some flow visualization pictures shall now be evaluated. It has often been argued that fluid markers producing time lines and streaklines in unsteady flow may be quite confusing and may easily be misinterpreted. The wellknown paper of Hama [5] illustrates this fact. With some care, however, useful information can be extracted from flow visualization pictures.

In Fig. 6, for example, it may be assumed that a fluid marker device like a hydrogen bubble wire moving parallel to the wall through the center of the cat's-eye, leaves a sheet of bubbles behind. Due to the differential rotational speed inside the cat's-eye the sheet is deformed into an S-like band. The visibility of the bubble sheet observed perpendicular to the wall is obviously best, where the sheet is folded. Therefore, a short time after the wave moved over the bubble wire we expect two bubble "fronts", one downstream and above, and another one below and behind the center of the cat's-eye. This picture is not changed very much by the fact that the bubble wire moves only with wave speed through the cat's-eye. The important point is that the bubble sheet with its rollingup-process is always trapped into the inner cat's-eye, and the two "front" lines give an indication of the position of the cat's-eye and the distortion of a wave front in the spanwise direction.

When the basic boundary layer is not completely two-dimensional but has slightly different mean velocity profiles in the spanwise direction the different shear rate which causes the separation of the folded bubble "fronts" should become visible. A wider separation indicates the spanwise position of a peak. The photograph in Fig. 7 shows a bubble front, folded by the Tollmien wave which runs toward the inactive fork-like probe. The center line of the fork is practically parallel to the flow direction. The forward and the faint rearward front are clearly more separated between the prongs than outside. The vorticity lines in this region form a bow which is bent forward and upward, as indicated in Fig. 1. This process - let's call it warping instability - starts with very small amplitudes at the splitting point and has (Fig. 7) developed into the top or hairpin vortex which can be seen by the dark bow between the probe. The top vortex has also been observed in earlier flow visualizations [6] or [7] . In Fig. 8 taken about one second later, the tail vortex to the right of the top vortex becomes visible. these two pictures have been selected to show that the relatively large Tollmien wave (the wave length equals nearly the width of the photograph) is still the dominant feature and the well developed top and tail vortices are local events which up to this state do not interact very much with the wave.

Obviously the top and tail vortex are at different wall distances and are connected by a band of less concentrated vorticity oriented obliquely in all three dimensions. If such a configuration moves over an active hydrogen bubble wire whose wall distance is between the top and tail vortex, as in Fig. 9, we have in a plane of symmetry through the center of the top vortex to expect a downward and rearward motion, whereas outside this plane the velocity components w' in the spanwise direction should change their sign in front of and behind the oblique vorticity.

Fig. 10 shows some bubble lines which start from a very thin wire, only visible by a row of small white dots which insulate single points on the wire. The grid lines on the wall have a mesh size of five centimeters. The bubble lines close to the wire can still be interpreted as a velocity distribution and show in the center the velocity defect typical for vortex tubes inside a boundary layer which are locally curved upward and forward.

The more distant time-lines in Fig. 10 indicate not only w'-components toward the center plane, but they are also three-dimensionally conveyed. The downward/rearward convection is strongest in the central region. Fig. 11 taken a little later and five centimeters downstream shows the much stronger deformation of the time-lines in the central region. The particles in the center now start an upward motion. On both sides of the center line the obliquely distributed vorticity pierces the bubble sheet which starts now to be folded due to the inward and downward motion in front and the upward and outward motion behind the stretched vortex tubes. Fig. 12 taken again five centimeters downstream shows the result in the center of the peak region. Due to the stretch which seems to concentrate to a small region near the center line there exists now a clearly visible top or hairpin vortex which is not only curved in the plane of Fig. 12 but also upwards. The faint bubble clouds in the neighborhood indicate that the environment is still filled with distributed vorticity.

Fig. 13 illustrates the same processes by a sequence of photographs and in addition shows the development of the tail vortex. The grid lines on the wall are the same as in Fig. 11-13. The double line, more clearly seen in the right pictures, marks the center of the test section. The top vortex develops at $z = -10$ cm and the tail vortex at $z = +8$ cm (Nearly one Tollmien wave apart). There is a time shift between these movie pictures as indicated by the connecting line in Fig. 13. Between top and tail vortex a diagonal crossing of two separate bubble clouds is visible. The first one which connects the top and the tail vortex marks the dominating vorticity which will transport the second cloud front into a diagonal position.

It can also be seen that the oblique vorticity rolls up to a real vortex only in the bows itself and that the spanwise extension is very small. The distributed vorticity outside the bow regions, however, will always produce u' fluctuations similar to the Tollmien wave. Klebanoff, Tidstrom and Sargent ([2] pp. 17 - 18) observed just such a result in their hot wire measurements and interpreted this as an argument against the vortex loop concept of Theodorsen [8] . This remark has probably veiled the physics of the dominating phenomena.

Conclusion:

On the basis of observations with hydrogen bubble photographs the following picture for the origin of the strong instability behind the splitting point of Tollmien waves has been proposed: beyond a certain threshold, characterized by a combination of wave amplitude and Reynolds number and inside the Tollmien wave at a phase position near the center of the cat's-eye there emerges a weak local vorticity peak. This peak does not move relative to the surrounding matter and therefore resembles an ordinary vortex tube. Beyond the threshold it will develop its own warping instability which splits the amplification rate of the Tollmien waves. The three-dimensional bending or warping of vortex tubes inside a boundary layer leads to different transport velocities or a convection which stretches the vortex tubes. The stretch in turn produces more convection and so, within the range of one or two wave lengths of the Tollmien waves the vorticity distribution is not only converted into an oblique orientation, but also concentrated into a top- and tail-vortex. The first one will be observed by a hot wire probe as a single spike. It is therefore concluded that the whole development between the splitting point and the appearance of a single spike is a continuous and coherent physical phenomenon.

Acknowledgment

I am indebted to Dr. Thorwald Herbert for many fruitful discussions about the above subject.

References

[1] T o l l m i e n, W.: Über die Entstehung der Turbulenz, 1. Mitt. Nachr. Ges. Wiss. Göttingen, Math. Phys. Klasse 21-44 (1929), NACA TM 609 (1931).

[2] K l e b a n o f f, P.S., T i d s t r o m, K.O., S a r g e n t, L.M.: The three-dimensional nature of boundary-layer instability. Journ. Fluid Mechanics 12 (1962) pp. 1-34.

[3] H a m a, F.R., W i l l i a m s, D.R., F a s e l, H.: Flow field and energy balance according to the spatial linear-stability theory of the Blasius boundary layer. IUTAM Symposium on Laminar Turbulent Transition, Stuttgart, Sept. 16-22, 1979.

[4] H a m a, F.R.: Progressive deformation of a perturbated line vortex filament. Physics of Fluids 6 (1963) p. 526.

[5] H a m a, F.R.: Streaklines in a perturbated shear flow. Physics of Fluids 5 (1962) p. 644.

[6] H a m a, F.R.:, N u t a n t, J.: Detailed flow field observations in the transition process in a thick boundary layer. Proc. Heat Transfer and Fluid Mech. Inst. (1963) p. 77.

[7] B r o w n, F.N.M.: See the wind blow. Univ. of Notre Dame (1971) p. 54.

[8] T h e o d o r s e n, Th.: The structure of turbulence. Proc. Second Midwestern Conf. Fluid Mechanics 1 (1952). 50 Jahre Grenzschichtforschung. Görtler-Tollmien (1955) Vieweg, Braunschweig.

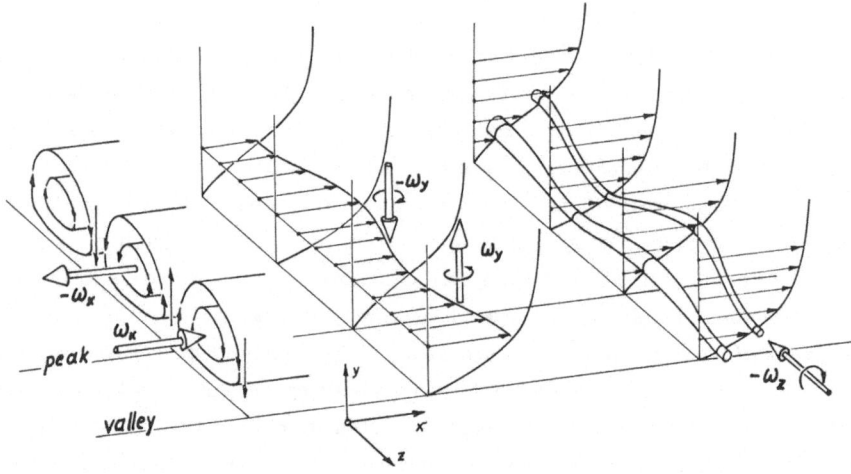

Fig. 1: Spanwise variation of velocity boundary layer profiles and vortex tubes due to steady longitudinal vortices.

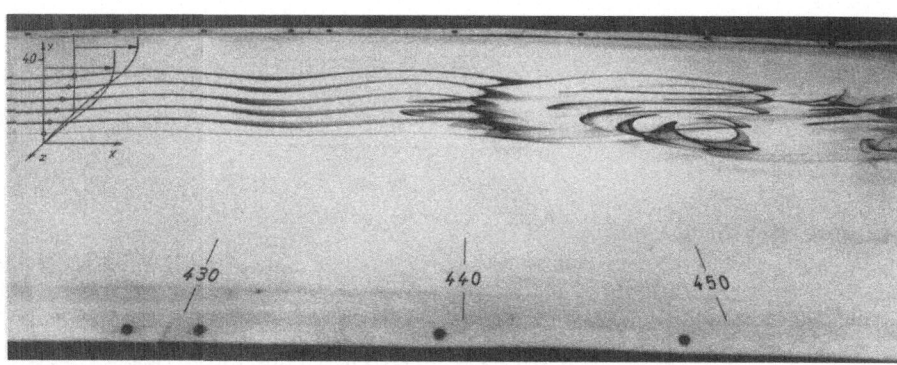

Fig. 2: Streaklines inside a flat plate boundary layer near the transition region.

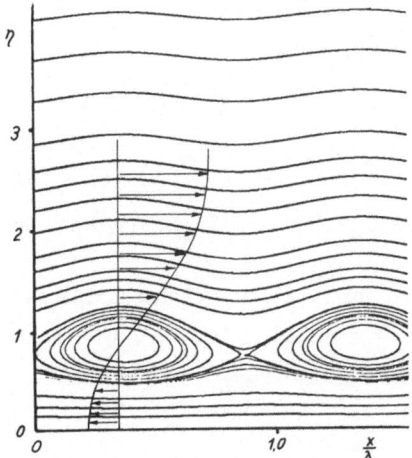

Fig. 3: Cat's-eye streamlines in a flat plate boundary layer at $Re_{\delta^*} = 700$. The velocity amplitude of the Tollmien wave $u'_{max}/u_\infty = 3\%$; λ = wave length.

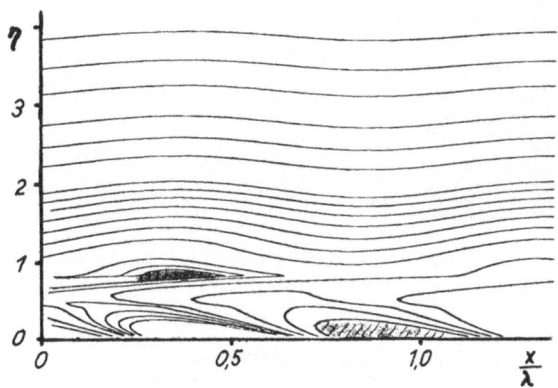

Fig. 4: Total velocity field ($Re_{\delta^*} = 700$) $(u')_{max}/U_\infty = 3\%$.

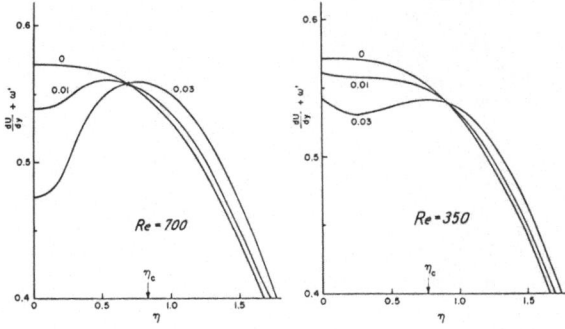

Fig. 5: Instantaneous total vorticity profiles with peak $x/\lambda = 0.3$, (numbers are $(u')_{max}/U_\infty$).

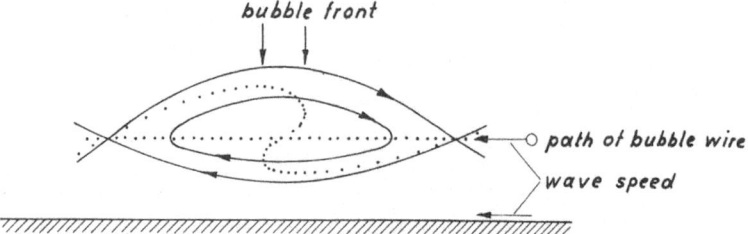

Fig. 6: Differential rotational speed near the center of the cat's eye folds up the hydrogen bubble sheet into two front lines.

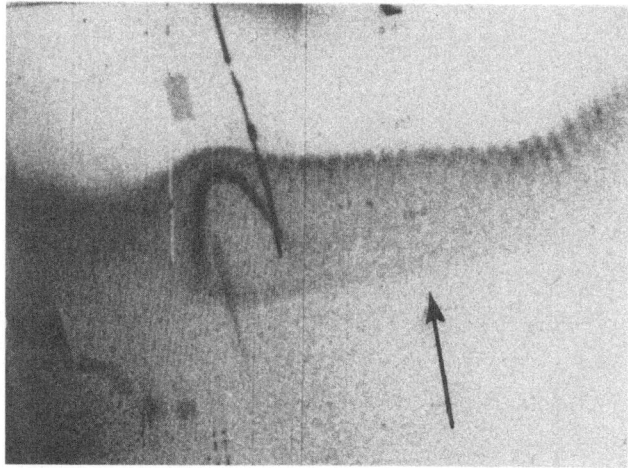

Fig. 7: Distoted front lines of the folded bubble sheet indicate a "peak" situation between the prongs of the probe. The dark bow is the developing top or hairpin vortex.

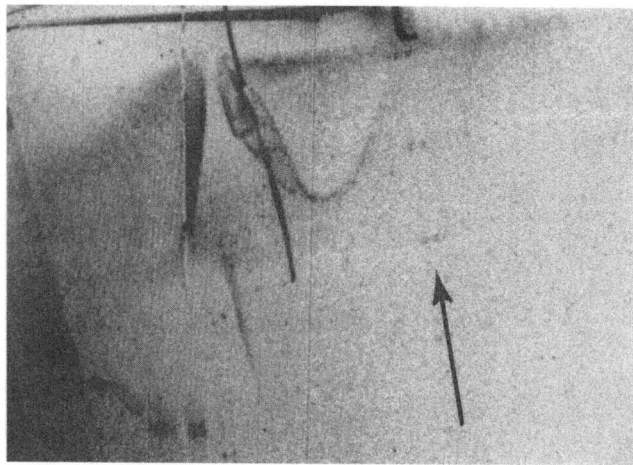

Fig. 8: This photograph taken about one second after Fig. 7 makes visible the tail vortex (to the right of the probe).

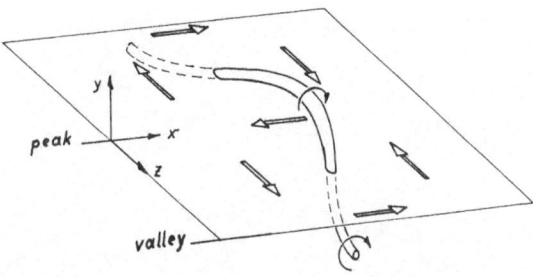

Fig. 9: Perturbation velocities associated with a warped vorticity distribution in a plane whose wall distance is between the top and tail vortex.

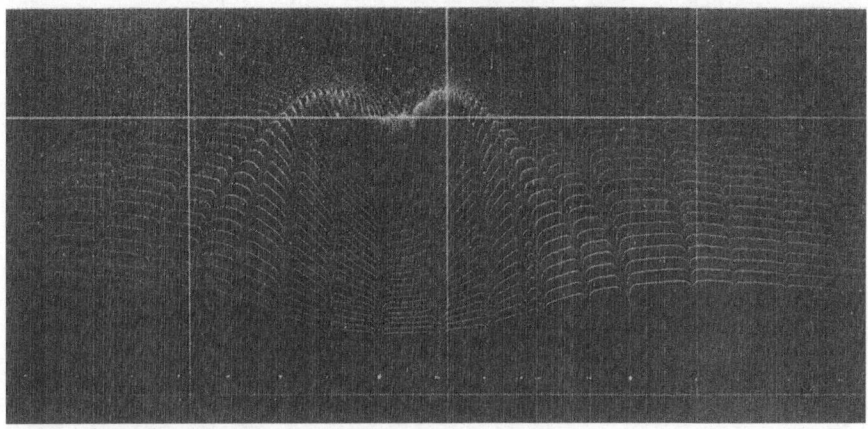

Fig. 10: Time-lines of hydrogen bubbles indicate the u' and w' components due to the warped vorticity distribution similar to Fig. 9. The reference lines on the wall have mesh size of 5 cm. U_∞ = 9.93 cm/s; δ^*= 1,3 cm.

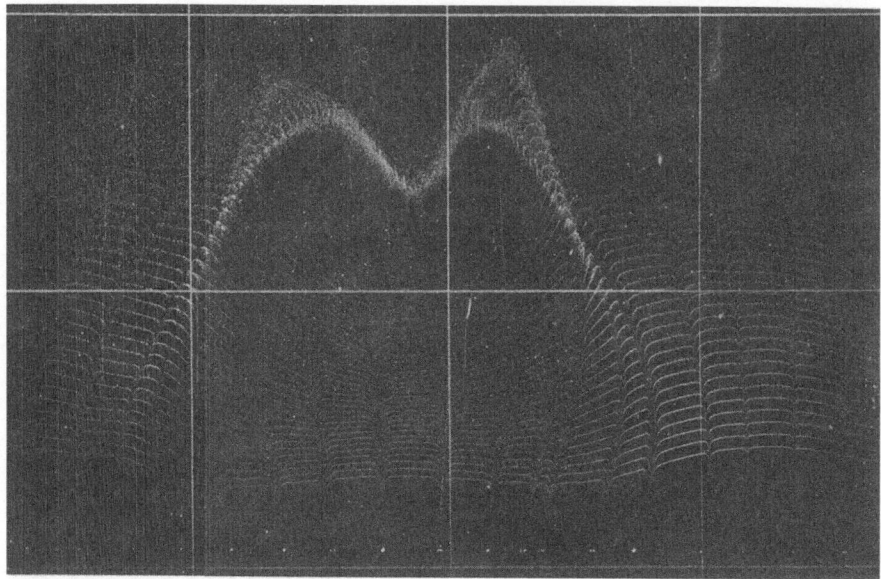

Fig. 11: Time-lines as in Fig. 10, taken 5 cm downstream at the same phase position, show the roll up of bubbles into a stretched and more concentrated oblique vorticity.

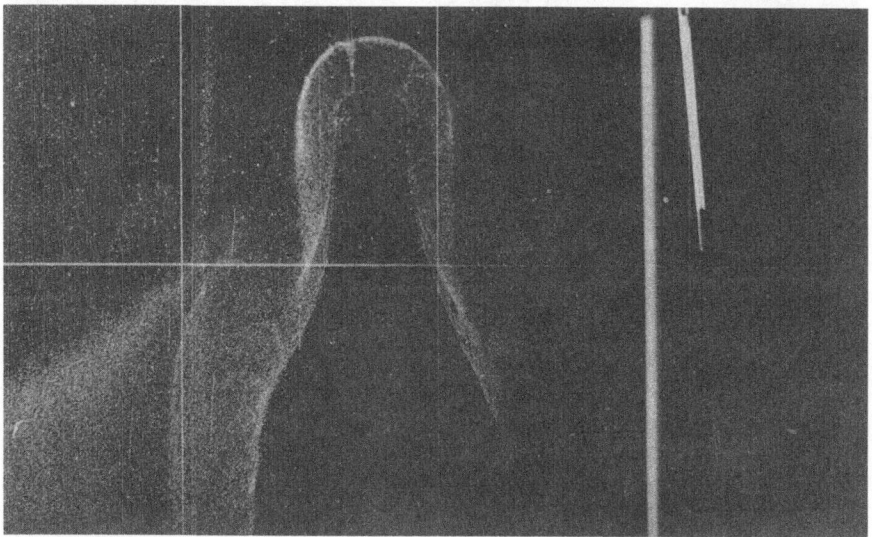

Fig. 12: Top or hairpin vortex becomes visible 5 cm downstream of Fig. 11.

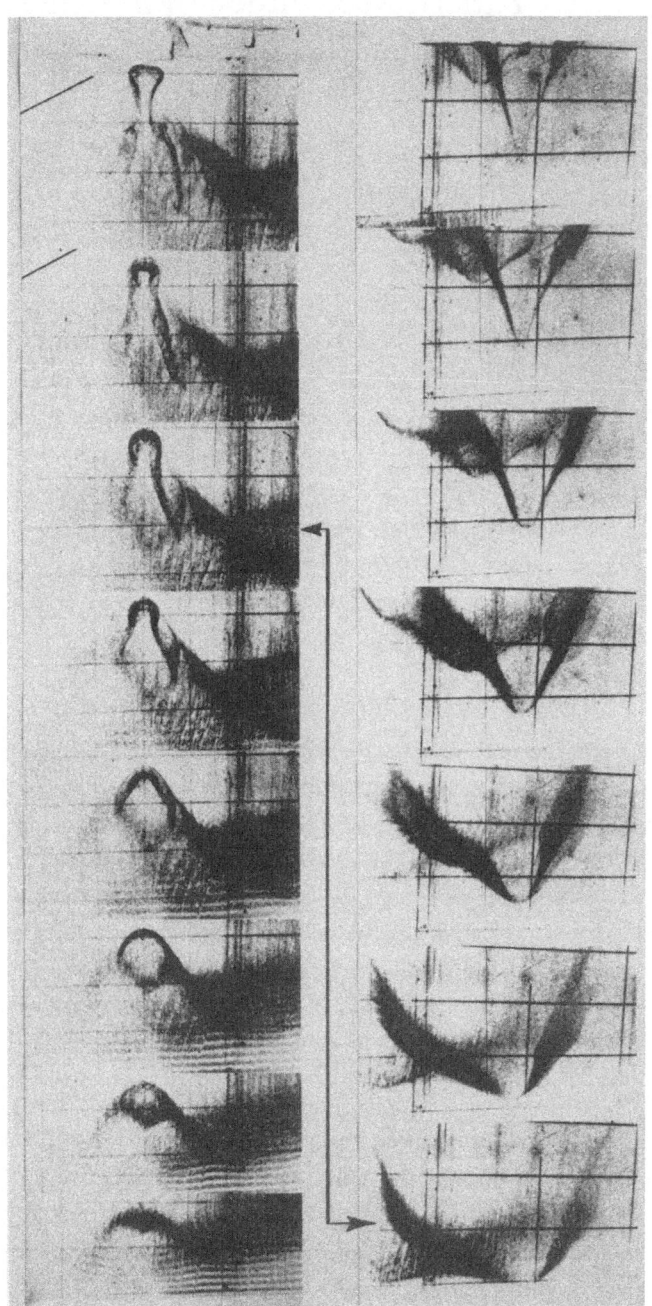

Fig. 13: Movie pictures of the development of top or hairpin vortex and tail vortex.

Pattern Recognition of Bounded Turbulent Shear Flows

H. Eckelmann
Göttingen

There is today no question that coherent structures exist in wall bounded turbulent shear flow. This is well documented by numerous publications in the last ten to fifteen years. A digital computer was used by applying simple criteria to recognize, extract and ensemble average coherent structures which repeatedly appeared in the mean stream component, u, of a turbulent channel flow. Simultaneously the two other velocity components, v and w, as well as the product uv and various vorticity components were also processed without applying any criteria on these signals. The results obtained provide insight into the structure of turbulence. Using simultaneously two probes at different locations our pattern recognition technique also provides additional information about the spacial extension of the coherent structures.

1. Introduction

A pattern recoginition technique for the analysis of turbulent velocity signals will be described and an application to turbulent channel flow data will be given. This technique has been introduced in two proceeding papers by Wallace, Brodkey and Eckelmann [1] and Eckelmann, Nychas, Brodkey and Wallace [2]. In a further investigation by Kastrinakis, Wallace, Willmarth, Ghorashi and Brodkey [3] the pattern recognition technique was applied to velocity and vorticity signals measured simultaneously at two different locations in a fully developed turbulent channel flow. Finally Eckelmann, Wallace and Brodkey [4] showed that the technique can also be used for tests of complex probe configurations that shall be used in turbulent flows. Here the patterns measured with the complex probe were compared with those obtained from a standard probe.

The pattern recognition technique is capable to extract the repetitive structure that exists in the velocity signal, U, between two maxima in the time derivative $\partial U/\partial t$ (U is the velocity component in mean flow direction). This structure, which is found in more than 65 % of the time, can be described as a gradual deceleration from a local maximum in the U component followed by a strong acceleration.

A probe being in the flow field will only occasionally intersect the center of such structures. In most of the cases the structure will be intersected near an edge. In addition these structures will pass by the probe at different stages of their lifetimes. For all these reasons wide distributions of both amplitudes and pattern lengths can be seen. The pattern recognition technique is able to detect the whole range of sizes. To obtain a meaningful average pattern their length is normalized. Furthermore the recognized patterns can be related to the visually observed coherent structures by Kline, Reynolds, Schraub and Rundstadler [5] and Corino and Brodkey [6].

2. Experimental Facility

The data analyzed with the pattern recognition technique were acquired in the oil channel and the low speed wind tunnel of the Max-Planck-Institut für Strömungsforschung. The details of the oil channel flow and the instrumentation have been described by Eckelmann [7] and by Wallace, Eckelmann and Brodkey [8] . The details of the wind tunnel flow and the used instrumentation can be found in the dissertation of Kastrinakis [9]. Some of the most important features of both facilities will be repeated here for the convenience of the reader.

Most of the data were obtained in a fully developed turbulent channel flow with a pure paraffin-based oil as the testing fluid. The Reynolds number of the flow based on the centre-line velocity of 21 cm/sec and the channel width of 22 cm was 7700. The kinematic viscosity v of the oil is 0.06 cm^2/sec. Under these conditions the wall region in this channel flow is greatly magnified. A measure for this magnification is the thickness of the viscous sublayer, which for this flow is about 3 mm for $y^+ = y \cdot u_\tau / v = 5$. In this formula y is the wall distance and u_τ the friction velocity.

A standard TSI model 1241 - 20 W X-probe with quartz coated film sensors was used to measure the U and v velocity components. Here $U = \bar{U} + u$ denotes the mean stream velocity component and u and v the fluctuating velocity components in mean stream direction and normal to the wall respectively. A self designed and built V-probe using 1 mm long 50 µm thick TSI model 10121 -20 W quartz coated sensors (see Kreplin and Eckelmann [10]) was used to measure U and w (w is the fluctuating velocity component in spanwise direction). An also self designed 5 - sensor probe with the same type of quartz coated sensors which was built by TSI (Model 1294 Be - 20 W see Eckelmann et al. [2]) was used to measure simultaneously the U, v, w, $\partial u / \partial y$ and $\partial u / \partial z$ velocity and vorticity components respectively. The assembly of all five sensors had a diameter of 4 mm ($d^+ = d \cdot u_\tau / \approx 7$). Pairs of constant-temperature anemometers and linearizers of the same kind were used to operate the various sensors belonging to one another. For all probes the overheat ratio was 1.01.

A wind tunnel with a 9 m long test section was also used. Here the probes were mounted at the exit of the test section where the two turbulent boundary layers of either wall side had grown together. The Reynolds number of the flow based on the centre-line velocity of about 62 cm/sec and on the channel width of 28 cm was 12500. Also in this flow the thickness of the viscous sublayer is relatively large (about 2.8 mm corresponds to $y^+ = 5$). The measurements were made with a pair of vorticity probes whose design was originally conceptualized by Kovasznay [11] and which were constructed by Kastrinakis [9]. Such probes are sensitive to the streamwise velocity component, U, and the streamwise vorticity component, $\omega_x = \partial w / \partial y -$

$\partial v/ \partial z$. The probe had a diameter of 2.8 mm ($d^+ \approx 5$). To operate the two probes a pair of constant-current anemometers and four high gain dc-amplifiers were necessary. The currents through the probes were 60 mA yielding only a moderate wire temperature.

The signals from the various probes were digitized and stored on the magnetic disk of a PDP-15 computer. For the oil channel data a rate of 50 samples/sec and for the wind tunnel data a rate of 500 samples/sec were used. The data were later transferred to a magnetic tape so that the analysis could be done on an UNIVAC 1108 computer.

3. The Pattern Recognition Technique

Fig. 1a shows a record of the u signal containing the characteristic pattern. This record is part of the raw digitized data with every second point shown. The characteristic pattern, a weak deceleration followed by a strong acceleration, can easily be recognized: it becomes even clearer when a short time smoothing is applied to the raw data (Fig. 1b). A similar pattern in the u signal was observed by Blackwelder and Kaplan [12] and Blackwelder and Eckelmann [13] at $y^+ = 15$ during the occurrence of a burst using the VITA detection technique. The skewness factor $S = [du/dt]^3 / ([du/dt]^2)^{3/2}$ is positive for all wall distances investigated (Fig. 2) which also confirms that the characteristic pattern occurs during much of the time. In [1] the pattern recognition technique is described in detail, whereas here only the key ideas are repeated for the convenience of the reader.

The first idea was the concept of using a short-time temporal average (here referred to as TPAV) and not the long time mean. TPAV is only taken over the length of a pattern. This made of course an iterative detection scheme necessary. With this scheme patterns, as shown in Fig. 3, could be recognized. Note that both patterns would have been missed if the long time average was used. TPAV is taken over the period from one maximum, $\partial u/ \partial t$, to the next (Fig. 1c).

It should be emphasized that the pattern recoginition technique does not need an arbitrarily chosen discriminator level. The decision whether a pattern should be accepted or rejected is only made by comparison of the absolute values of the deceleration and acceleration slopes of the smoothed u signal. To be an accepted pattern $|(\partial u_s/ \partial t)_{min}| < |(\partial u_s/ \partial t)_{max}|$ has only to be fulfilled. The smoothed u signal was used for this criterion so that the pattern recognition technique will not recognize jitter in the u signal. The smoothed signal, however, was only used for the recognition process itself; when the ensemble averages were formed the raw data were used.

The lengths of the recognized patterns vary over quite an extensive range, 1 : 25 (Fig. 4). To see the characteristics of the patterns clearly they have to be normalized to an arbitrary length. Otherwise short patterns will average into only some part of a larger pattern. The time interval for this normalization was taken as 120 points. Thus longer patterns had to be compressed and shorter ones had to be stretched. This normalization was done with the raw data u signal as well as with the simultaneously measured other velocity or vorticity component. Only the recognition process itself was done with the smoothed data. The u signal and any other signal measured simultaneously with it were normalized and ensemble averaged the same way.

The computer flow diagram for the pattern recognition technique is shown in Fig. 5. The single steps can easily be followed with the help of the notations given in Figs. 1b and 1c.

4. Results

In Fig. 6 the normalized and ensemble averaged u, v, w and uv patterns measured in the oil channel at a wall distance of $y^+ = 15$ are shown. The 5-sensor probe described by Eckelmann et al. [2] was used. This probe enabled us to measure simultaneously all three velocity- and two vorticity components at one location. The u-curve is an ensemble average of about 1000 recognized u-patterns which occur in about 65 % of the time at this location. The shape of the u pattern is only a result of the criterion that the maximum positive slope in the u pattern be larger in absolute value than the maximum negative slope. This quality of the u velocity component is observed over most of the channel width. Randomly generated signals which were also investigated with the pattern recognition technique by Wallace et al. [1] did not show these characteristics.

The v and w patterns of Fig. 6 were obtained without applying any conditions to the raw data. However, when the u pattern was normalized the same normalization procedure was applied to the v and w signal before building the ensemble average. The general shape of the u pattern is predetermined as described above, and the v pattern must be generally out of phase with u since it is well known from previous results (e.g. Eckelmann [14]) that these two signals have a correlation coefficient of about -0.4 over most of the channel halfwidth. Between the u and w signals in the average the correlation is zero; therefore an ensemble averaged w pattern should not exist. The measured w pattern, however, shows a small amplitude along the time axis. This indicates that the motion, in all its phases, has little preferred direction in the spanwise direction. This can be caused by a small misalignment of the probe in the spanwise direction during the measurements. Detailed features of the uv patterns cannot be predicted. In Fig. 6 one sees positive and negative pulses in the v pattern which are largely out of phase with such pulses in the u pattern resulting in the large negative pulses in the uv pattern. Such pulses are the times when Reynolds stress is being produced and are associated with the ejection and sweep events in the event cycle described by Corino and Brodkey [6] . They characterize ejection events by an energetic motion away from the wall of coherent regions of low streamwise velocity. Sweep events have a streamwise velocity higher than the local mean, and move parallel to or slightly towards the wall. Because the u and v patterns are not completely opposite in phase, a small positive pulse in the uv pattern results.

In [1] the slope comparison criterion was varied by requiring that acceptable patterns have $|(\partial u_s / \partial t)_{min}| < a \; |(\partial u_s / \partial t)_{max}|$ with a being 1, 1/2 and 1/3. The result was that the essential character of the patterns was not changed, although the characteristic features of the patterns were accentuated when stronger criteria were used.

To show further characteristics of the coherent-motion event sequence, Wallace et al. [1] incorporated an analysis technique first used by Wallace et al. [8] and Willmarth and Lu [15] , namely that of truncating the u and v signals about their zero levels to obtain four classification criteria based on the pair of signs of u and v. For more details it is referred to this publication.

In a further investigation Eckelmann et al. [2] studied the vorticity dynamics and turbulence production occurring during bursting as detected by the pattern recoginition technique. This investigation was made possible because the 5-sensor probe measured directly $\partial u/\partial y$ and $\partial u/\partial z$. With the gradients in the x-direction which were calculated using Taylor's hypothesis $[\partial/\partial x = -(1/U)\,\partial/\partial t]$ the two instantaneous signals of the vorticity components ω_z and ω_y could be obtained. The pattern recognized vorticity component ω_z for a wall distance of $y^+ = 30$ is shown in Fig. 7. The pattern recoginized gradient $\partial U/\partial y$ at the same y^+ position (see Eckelmann et al. [2]) is very similar to the ω_z pattern of Fig. 7, but 180 degrees out of phase. This means that $\partial v/\partial x$ contributes only little to the spanwise vorticity component. A similar behaviour can be observed with the ω_y pattern shown in Fig. 8 also for a wall distance of $y^+ = 30$. The pattern recognized signal bears remarkable similarity to the pattern recognized u signal; also 180 degrees out of phase. If we make the same assumption that the gradient, $\partial w/\partial x$, contributes little to the normal vorticity component, the gradient $\partial u/\partial z$ is very similar to the velocity pattern. More details about the pattern recognized gradients $\partial u/\partial y$ and $\partial u/\partial z$ will be published by Eckelmann and Wallace elsewhere.

Up to this point all measurements reported here were made in our oil channel using hot-film probes. In our wind tunnel Kastrinakis, Wallace, Willmarth, Ghorashi and Brodkey [3] used a pair of vorticity probes whose design was originally proposed by Kovasznay [11] and which are sensitive to both the U component of the velocity and the streamwise vorticity, ω_x. Details about the nature of this probe can be found in [9] . When using two probes at different spanwise locations, either u signal pattern can be recognized and the simultaneously occurring other signals from the probe can be averaged ensembly. Fig. 9 shows the u patterns normalized with the rms value of the streamwise velocity fluctuations, u, obtained from each of the two probes (A and B). Both probes were located at $y^+ = 15$ and the u patterns are quite similar to each other and to the pattern at the same y^+ position measured in the oil channel flow (Fig. 6 top). The comparison of the u patterns measured in the two different flows and at different Reynolds numbers, shows that this pattern is a distinguishing characteristic of the flow, since the two flow systems were considerably different. Finally Fig. 10 gives the ω_x pattern obtained from probe A with the u pattern recognized from the same probe.

Here the two patterns are also normalized with the rms values of u and ω_x respectively. Symmetry considerations require that the ω_x pattern occurring simultaneously with the recognized u pattern from the same probe should, on the average, be zero over the entire time interval. Fig. 10 shows that this is practically the case except at those normalized times where the u pattern shows a strong gradient $\partial u / \partial t$.

5. References

[1] W a l l a c e, J.M., B r o d k e y, R.S., E c k e l m a n n, H.: Pattern recognized structures in bounded turbulent shear flows. J. Fluid Mech., 83 (1977), pp. 673-693.

[2] E c k e l m a n n, H., N y c h a s, S.G., B r o d k e y, R.S., W a l l a c e, J.M.: Vorticity and turbulence production in pattern recognized turbulent flow structures. The Physics of Fluids, 20 (1977), pp. 225-231.

[3] K a s t r i n a k i s, E.G., W a l l a c e, J.M., W i l l m a r t h, W.W., G h o r a s h i, E., B r o d k e y, R.S.: On the mechanism of bounded turbulent shear flows. Lecture Notes in Physics 75, 175, Springer Verlag Berlin, Heidelberg, New York (1978).

[4] E c k e l m a n n, H., W a l l a c e, J.M., B r o d k e y, R.S.: Pattern recognition, a means for detection of coherent structures in bounded turbulent shear flows. Proc. of the Dynamic Flow Conference, Marseille and Baltimore, (1978), pp. 161-172.

[5] K l i n e, S.J., R e y n o l d s, W.C., S c h r a u b, F.A., R u n d s t a d l e r, P.W.: The structure of turbulent boundary layers. J. Fluid Mech., 30 (1967), pp. 741-773.

[6] C o r i n o, E.R., B r o d k e y, R.S.: A visual investigation of the wall region in turbulent flow. J. Fluid Mech., 37 (1969), pp. 1-30.

[7] E c k e l m a n n, H.: The structure of the viscous sublayer and the adjacent wall region in a turbulent channel flow. J. Fluid Mech., 65 (1974), pp. 439-459.

[8] W a l l a c e, J.M., E c k e l m a n n, H., B r o d k e y, R.S.: The wall region in turbulent shear flow. J. Fluid Mech., 54 (1972), pp. 39-48.

[9] K a s t r i n a k i s, E.G.: Experimentelle Untersuchungen der Längsschwankungen des Geschwindigkeitsvektors und der Rotation des Geschwindigkeitsvektors in einer ausgebildeten turbulenten Kanalströmung. MPI für Strömungsforschung, Bericht 5/197, Universität Göttingen.

[10] K r e p l i n, H.-P., E c k e l m a n n, H.: Propagation of perturbations in the viscous sublayer and adjacent wall region. J. Fluid Mech., 95 (1979), pp. 305-322.

[11] K o v a s z n a y, L.S.G.: Turbulence Measurements. In: "High Speed Aerodynamics and Jet Propulsion", 9, "Physical Measurements in Gas Dynamics and Combustion", Princeton University Press, Princeton, New Jersey (1954), pp. 211-285.

[12] B l a c k w e l d e r, R.F., K a p l a n, R.E.: On the wall structure of the turbulent boundary layer. J. Fluid Mech., 76 (1976), pp. 89-122.

[13] B l a c k w e l d e r, R.F., E c k e l m a n n, H.: Streamwise vortices associated with the bursting phenomenon. J. Fluid Mech., 94 (1979), pp. 577-594.

[14] E c k e l m a n n, H.: Experimentelle Untersuchungen in einer turbulenten Kanalströmung mit starken viskosen Wandschichten. Mitt. MPI für Strömungsforschung, Göttingen, Heft 48 (1970).

[15] W i l l m a r t h, W.W., L u, S.S.: Structure of the Reynolds stress near the wall. J. Fluid Mech., 55 (1972), pp. 65-92.

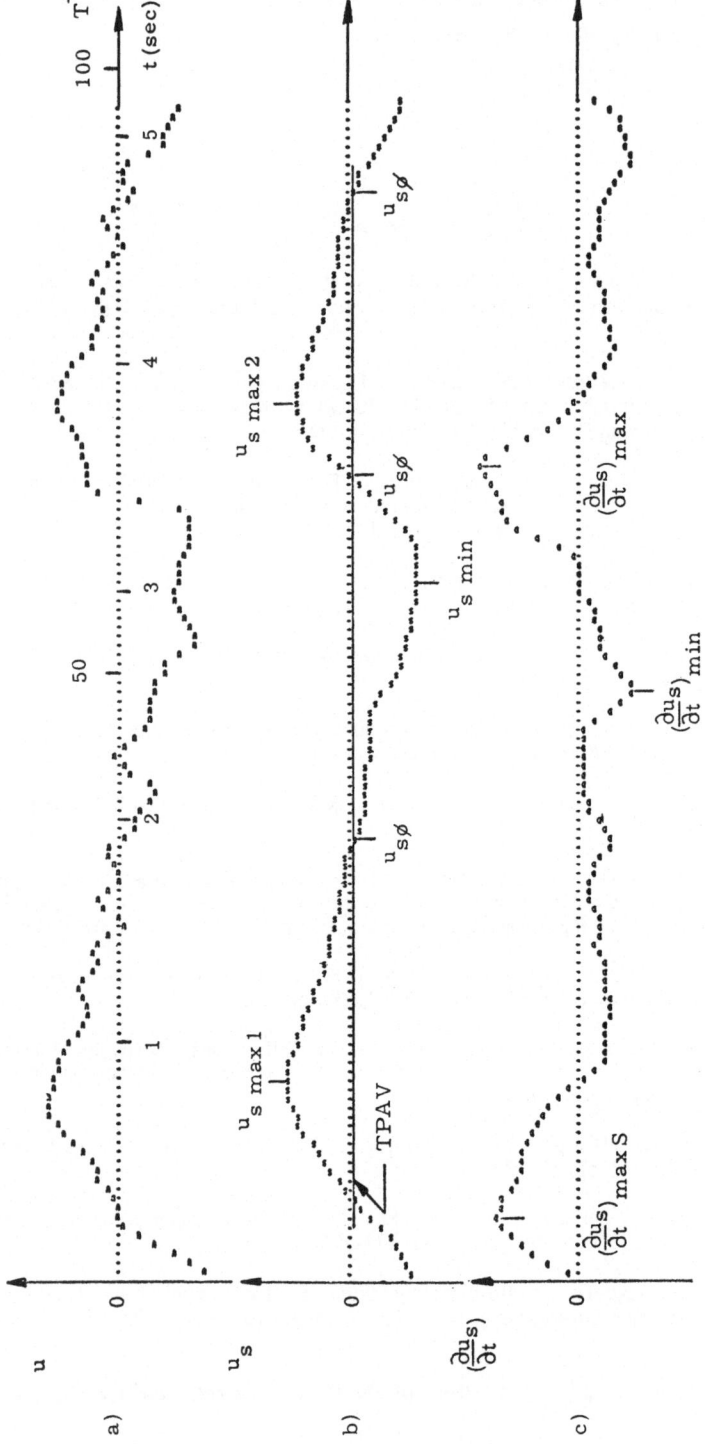

Fig. 1. a) u signal (raw data) digitized with every second point shown
b) Smoothed u Signal
c) Time derivative of the smoothed u signal.

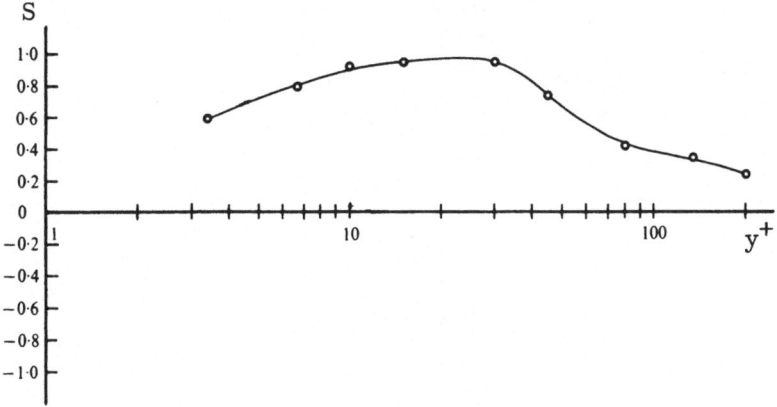

Fig. 2. Skewness factor of ∂u/ ∂t over the channel half-width.

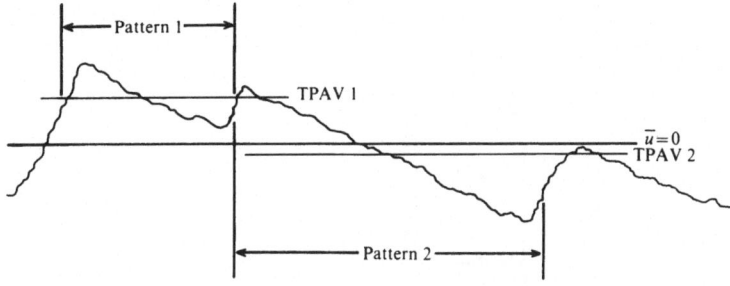

Fig. 3. Sketch illustrating the use of TPAV in the recognition program.

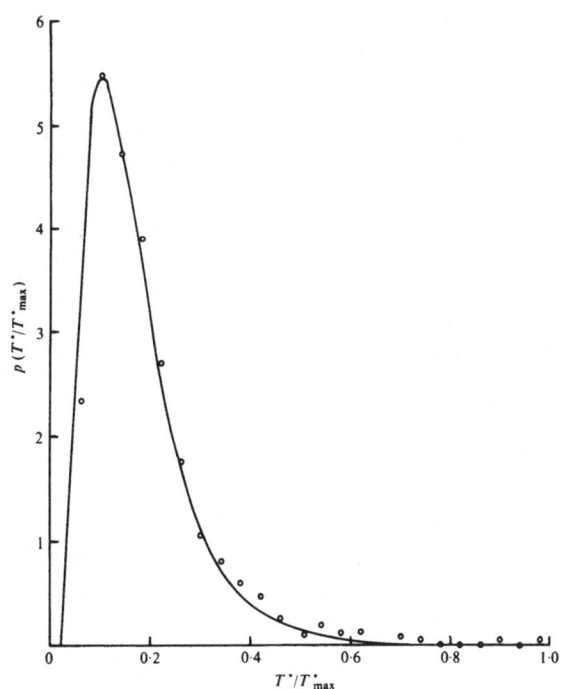

Fig. 4. Probability density distribution of the normalized pattern length.

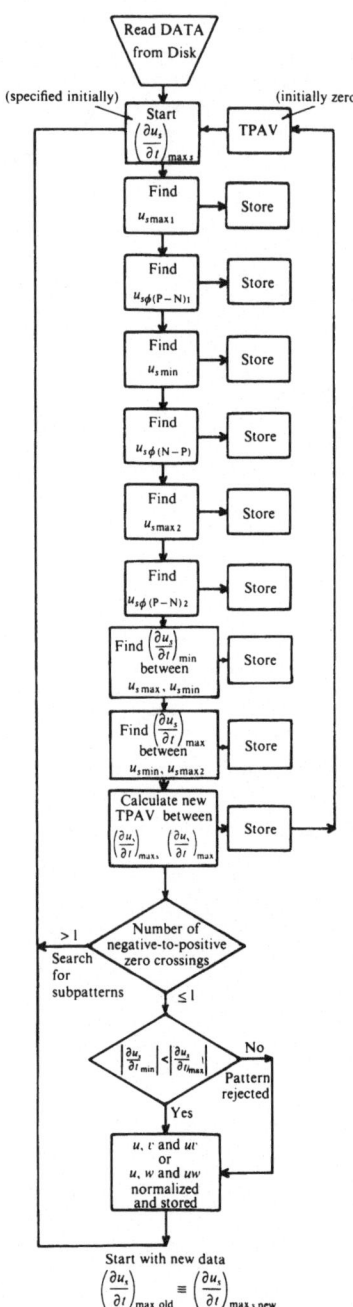

Fig. 5. Flow diagram showing the operation of the recognition program. For explanation of the symbols see Fig. 1.

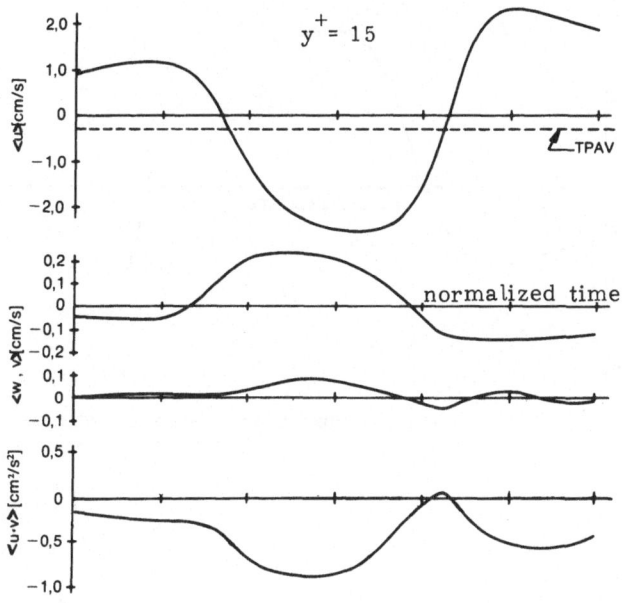

Fig. 6. Normalized and ensemble averaged u, v, w and uv patterns.

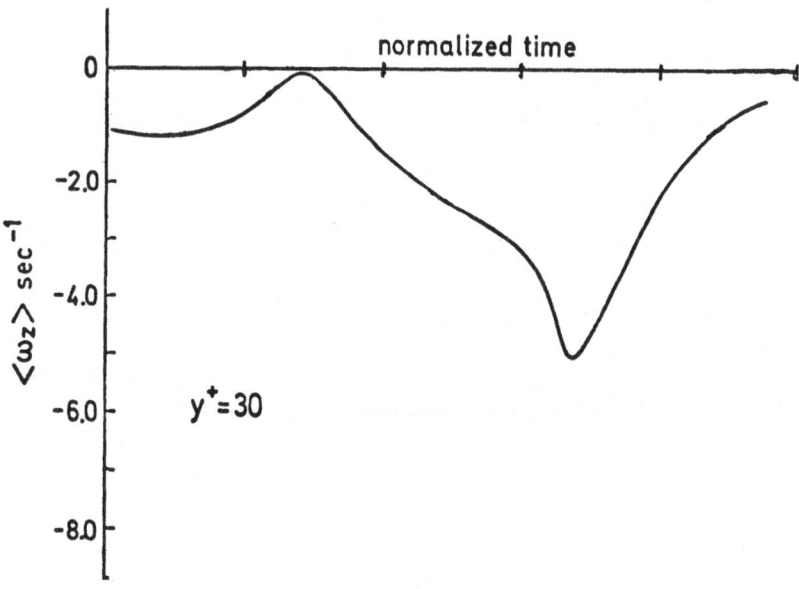

Fig. 7. Normalized and ensemble averaged ω_z patterns.

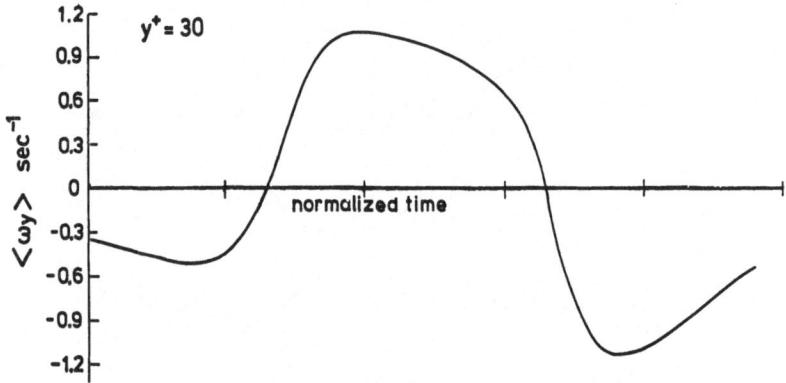

Fig. 8. Normalized and ensemble averaged ω_y patterns.

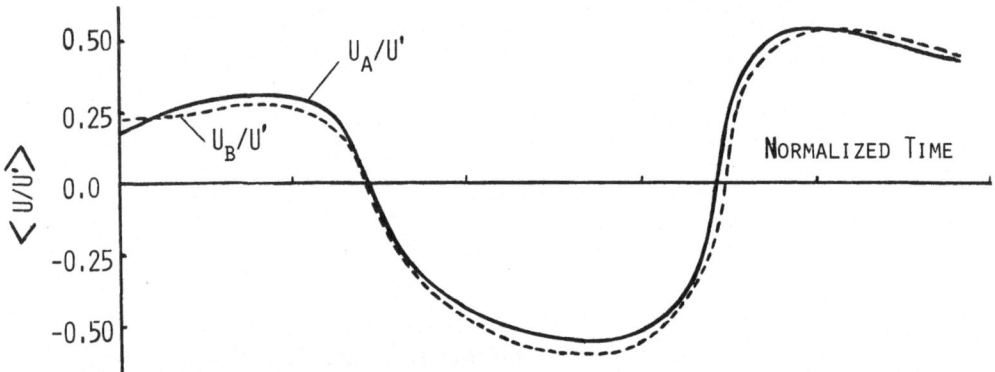

Fig. 9. Pattern recognized streamwise velocity component normalized
with rms values: solid line recognized on probe A; dashed line recognized
on probe B from [3] .

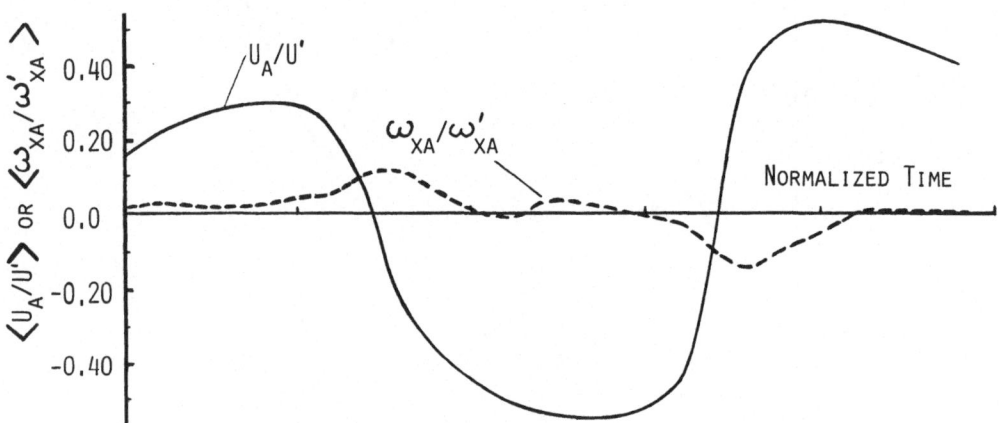

Fig. 10. Simultaneously averaged vorticity component ω_x from
same probe as pattern recognized u velocity from [3] .

Biological Flow in Deformable Vessels

Y. C. Fung
San Diego

It is, indeed, a great honor and pleasure for me to participate in the inauguration of the Aerodynamisches Institut's reconstructed Building. This is the old home of Dr. Theodor von Kármán. This is the place where his colleagues and successors worked and continue to work. I felt an affinity to you through von Kármán, because although I was not old enough to have worked directly with him, I did know him personally. I studied and worked in his American home at the California Institute of Technology in Pasadena for 20 years, and have always held him as a model to emulate. So, it is, indeed, a pleasure to be here with you this afternoon.

Ladies and gentlemen, this afternoon I would like to speak with you on biomechanics. Biomechanics is another tradition of the Aerodynamisches Institut. Professor Naumann and Professor Zeller have led the way and made important contributions to biomechanics, and Professor Krause is quickly establishing here a leading institution on biomechanics research. I know that early in 1980 the Medical Faculty of the RWTH Aachen honored Professor Naumann with a honorary degree in medicine. Congratulations, Professor Naumann! The Medical Faculty should be congratulated also for its astuteness in recognizing the contributions of an outstanding fluid mechanicist and the science of fluid mechanics to medicine, setting an example for other countries to follow.

Since this audience is one that has a general interest in fluid mechanics but not particularly specialized in biomechanics, I would like to interpret my title liberally and make some general comments on biomechanics. My paper may be better described by an alternative title: ON BIOMECHANICS IN GENERAL, AND ON PULMONARY BLOOD FLOW IN PARTICULAR.

Biomechanics is an expression of man's interest in him/herself. Man's interest in man has always been very strong. Thus it is not surprising that the pioneers of mechanics, people like Aristotle, Galileo, Descartes, Hooke, Euler, Helmholtz, Young, etc. wrote pioneering papers on biomechanics. But as a vast territory bordering on biological and physical sciences, the field of biomechanics was only sparsely populated for a long time. Only recently has the population grown. Today it promises to be a thriving field. It seems, therefore, worthwhile to pause and survey briefly the present status and future prospects, and examine the key areas that must be worked on in order to bring us closer to the end period of the rainbow.

Objectives of Biomechanics

Workers on biomechanics may aim at one of the following objectives:

(a) Scientific curiosity, (b) Clinical applications, (c) Health care delivery, (d) Industrial developments. Much can be said about each objective. To illustrate the contents of biomechanics with respect to these objectives, consider the science of physiology. Topics like blood flow, ventilation, water movement, ion transport, diffusion, etc. are such obvious mechanical events that a quantitative understanding can hardly be achieved without a good mechanical analysis. Old physiology avoids mathematical analysis, it is qualitative and hard to understand. New development of biomechanics gives physiology a quantitative approach to make it easier to understand. The motivation to study biomechanics in physiology is usually scientific curiosity. But in the course of such a study results with clinical apllications often appear. For example, in order to do the biomechanical analysis one has to know the mechanical properties of biological fluids and solids. Hence the science of biorheology is developed. Once these rheological properties are known, they can be used in diagnosis or treatment, because disease is nothing but deviation from the normal, and measurements of the mechanical properties of the tissues to decide whether the tissues are normal or not can be very helpful for diagnosis. Thus instruments of biorheology can become tools of medicine and surgery.

In association with biomechanical analysis is the need to know the geometry of an organ, such as the shape of the heart, the size of the blood vessels, the details of the bone structure, etc. Hence the frontiers of biomechanics overlap with the frontiers of anatomy, and there results a lot of activities such as the use of computer to determine the geometry of the heart and lung in vivo, biplane x-ray photography, etc. Furthermore, biomechanical analysis requires a clear statement of boundary conditions, which for internal organs are often rather difficult to determine. Careful and ingenious observations are often required to gain information, and the observational methods often become practical tools of the clinic.

Recent advances in cardiovascular surgery can illustrate the direct relation between basic research and industrial development. With the advent of prosthetic heart valves, left ventricle assist pumps, heart-lung machines, hemodialysis machines, total heart replacement, etc., an industry of considerable size has sprung up. These devices are being improved continuously with the help of detailed biofluid mechanical studies.

Surgery is an art of inflicting an injury according to a design. Healing of wounds is a great event that nonliving machines cannot do. The use of artificial internal organs, implantable artificial prosthesis, bone repair, joint replacement, artificial limbs, orthodontics, and the treatment of malocclusion, dental caries, etc. require basic biomechanical studies. Improvement in basic understanding of the biomechanics of these devices will bring improvements to health delivery.

Other biomechanics problems may be more closely associated with society. Problems involved in safety in highway driving, aircraft flying, occupational hazards such as work in mines, factories, farms, sports, etc. may range from locomotion, impact, external and internal injuries, short and long term effects on the lung, heart, eyes, ears, bones, muscles, nerves, etc. Design of wheelchairs for the handicapped and beds for the bedridden calls for ingenuity in mechanical and electric designs. Design of monitoring instruments of blood pressure and ECG for use in hospital or at home is important. Instruments to be used by patients at home but with data transmitted to a clinic hold great promise. Instruments of this kind may be effective in reducing medical costs. One of the greatest aspirations of bioengineering of the future is to lower the health care costs to a reasonable degree while improving the quality of health delivery. All countries are faced with the problem of runaway health care costs. Engineers can help bring the costs down. Up to this point bioengineering has helped physicians to make medicine and surgery more sophisticated but more costly. The next stage should be initiated by government and society to reward bioengineering developments that bring the costs down. Improved understanding of biomechanics should play a central role in this development.

Methods for Biomechanics

The methods for biomechanics are not very different from those of any other branches of mechanics. Advance in other branches of fluid mechanics influences biofluid mechanics. However, biomechanics does have a character of its own. It has an overwhelming concern about the container. We speak of blood flow, we are concerned about blood vessels. We study the flow in the heart, we may be actually worrying about the health of the heart muscle. We study wave propagation in arteries, our objective might be the detection of signs of diseases in the pulsation. Virtually all significant biofluid mechanical problems involve a significant participation of the wall. Walls are usually solid. Hence biofluid mechanics is practically always coupled with biosolid mechanics.

In general, to formulate any problem in mechanics one needs to know the geometrical configuration, the materials of construction, the boundary conditions, and the field equations. In biomechanics, geometrical determination means anatomy and histology, material determination means biorheology, biochemistry, and biophysics. The proper understanding of boundary conditions is crucial, and sometimes the most difficult. But generally speaking, it is the material property that characterizes each special field of mechanics. Thus, Hooke's law characterizes the classical theory of elasticity; Newton's law of viscosity characterizes the classical theory of fluid dynamics. Then what replaces these laws in biomechanics? To provide an answer, I shall present a survey of the constitutive equations for living tissues.

The Constitutive Equations for Living Tissues

There are three kinds of constitutive equations that must be known. First, those describing stress-strain relationship in normal state of life (a dynamic state of equilibrium which physiologists call homeostasis). Second, those describing transport of water and other substances in tissues. Third, those describing growth or resorption of tissues in response to long term changes in state of stress and strain. Of these three, the third is the most fascinating; but about which there is very little quantitative information except for the bone. The second turned out to be very complex because living tissues are nonhomogeneous. Mass transport in tissues being a molecular phenomenon, it is often accentuated by the nonhomogeneity at the cellular level. At the present time, we have a reasonable grasp only of the first item, the stress-strain relationships of living tissues in normal physiological conditions. On this subject I would like to present a few explicit examples.

Blood

Blood may be considered as an incompressible non-Newtonian fluid in ordinary circumstances. If blood is tested in a Couette viscometer or a cone-plate viscometer, it is found that in steady flow the coefficient of viscosity is higher if the shear gradient is smaller. The shear stress τ is related to the shear strain rate $\dot{\gamma}$. The relationship is approximated by Casson's equation [1]

$$\sqrt{\tau} = \sqrt{\tau_y} + \sqrt{\eta\dot{\gamma}} \tag{1}$$

where τ_y and η are constants. τ_y has the signifiacance of a yielding stress. η is called Casson's coefficient of viscosity. This relation reflects the deformability of red blood cells in shear flow, and the ability of red cells to stick to each other and form a spatial structure when the shear gradient decreases.

Generalizing equation (1) to a stress-strain law, we may write [2]

$$\sigma_{ij} = - p\delta_{ij} + 2\mu \ (J_2) \ V_{ij} \tag{2}$$

where σ_{ij} is the stress tensor, V_{ij} is the strain rate tensor, v_i is the velocity vector, p is the pressure, δ_{ij} is the Kronecker delta, and μ is a function of the strain rate invariant J_2:

$$\mu(J_2) = \left[(\eta J_2)^{1/4} + 2^{-3/4} \ \tau_y^{1/2} \right]^2 \ J_2^{-1/2} \ , \tag{3}$$

where

$$J_2 = \frac{1}{2} \ V_{ij} \ V_{ij}, \quad V_{ij} = \frac{1}{2} \ (v_{i,j} + v_{j,i}) \tag{4}$$

If τ_y is taken to be a yield stress literally, then these equations must be supplemented by a yield rule, and an elastic stress-strain relationship below yielding. But it is quite possible that the concept of yielding is only a mathematical idealization, and the real process is much more complex. Equation (2) does not account for the normal stress due to shear, nor other viscoelastic features of blood which exist but whose significance has not yet been demonstrated.

Red Blood Cells

Red blood cells are filled with hemoglobin solution, which in normal blood is a Newtonian fluid with a viscosity about 6 cp. In flowing through capillary blood vessels the red cells are severely deformed. When stationary, however, the cells assume a regular shape of a biconcave circular disk. Human red cells have an average diameter of about 7.8 μm, a thickness of about 2 μm, and a wall (cellmembrane) thickness of about 0.005 μm.

It is easy to demonstrate theoretically that the biconcave disk shape (a disk with a reduced thickness at the center) assumed by the cell in static equilibrium in a fluid implies that the static pressure in the interior of the cell is approximately equal to the pressure outside the cell (see [2] for details and bibliography). It follows that the cell membrane is unstressed in the normal condition. A liquid-filled biconcave disk can be deformed into a large variety of other shapes through isometric transformation. Such a transformation changes one surface into another surface without tearing and without stretching; hence without strain, and without stress. Thus the red cells may be considered as having been designed (by having a biconcave disk as its natural shape) to withstand a variety of large deformations without inducing any membrane stresses. This remarkable flexibility without being stressed may be the secret of the durability of the red cells in their strenuous courses through the microcirculation.

By testing red blood cells in the off-design condition which causes change of metric and introduces membrane strain and stress, one can determine the mechanical properties of the cell membrane. Through many years of research (see [2, 3] for bibliography) we now know that red blood cell membranes are deformable, can be stretched uniaxially several hundred percent, but its area cannot be changed more than three or four percent without breaking it. The constitutive equations for the cell membrane may be written as follows [2], modified from Evans and Skalak, [3] :

$$N_1 = K (\lambda_1 \lambda_2 - 1) + \mu (\lambda_1^2 - \lambda_2^2) \quad , \tag{5a}$$

$$N_2 = K(\lambda_1 \lambda_2 - 1) - \mu (\lambda_1^2 - \lambda_2^2) \tag{5b}$$

Here N_1, N_2 represent the principal stress resultants in the cell membrane (in the sense of Kirchhoff), λ_1, λ_2 represent the principal stretch ratios. The product $\lambda_1 \lambda_2$ obviously represents the area ratio, and $\lambda_1 \lambda_2 - 1$, the change of area. The principal Green's strains are

$$E_1 = \frac{1}{2} (\lambda_1^2 - 1), \quad E_2 = \frac{1}{2} (\lambda_2^2 - 1) \tag{6}$$

Hence, by Mohr's circle construction one recognizes at once that the maximum shear strain is

$$\gamma_{max} = \frac{1}{2} |E_1 - E_2| = |\lambda_1^2 - \lambda_2^2| \tag{7}$$

whereas $N_1 - N_2$ is twice the maximum shear stress resultant. Thus, by adding and subtracting (5a) and (5b) we see that

The mean membrane stress resultant $= K (\lambda_1 \lambda_2 - 1)$ \hfill (8)

The maximum membrane shear resultant $= \mu |\lambda_1^2 - \lambda_2^2|$. \hfill (9)

Hence K is an elastic modulus for cell membrane area change, μ is an elastic modulus relating shear stress and shear strain. The unique feature of red cell membrane is that K is about four orders of magnitude larger than μ :

$$K \sim 10^4 \, \mu \tag{10}$$

If the cell membrane material is incompressible, this result says that while the membrane can be distorted, its thickness remains essentially constant.

Blood Vessels

All living tissues are pseudo-elastic; that is, they are not elastic, but under periodic loading a steady-state stress-strain relationship exists which is not very sensitive to strain rate. The loading and unloading branches can be treated separately, and a pseudo-elastic potential can be introduced to describe the stress-strain relationship in either loading or unloading. The elastic potential $\rho_o W$ is a function of the Green's strain components E_{ij}. The partial derivative of $\rho_o W$ with respect to E_{ij} gives the corresponding stress S_{ij} (Kirchhoff stress). W is defined for unit mass of the vessel wall, ρ_o is the density in the initial state, hence $\rho_o W$ is the strain energy per unit initial volume. For arteries and veins the pseudo strain energy functions are of the form [4]

$$\rho_o W^{(2)} = C \exp \left[a_1 E_1^2 + a_2 E_2^2 + 2a_4 E_1 E_2 \right] \tag{11}$$

Here the superscript (2) over $\rho_o W$ signifies that this is a two-dimensional approximation, treating the arterial wall as a membrane, ignoring the radial stress, with subscript 1 referring to the circumferential direction, and 2 referring to the longitudinal direction. Comparison of (11) with other proposed strain energy functions such as polynomials is discussed in detail in [4]. A polynomial strain energy function proposed by Patel and Vaishnav [5], is especially well-known.

Skin

The skin is an anisotropic material. Treated as a membrane, the two-dimensional pseudo-strain energy function for the skin is [6, 7].

$$\rho_o W^{(2)} = \frac{1}{2} \left[\alpha_1 E_1^2 + 2 \alpha_4 E_1 E_2 + \alpha_2 E_2^2 \right]$$

$$+ \frac{1}{2} C \exp \left[a_1 E_1^2 + a_2 E_2^2 + 2a_4 E_1 E_2 + E_1 E_2 (\beta_1 E_1 + \beta_2 E_2) \right] \tag{12}$$

where C, a's, α's, and β's are constants. The β terms may be omitted without significant sacrifice of accuracy.

The Lung Tissue

The lung is a highly compressible soft tissue analogous to a foam rubber. In air-inflated lung the mechanical properties are dominated by surface tension that acts on the moist interface between the interalveolar septa and air. If the surface tension is eliminated then we obtain the tissue component of the stress-strain relationship. In laboratory experiments the surface tension can be reduced to a negligible amount by filling the airway and alveoli with saline. Then the stress-strain relationship of the lung tissue is due to the elasticity of the interalveolar walls. If we have determined the geometric structure of these interalveolar walls, we can then construct the stress-strain relationship of the lung tissue on the basis of the elasticity of the walls. On assuming that the interalveolar wall obeys (11), the constitutive equation of the lung tissue can be derived. The theoretical constitutive equation has been derived in two forms. In [8], the tissue is assumed anisotropic, and we obtain

$$\rho_o W = C \exp \left[a_1 E_1^2 + a_2 E_2^2 + 2a_4 E_1 E_2 \right]$$

$$\text{(13)}$$

$$+ \text{ symmetric terms by permutation,}$$

where C, a_1, a_2, a_4 are material constants. In [9] isotropy is assumed and we have

$$\rho_o W = \frac{\acute{c}}{2D} \exp(\alpha I_1^2 + \beta I_2) \tag{14}$$

where $\acute{c}$, α, β are material constants whereas D is the average diameter of the alveoli. I_1, I_2 are strain invariants:

$$I_1 = E_1 + E_2 + E_3 \doteq E_{11} + E_{22} + E_{33} \tag{15a}$$

$$I_2 = E_1 E_2 + E_2 E_3 + E_3 E_1$$

$$= E_{11} E_{22} + E_{22} E_{33} + E_{33} E_{11} - E_1^2 - E_{23}^2 - E_{31}^2 . \tag{15b}$$

Here E_{ij} is the Green's strain tensor, E_1, E_2, E_3 are the principal strains.

Comparison between (13) and (14) and their application to experimental results show that they do not differ very much; but that in general (13) fits the experimental data better. Extensive experimental results are reported in [8], and (13) and (14) have been shown to be able to fit the experimental results quite well, [9]and[10] .

Experimental results on lung tissue subjected to triaxial loading have been published by Hoppin, Lee and Dawson, [11];and on the basis of these data Lee and Frankus [12] have formulated empirical strain energy function in the form of a polynomial. Good agreement was obtained.

Other Biological Tissues, Muscles

In the above we presented some typical constitutive equations of biomaterials. These examples convey the impression that biomaterials are complex; but there is also a unifying simplicity. For example, most of the soft tissues are exponentially stiffening. For certain purpose of biomechanical analysis it is sufficient to know these general features. For other purposes, especially for clinical application, it is necessary to know the details of the mechanical properties.

Of all the constitutive equations in biomechanics, that of the muscle must be one of the most important. Without a constitutive equation for the active contraction of the heart muscle we cannot build a rigorous theory of cardiology. Without the constitutive equations for the vascular smooth muscle, we cannot develop a mathematical theory of circulation regulation and control. Similarly, for theories of other internal organs we need the constitutive equations of their smooth muscles. Yet reliable constitutive equations for these muscles are not yet available! There is a huge literature, a large number of investigators, a lot of controversy, but no unchallenged success. See, for example, [2] and [13] . Hence, looking toward the future, I would say that the clarification of the constitutive equations of the muscles which describe the contraction process in terms of stress and strain as a function of time after stimulation and history of deformation, holds the key to future development of biomechanics.

Bone

The bone obeys Hooke's law of elasticity and fails at a strain so small that the classical linearized theory of elasticity is sufficiently accurate for most problems concerning the stress distribution in the bone. Yet living bone is a biological material. It has blood circulation. It responds to stress and strain by growth or resorption (i.e. decrease in tissue mass). An astronaut in weightless condition loses calcium in his bone. Appropriate stress from normal exercise maintains bone metabolism. Overstress causes resorption, as one sees sometimes when broken bones are fixed with nuts and bolts which are tightened too much. Studies of this nature lead to constitutive equations of the third kind mentioned earlier. Other tissues probably behave the same way; but quantitative information is lacking. This, then, is another important frontier for future biomechanics research.

The Importance of Biological Observation

In the preceding sections we have described the methods for biomechanics and the mechanical properties of some typical biological materials. With the constitutive equation and the usual laws of conservation of mass, momentum, and energy, field equations are derived and boundary-value problems can be formulated. Most meaningful problems in physiology, however, are very complicated for the following reasons:

> a. Very complex geometry, and irregular boundary shape.
> b. Finite deformation, making it necessary to define finite strains and use nonlinear strain-displacement relationship.
> c. Nonlinear stress-strain relationship or nonlinear viscosity or viscoelasticity.

    d. Wide range of Reynolds number, e.g. in blood flow in man the Reynolds number ranges from several thousand to 10^{-3}.

    e. In periodic phenomenon a wide range of Stokes number (or its square root, the Womersley number). Again, e.g., in blood flow in man the Stokes number ranges from 200 to 10^{-4}.

In face of such complications one usually thinks of powerful numerical methods: such as the finite elements or the finite differences methods, and use a large computer as a laboratory. There is no doubt that future theoretical approaches need the computer. But one should also realize that problems of biomechanics are usually much more complex than what we normally encounter in aeronautics or mechanical engineering, in which air or water or its like is the working fluid, and the walls are usually rigid. In biomechanics it is yet too early to rely on the brute-force method to get any useful results. At this stage of development it is extremely important to make biological observations in order to search for permissible simplifications. Significant advances are made when appropriate simplifications are obtained.

In the following I shall present two examples with which I am familiar in order to illustrate the point. In the first example we consider the blood flow in the lung, with the objective to determine the relationship between the volume rate of flow and the pressures in the pulmonary arteries and veins. In the second example we consider the so-called "water-fall" phenomenon in the lung: a condition in which the flow rate will not increase even though the downstream pressure in pulmonary vein continues to decrease while the upstream pressure remains constant. In fact the flow decreases slightly when the venous pressure decreases. In neither case is the flow proportional to the difference between the inlet pressure and exit pressure, showing the existance of a strong nonlinear phenomenon.

Blood Flow in the Lung

Our lung is the organ to oxygenate the blood. The venous blood of peripheral circulation is pumped by the right ventricle of the heart into the pulmonary artery. The artery divides and subdivides again and again into smaller and smaller blood vessels. The smallest blood vessels are called the capillaries, which have very thin walls (about 1 μm thick) across which the blood exchanges oxygen and CO_2 with the air that we breathe. The oxygenated blood then flows into the veins, and finally into the left atrium of the heart. The last generation of pulmonary artery before entering the capillaries is called the arteriole; the first generation following the capillaries is called the venule. Between the arterioles and venules the blood-gas exchange takes place in the capillaries. The capillary blood vessels of the lung are shown in Fig. 1 which shows two views of the interalveolar septa (i.e., sheets of capillary blood vessels) of the cat, (which is quite similar to the human lung). These photographs were obtained by perfusing the blood vessels with a liquid silicone rubber which was catalized to solidify in the blood vessels after the lung was perfused an equilibrated with a controlled static pressure in the blood and in the air space. No red blood cells are seen. The large blank spaces are the air pockets of the alveoli. The small connected channels are the capillary blood vessels.

As a structural material, the lung is not unlike a foam rubber. The capillary blood vessels lie in the foamed walls which are called interalveolar septa. Each unit of air space bounded by the capillary blood vessels is called an alveolus. A human lung has about 300 million alveoli, with a blood-gas exchange area of between 50 and 100 square meters.

The capillary blood vessels in the lung are organized into tightly knit networks in the interalveolar septa. The left side of Fig. 1 shows a plane view of the net work. The right side shows a side view. These vessels may be looked upon in an overall manner as a thin sheet of blood bounded by two membranes which are interconnected by an array of avascular posts [14, 15] . The posts provide elasticity against distension of the sheet. But the major part of the distinsibility of the sheet against the transmural pressure (blood pressure minus alveolar gas pressure) comes from the tension in the stretched bounding membranes, in a manner directly analogous to the rigidity of a musical drum against lateral load (Fung and Sobin, [16]). In the case of a drum the tension times curvature equals the lateral pressure, so that the load-deflection relationship is linear. The same is the case of the pulmonary alveolar septa, within certain limits. Measurements of the mean sheat thickness of the vascular space (i.e. blood space) in pulmonary interalveolar septa of the cat [17] show that when the blood pressure changes the mean sheet thickness changes with the transmural pressure. When the transmural pressure is positive the thickness h increases linearly with increasing pressure according to the formula

$$h = h_o + \alpha \ (p - p_{alv})$$ (16)

where p is the static pressure of the flowing blood and p_{alv} is the alveolar gas pressure. When the transmural pressure is negative it is a good approximation to take the thickness h as zero. When $p - p_{alv}$ is large the posts and membranes cannot extend further, and the average thickness h tends to a limiting constant. The experimental results for the cat lung are shown in Fig. 2. Those for man and dog are similar, except that the values of α and h_o do vary.

On the other hand, the geometry and dimensions in the plane view do not change with blood pressure. The capillaries in the lung are distensible with respect to blood pressure in the direction perpendicular to the sheet but are indistensible in the plane of the sheet. The dimensions in the plane of the sheet can be changed only by inflation of the lung, not by blood pressure [15, 16, 17] .

Although blood flow in the capillary blood vessels of the frog lung was discovered by Marcello Malpighi (1628-1694) and communicated by him in his letters addressed to Borelli in 1661, the pressure-flow relationship remained unknown until recently. This relationship is important because we want to know the answers to the following type of questions: How much resistance do these blood vessels offer to the flow? In what situation the high resistance in these vessels will cause hypertension (high blood pressure) in the pulmonary artery? In what way a high end-diastolic pressure in the heart's left ventricle will cause trouble to the lung? In what condition an excessive amount of water will leave the blood vessels and enter the alveolar air space, causing edema? Would the oxygenation of the blood be complete? What happens if some part of the lung collapses (atelectasis)? What is the effect on blood flow by acceleration of the body (as in sports or in accidents), or in low gravitational field as in space flight? How long does it take an average red blood cell to go through the lung, which is measureable and is useful clinically? How much is the blood volume in the lung (which is measureable by indicator dilution method, or CO inhalation, etc.)? What is the dynamic impedance of the lung (flow/pressure fluctuations) as seen by the blood at the pulmonary artery?

To formulate a theory to deal with these problems requires considerable preparation. First, the anatomy of these vessels must be studied [15, 17, 18]. Then the geometry of the red blood cells [19], and the flexibility of the red blood cells are studied [20, 3]. With this information the apparent viscosity of blood flowing through an alveolar sheet is measured by model testing [21] and compared with theoretical analysis [21, 22]. Then the elasticity of the connective tissues which make up the interalveolar septa (i.e. the alveolar walls) is considered [23] and the distensibility of the sheet with respect to blood pressure is measured [16, 17, 24].

Having completed these preliminary studies, we can assemble the pieces together to look at the sheet flow in the lung. The flow is a generalized two-dimensional flow. The principal variable is the sheet thickness h. If we use a set of rectangular Cartesian coordinates x, y to denote a location in the sheet, and U, V to denote the local average velocity components in the directions of x and y, then the equation of continuity is:

$$\partial (hU)/\partial x + \partial (hV)/\partial y = efflux + \frac{\partial h}{\partial t} , \tag{17}$$

Here "efflux" means the rate of change of mass across the surface of the sheet. The efflux is zero at a steady state; but is proportional to the rate of change of thickness, $\partial h/\partial t$, if pulsatile flow is considered, and is proportional to the fluid transfer into interstitial space if filtration is considered.

Now, according to our preliminary investigations, [21], the pressure-flow relationship is

$$(U, V) = \mu fh^2 (\partial p/\partial x, \partial p/\partial y) \tag{18}$$

where μ is the apparent viscosity of the blood, and f is a numerical factor (about 4) which depends on the geometric details of the sheet. Combining these equations and (16) we obtain

$$(\frac{\partial^2}{\partial x^2} + \frac{\partial^2}{\partial y^2}) \, h^4 = 4 \, \mu f \alpha \, [\, \frac{\partial h}{\partial t} + \frac{2K}{\rho \alpha} \, (h - h^*) \,] \tag{19}$$

The last term in (19) applies when Starling's law of filtration is assumed. K is the permeability of the blood-tissue interface, ρ is the blood density, and

$$h^* = h_o + (p^* - p_{alv}) \tag{20}$$

$$p^* = \sigma \, (-\pi_{tissue} + \pi_{blood}) + P_{tissue}$$

σ is the reflection coefficient, π is the osmotic pressure. The subscripts "tissue" and "blood" refer to the interstitial space and blood respectively.

These equations apply when the blood pressure is greater than the alveolar gas pressure. If $p < p_{alv}$, then $h = 0$. If $p - p_{alv}$ becomes larger than a certain value, say, 20 cm H_2O, then h tends to a limiting constant value.

With the basic differential equation known, we can find the thickness distribution h(x,y,t) for appropriate boundary conditions. From h(x,y,t) we can find the flow velocity U, V and pressure distribution p(x,y,t).

By integrating the basic equation along a streamline and summing up the flow over an entire sheet, we obtain the following pressure-flow relationship [24] in case of steady flow without fluid loss across the wall. If the flow per unit width of sheet is Q, then

$$\dot{Q} = (h_{art}^4 - h_{ven}^4)/C \tag{21}$$

where C is a constant, and h_{art} and h_{ven} are sheet thickness at the arteriole and venule, respectively:

$$h_{art} = h_o + \alpha \, (p_{art} - p_A)$$

$$h_{ven} = h_o + \alpha \, (p_{ven} - p_A) \tag{22}$$

$$\frac{1}{C} = \frac{\text{area of sheet}}{(\text{visc.coeff}) (\text{mean path length})^2 (\text{compliance } \alpha)} \tag{23}$$

Equation (21) shows that the flow is proportional to the difference in the 4th power of the thickness at the arteriole minus the 4th power of that at the venule. The 4th power is a powerful factor. If h_{ven} is $(1/2) \, h_{art}$, then h_{ven}^4 is only 1/16 of h_{art}^4, and becomes quite unimportant in determining the flow Q. For this reason we see that when the pressure in pulmonary vein is low, the controlling factor for perfusion is pulmonary arterial pressure.

Equation (21) relates the pulmonary blood flow in each sheet to the pressures in the arteriole and venule. Summation of the flow over different regions of the lung gives the total blood flow in the lung as a function of blood pressure distribution in the lung. Adding to this the flow-pressure relationships in the arteries and veins, we can derive an expression relating the total pulmonary blood flow to the pressures in the pulmonary artery and vein. The final result is in such a form that can be tested by animal experiments. So far the result has been checked against experiments in pressure-flow relationship, pulmonary blood volume, the blood transit time through the lung and the regional distribution of blood flow, and was found to be reasonably satisfactory [25] . Other predictions of the theory, such as the input impedance of the capillary bed [26] and the water flow in the interstitium [27] , still await experimental confirmation.

The confirmation by physiological experiments is an important step in the process, because the ad hoc simplifying assumptions introduced in the mathematical analysis are so far-reaching that their effects are not easy to estimate. With reasonable confirmation from in vivo experiments, one obtains an assurance that the approach is on the right track.

The Water-Fall Phenomenon

John West [28] defines regional distribution of blood flow in the lung in three zones:

> Zone 1 - alveolar pressure > arterial pressure > venous pressure
> Zone 2 - arterial pressure > alveolar pressure > venous pressure
> Zone 3 - arterial pressure > venous pressure > alveolar pressure

In zone 1, the capillary blood vessels are collapsed (see Fig. 2) and there is very little flow. In zone 3 all capillaries are wide open and (21) applies. The question is zone 2. In zone 2 the pressure in the pulmonary vein is smaller than the alveolar pressure, but how about the venular pressure? The venular pressure differs from the vein pressure by the sum of the hydrostatic head and the loss which is equal to flow times resistance. In collapsible tubes subjected to negative transmural pressure the resistance can vary a great deal. Hence, the blood flow in zone 2 condition needs more careful consideration.

Because of hydrostatic head difference, in a man in upright position the apex of the lung is usually in zone 1 condition, the bottom is in zone 3, leaving the middle in zone 2.

Permut et al. [29] have experimented on dog's lung in zone 2 condition, and their results are shown in Fig. 3. The figure shows three curves. For each curve the pulmonary arterial pressure is fixed. On continually decreasing the venous pressure, it is seen that flow increases at first, then reaches a plateau when the venous pressure becomes equal to the alveolar gas pressure. Then on further decrease of venous pressure the flow decreases somewhat, but essentially remains constant.

This is analogous to the water flow in a river. Whereas normally the flow rate varies directly with the rate of drop of the elevation of the riverbed, a limit is reached when the drop is so large that the river becomes a waterfall. Any further lowering of the bed below the waterfall has no effect on the total rate of flow. The pulmonary phenomenon in zone 2 is similar; hence it is called the "waterfall" phenomenon [29, 30] . It is also analogous to the phenomenon of sluicing or the sonic section in laval nozzles used in steam turbines, or a supersonic wind tunnel used in aeronautics. If the speed of sound that figures in the wind tunnel problem is replaced by the pulse (progressive) wave speed in the blood vessel, then in a one-dimensional approximation the mathematical problems of the three cases - blood flow, wind tunnel, and river, turn out to be identical. They are governed by the same differential equations, only the interpretations of the mathematical symbols are different.

Lest our aeronautical friends will think the problem too elementary, we must remind them that the pulmonary problem begins where they leave off. Whereas the wind tunnel has rigid walls, in the lung we have collapsile tubes. The shape of the wall is unknown. Our problem is: where is the sonic throat, if any?

The problem has attracted many research workers (see Bibliography in[31, 32]) because similar questions occur in many other applications such as the blood flow in vena cava, Korotkov sound in the artery when it is compressed-as in blood pressure measurement, the urine flow in both male and female urethra, the Starling resistor used in physiological experiments and in artificial heart-lung machines, etc. Shapiro [31] and Pedley et al. [32, 33] have presented excellent mathematical analysis. Mathematically it is a very complex problem.

The lung problem appears to be even more complex than these other problems because we have to deal with a complex system of branching tubes whereas others need to consider only a single tube.

To deal with the lung problem, Dr. Sidney Sobin and I decided to measure the elastic behavior of pulmonary veins under negative transmural pressure (i.e., the difference between the static pressure of the blood and the alveolar gas pressure). We perfused the lung of a cat with liquid silicone rubber. The cat was anestetized and killed by an overdose of Numbutal and kept on the respirator with its chest closed. When good perfusion was obtained, the respirator was turned off, the trachea was opened to atmosphere, the perfusion was stopped, and the liquid pressure in the pulmonary blood vessels was reduced to -17 cm H_2O (below atmospheric). The liquid silicone rubber was precatalyzed and it solidified in about 30 minutes. After solidification of the rubber the cat was put in a refrigerator for four weeks to let the polymer harden and develop its full strength. The chest was then opened and the lungs taken out. The excised lungs were then hung in jars of concentrated NaOH to remove the tissue.

With very gentle handling and several changes of the NaOH solution a clear cast of the blood vessels was obtained.

Since the silicone rubber occupied the vascular space, any vessel which was collapsed and had insignificant luminal area would have no rubber in it and would have been dropped off in the corrosion process. All capillary blood vessels disappeared: they were all collapsed. This is consistent with our previous finding that the pulmonary alveolar sheet thickness is reduced to zero when the transmural pressure falls below 1 cm H_2O (see Fig. 2). What is left on the arterial and venous trees are vessels that remained patent under the negative transmural pressure.

This cast answers the question of stability of the veins positively: the larger veins do remain open at -17 cm H_2O.

Our interest is then shifted to the smallest twigs on the tree, because it is at the tips of these twigs that the pulmonary microcirculation emerges and flows into the vein. In the zone 2 flow condition these are the places where sonic throats are anastomosed to the venous tree.

We measured the width of the tips of the twigs of the venous tree and obtained a histogram. The mean width (call it "diameter") of the tips was found to be 24.46 ± 9,38 μm when the transmural (blood minus alveoli) pressure was -17 cm H_2O. This was the average diameter of the smallest vein which remains patent under negative pressure.

We then repeated the experiment with another cat at a transmural pressure of negative 2 cm H_2O. We were surprised to find that the mean width of the smallest veins which remain patent was 24.22 ±8.23 μm. A third cat was done with the same negative 2 cm H_2O transmural pressure, and the mean width of the smallest open veins was found to be 29.30 ± 7.37 μm. The difference is attributable to individual difference between animals. We must conclude then that the smallest veins in the cat's lung, with diameter of the order of 25 μm, remains patent under a negative pressure from -2 to -17 cm H_2O.

We have found earlier [34] that the smalles arteriole or venule immediately next to the capillary sheets in the cat's lung has a diameter of 15 μm. We have studied the relative position of arterioles, venules, and capillaries in the lung parenchyma [35]. With the patency measurement named above, we now seek the reason for such an interesting result. We found the reason to be tethering. An example is shown in Fig. 4. Here a venule is seen to be tethered by three interalveolar septa, positioned roughly at 120° apart. There are tensions in these interalveolar septa, due to stretching of the tissue beyond its no-load size, and due to the interfacial tension between the moist surface and the gas in the alveoli. The tension tends to keep the venule open, over a broad range of negative transmural pressures.

The relationship between the terminal venule and the interalveolar septa has been reported by Sobin [36]. From a large number of histological observations, we present a sketch of the smallest veins in Fig. 5. The venule with a diameter of the order of 25 μm is shown tethered by three interalveolar septa. A short, small branch extends over a capillary network and drains the arterialized blood into the vein. The draining points are fairly sparsely posotioned. Neighboring draining points are spaced at a distance of the order of 550μm, [35].

We have also collected some information about the elasticity of the venules and veins. In [34] we presented the elasticity of the smallest venules with respect to positive transmural pressure. In [37], Dr. Yen and Ms. Foppiano from our laboratory presented the distensibility of the pulmonary veins of the cat with respect to both positive and negative transmural pressures.

With these pieces of information, we can solve the venous "water-fall" problem rather simply. The general idea is illustrated in Fig. 6. Here an elastic vessel is shown draining into a vein. When there is no flow, we assume that the vessel is straight and has a uniform diameter. When flow exists as shown in the middle drawing, the pressure decreases as the flow proceeds downstream, because the pressure head is used up to overcome friction. The decrease of pressure reduces the diameter of the elastic tube. The reduced diameter increases the flow speed and the rate of dissipation of energy, and as a consequence of this feedback the contour of the tube wall becomes nonuniform. The diameter decreases at first rather slowly, then it shrinks faster and faster toward the exit end.

The third drawing shows the limiting case. At this condition there is a sonic throat and the flow becomes maximal. Any further decrease of p_v will not increase the volume rate of flow. The waterfall condition is reached.

Our negative transmural pressure study tells us that the vertical tube of Fig. 6 is the 25 μm venule, and the horizontal converging tube represents the interalveolar sheet and the smallest venule. The throat occurs in the smallest venule. Quantitatively the maximum flow achievable in zone 2 condition is given by (21) with the h_{ven}^4 term neglected in comparison with the first term, h_{art}^4. Hence the maximal flow that can be obtained for a given arteriolar pressure is

$$\dot{Q}_{max} = [h_o + \alpha(p_{art} - p_A)]^4 /C \qquad (24)$$

This result was given in [24] on the basis of (21) and under the assumption that the larger pulmonary veins will remain patent under negative transmural pressure. In [24], it is not known which particular generation of the venule can remain open at negative transmural pressure, and consequently the "mean path length" in (23) (the average distance between the juncture of the arteriole and capillary sheet and the location of the throat on the venule) cannot be determined. Our new results settle these questions definitively.

With the question of patency and throat location settled, and the elasticity of the pulmonary veins under positive and negative transmural pressure known, the entire blood flow problem from pulmonary artery to left atrium can be analyzed quite rigorously. The flow in the throat region has small Reynolds number and Stokes number (both ≪1, the Stokes number is the square of Womersley's number, or the ratio of transient inertia force to viscous force), and the flow speed is smaller than the progressive flexural wave speed on both sides of the throat. Thus the difficulties elaborated by Shapiro [31] do not arise, and a complete solution can be obtained with ease.

The features shown in Fig. 3 are now clear. As the pulmonary venous pressure decreases while the arterial pressure is held fixed the flow increases until a maximum is reached (as given by (24)), at which the flow velocity at the narrowest section reaches the speed of the flexural wave in the blood vessel. Flow cannot exceed this maximum. As the pulmonary venous pressure is decreased further the transmural pressure in the veins becomes more negative and their lumen become smaller; the resistance to flow increases rapidly so that the sucking pressure at the throat in fact decreases. The flow at the throat then become subsonic, showing the decline of flow with further decrease of pulmonary venous pressure.

Conclusion and Summary

The two examples concerning blood flow illustrates the coupling of fluid and solid mechanics which is a usual feature of biomechanics. Future progress in biomechanics needs new knowledge on the mechanical properties of biological fluids and solids. We emphasized especially the need for constitutive equations for the active contraction of muscles in single twitches. The mathematical description of growth and resorption of tissues as functions of stress and strain, and of mass transport in the tissues are the other two frontiers to be conquered. On the methodology side, we believe that numerical analysis will be the order of the day, but simplifications should be sought through biological observations.

Acknowledgement

The support of the National Science Foundation and the USPHS National Institute of Health - National Heart, Lung and Blood Institute on my research on cardiopulmonary dynamics is gratefully acknowledged.

References

[1] C o k e l e t, G.R.: The rheology of human blood. In Biomechanics: Its Foundations and Objectives, (Fung, Perrone, and Anliker, eds.), Prentice-Hall, Englewood Cliffs, New Jersey (1972), pp. 63-103.

[2] F u n g, Y.C.: Biomechanics: Mechanical Properties of Biological Materials, Springer-Verlag, New York. In press.

[3] E v a n s, E.A., S k a l a k, R.: Mechanics and Thermodynamics of Biomembranes. Critical Reviews in Bioengineering, CRC Press, Boca Raton, Florida, 3, (1979).

[4] F u n g, Y.C., F r o n e k, K., P a t i t u c c i, P.: On pseudo-elasticity of arteries and the choice of its mathematical expression. American Journal of Physiology, 237, (1979), pp. H620-H631.

[5] P a t e l, D.J., V a i s h n a v, R.N.: The rheology of large blood vessels. In Cardiovascular Fluid Dynamics, (D.H. Bergel, ed.), Academic Press, New York, 2, (1972), pp. 1-62.

[6] L a n i r, Y., F u n g, Y.C.: Two-dimensional mechanical properties of rabbit skin - II. Experimental results. Journal of Biomechanics, 7, (1974), pp. 171-182.

[7] T o n g, P., F u n g, Y.C.: The stress-strain relationship for the skin. Journal of Biomechanics, 9, (1976), pp. 649-657.

[8] V a w t e r, D.L., F u n g, Y.C., W e s t, J.B.: Elasticity of excised dog lung parenchyma. Journal of Applied Physiology, 45, (1978), pp. 261-269.

[9] F u n g, Y.C., T o n g, P., P a t i t u c c i, P.: Stress and strain in the lung. Journal of Engineering Mechanics, 104, (1978), pp. 201-223.

[10] V a w t e r, D.L., F u n g, Y.C., W e s t, J.B.: Constitutive equation of lung tissue elasticity. Journal of Biomechanical Engineering, Transactions of ASME, 101, (1979), pp. 38-45.

[11] H o p p i n, F.G., Jr., L e e, G.C., D a w s o n, S.V.: Properties of lung parenchyma in distortion. Journal of Applied Physiology, 39, (1975), pp. 742-751.

[12] L e e, G.C., F r a n k u s, A.: Elasticity properties of lung parechyma derived from experimental distortion data. Biophysical Journal, 15, (1975), pp. 481-493.

[13] B r a d y, A.J.: Mechanical properties of cardiac fibers. In handbook of Physiology, Sec. 2, The Cardiovascular System, Vol. 1, Heart (R.M. Berne, ed.), American Physiological Society, Bethesda, Maryland, (1979), pp. 461-474.

[14] F u n g, Y.C., S o b i n, S.S.: Theory of sheet flow in the lung alveoli. Journal of Applied Physiology, 26, (1969), pp. 472-488.

[15] S o b i n, S.S., T r e m e r, H.M., F u n g, Y.C.: Morphometric basis of the sheet-flow concept of the pulmonary alveolar microcirculation in the cat. Circulation Research, 26, (1970), pp. 397-414.

[16] F u n g, Y.C., S o b i n, S.S.: Elasticity of the pulmonary alveolar sheet. Circulation Research, 30, (1972), pp. 451-469.

[17] S o b i n, S.S., F u n g, Y.C., T r e m e r, H., R o s e n q u i s t, T.H.: Elasticity of the pulmonary interalveolar microvascular sheet in the cat. Circulation Research, 30, (1972), pp. 440-450.

[18] R o s e n q u i s t, T.H., B e r n i c k, S., S o b i n, S.S., F u n g, Y.C.: The structure of the pulmonary interalveolar microvascular sheet. Microvascular Research, 5, (1973), pp. 199-212.

[19] E v a n s, E., F u n g, Y.C.: Improved measurements of the erythrocyte geometry. Microvascular Research, 4, (1972), pp. 335-347.

[20] F u n g, Y.C.: Theoretical considerations of the elasticity of red cells and small blood vessels. Federation Proceedings, 25, (1966), pp. 1761-1772.

[21] Y e n, R.T., F u n g, Y.C.: Model experiments on apparent blood viscosity and hematocrit in pulmonary alveoli. Journal of Applied Physiology, 35, 4, (1973), pp. 510-517.

[22] F u n g, Y.C.: Studies on the blood flow in the lung. In Proceedings of the Second Canadian National Congress of Applied Mechanics, Waterloo, Canada, May 20-23, (1969), pp. 433-454.

[23] F u n g, Y.C.: Biorheology of soft tissues. Biorheology, 10, (1973), pp. 139-155.

[24] F u n g, Y.C., S o b i n, S.S.: Pulmonary alveolar blood flow. Circulation Research, 30, (1972), pp. 470-490.

[25] F u n g, Y.C., S o b i n, S.S.: Pulmonary alveolar blood flow. Bioengineering Aspects of Lung Biology, (J.B. West, ed.) Marcel Dekker Inc., New York, (1977), pp. 267-358.

[26] F u n g, Y.C.: Theoretical pulmonary microvascular impedance. Annals of Biomedical Engineering, 1, (1972), pp. 221-245.

[27] F u n g, Y.C.: Fluid in the interstitial space of the pulmonary alveolar sheet. Microvascular Research, 7, (1974), pp. 89-113.

[28] W e s t, J.: Ventilation/blood flow and gas exchange. 2nd Ed., F. Davis, Philadelphia (1970).

[29] P e r m u t t, S., B r o m b e r g e r-B a r n e a, B., B a n e, H.N.: Alveolar pressure, pulmonary venous pressure and the vascular waterfall. Medica Thoracic, 19, (1962), pp. 239-260.

[30] D a w s o n, S.V., E l l i o t t, E.A.: Wave-speed limitation on expiratory flow – a unifying concept. Journal of Applied Physiology, 43, (1977), pp. 498-518.

[31] S h a p i r o, A.H.: Steady flow in collapsible tubes. Transations of the ASME, Journal of Biomechanical Engineering, 99 (Ser. K), (1977), pp. 126-147.

[32] P e d l e y, T.J.: Mechanics of the arterial blood flow. Oxford University Press, London (1980).

[33] W i l d, Rosemary, P e d l e y, T.J., R i l e y, D.S.: Viscous flow in collapsible tubes of slowly-varying elliptical cross-section. J. Fluid Mech. 81, (1977), pp. 273-294.

[34] S o b i n, S.S., L i n d a l, R.G., F u n g, Y.C., T r e m e r, H.M.: Elasticity of the smalles noncapillary pulmonary blood vessels in the cat. Microvascular Research, 15, (1978), pp. 57-69.

[35] S o b i n, S.S., F u n g, Y.C., L i n d a l, R.G., , T r e m e r, H.M., C l a r k, L.: Topology of pulmonary arterioles, capillaries, and venules in the cat. Microvascular Research, 19, (1980), pp. 217-233.

[36] S o b i n, S.S.: The architecture and function of the microvasculature. In Biomechanics, Proc. ASME Meeting, Nov. 30, 1966. ASME, New York (1966), pp. 132-150.

[37] Y e n, R.T., F o p p i a n o, L.: Elasticity of small pulmonary veins in the cat. Journal of Biomechanical Engineering. In Press.

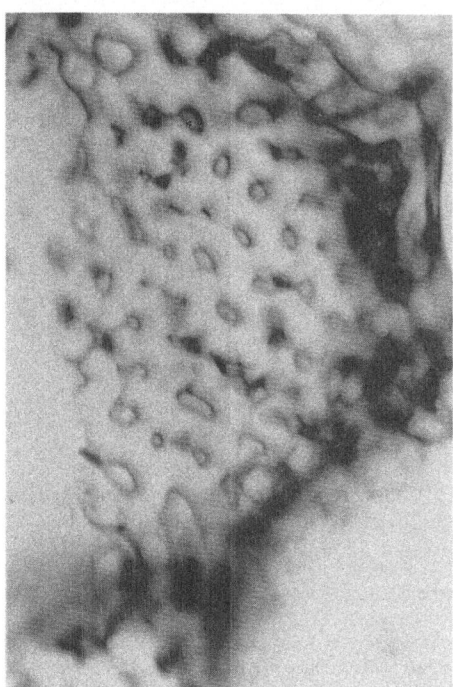 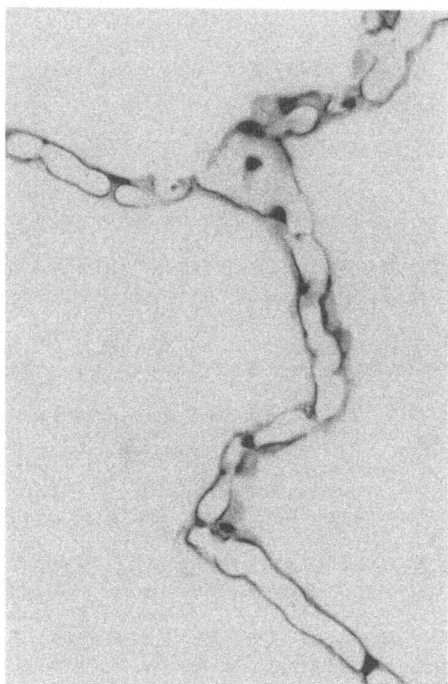

Fig. 1: <u>Left</u> - A photograph of an interalveolar septum when it is seen on its plane. Cat's lung. The interalveolar septa are filled with capillary blood vessels. The capillary blood vessels are seen to form a closely knit network. On this plane view, the average width of these capillary blood vessels is about 7 μm. The large blank spaces are the alveolar gas space. <u>Right</u> - A photograph of several interalveolar septa when they are viewed in cross section. The thickness of the capillary blood vessel sheet varies with the transmural pressure (blood pressure minus alveolar gas pressure), and is about 6 μm here. The gas in the alveoli is separated from the capillary blood by a membrane which is only about 1 μm thick. From Fung and Sobin [14].

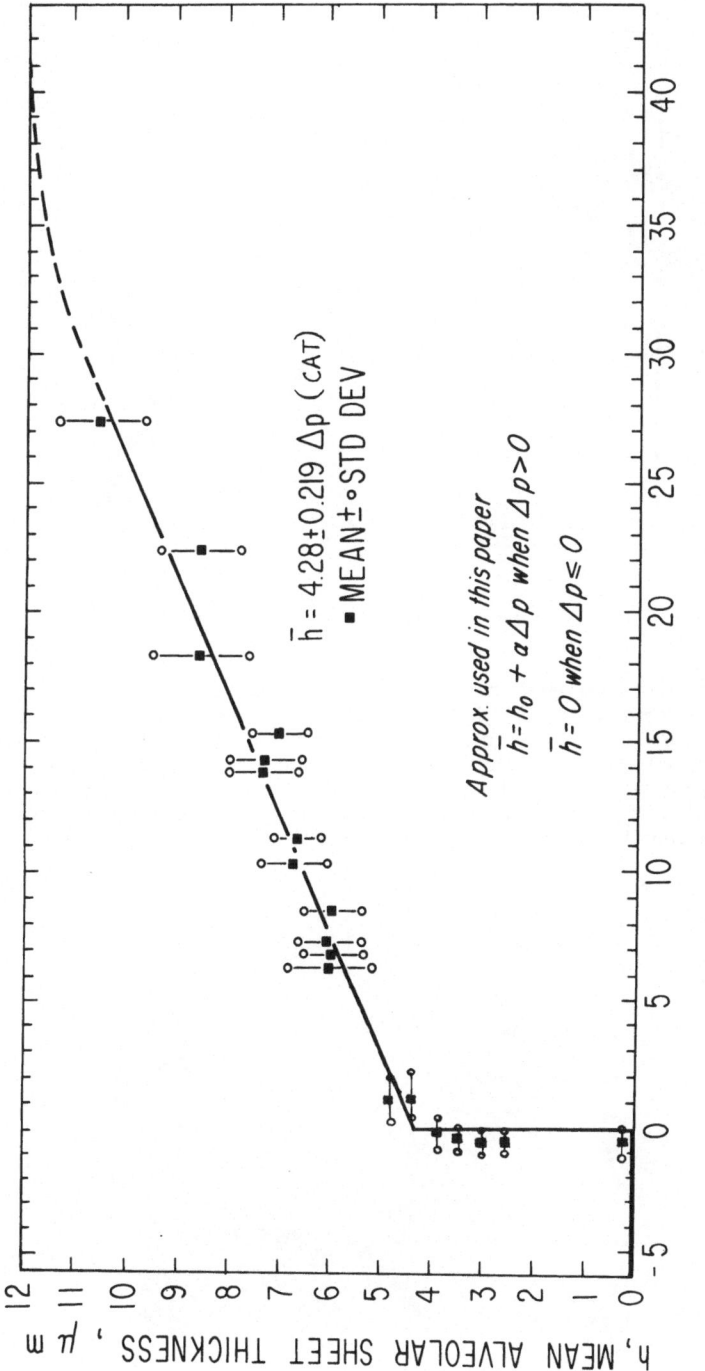

Fig.2: The variation of the thickness of a pulmonary alveolar sheet with the transmural pressure. Data points are for the lung of the cat. The data are summarized by (16) for $0 < (p - P_{alv}) < 28$ cm H_2O. From Fung and Sobin [16], and Sobin et al. [17].

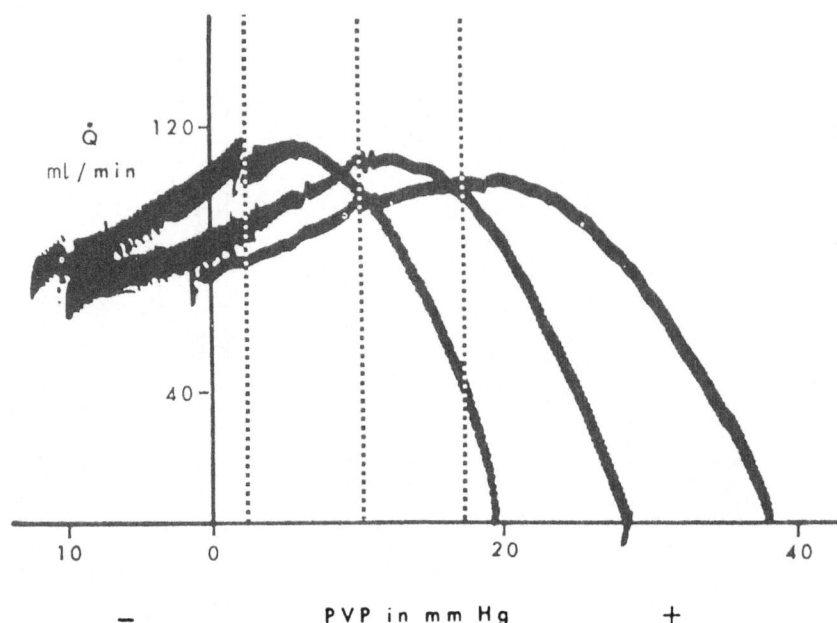

Fig. 3: Example of pressure-flow relationships under zone 2 conditions. The horizontal axis shows the pulmonary venous pressure (PVP). The pressure in the pulmonary artery (PAP) was fixed while PVP was varied. As PVP decreased, there was no flow ($\dot{Q}$) until PVP was smaller than PAP. Then $\dot{Q}$ increased with increasing (PAP - PVP). The flow reached a maximum when the venous pressure was about equal to the alveolar pressure (indicated by the vertical broken lines in each case). With further decrease in PVP the flow declined. The reason for this is the object of research to be reported below in this paper. From Permutt et al., 1962 [29].

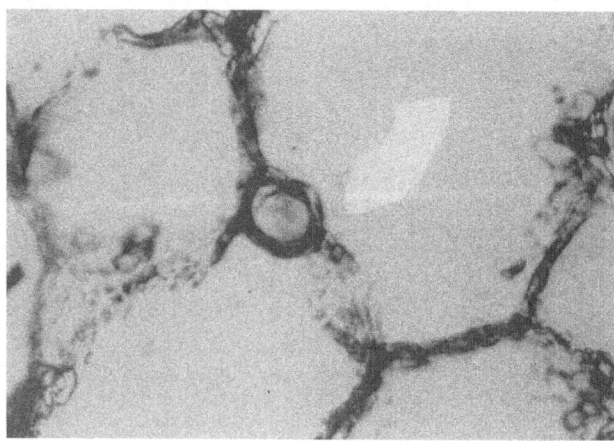

Fig. 4: Photograph of a venule in the cat's lung. Three interalveolar septa are attached to this venule. In an inflated lung these interalveolar septa are stretched and there are tension in the tissue. In addition, there are surface tension acting on the surface of the septa. Thus the venule is pulled by three membranes located roughly at 120° around its circumference. These septa tends to keep the venule open even when the internal (blood) pressure in the venule becomes smaller than the external gas pressure.

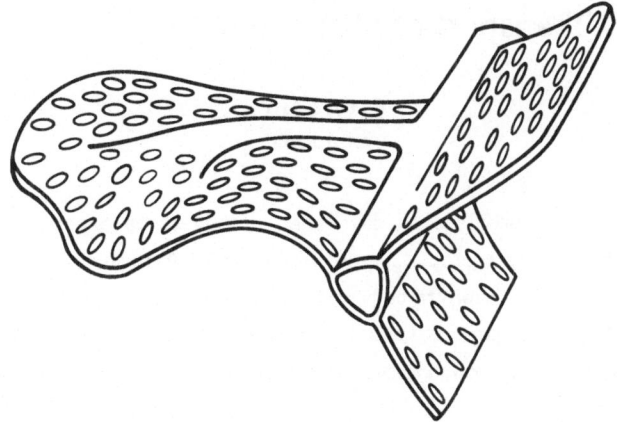

Fig. 5: A sketch of the relationship between the venules and the capillary blood vessels in the interalveolar septa in the lung. The terminal venule drains blood from the capillaries in the septa and passes it to the venule which is tethered by three septa as photographed in Fig. 4.

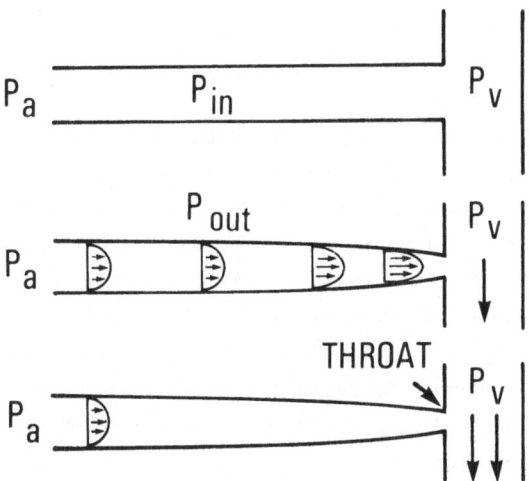

Fig. 6.: The concept of the "vascular waterfall" in blood flow in the lung in zone 2 condition. See text for explanation.

Steady Transport of Material in the Artery Wall

C. G. Caro
London

It is a great honour and privilege to have been invited to take part in the inauguration ceremonies for the reconstructed laboratories of the Aerodynamisches Institut. It is, furthermore, a tremendous pleasure to be back in Aachen. My contact with the Institute extends over many years. During this period I have grown increasingly to admire its many contributions to biofluid mechanics, have collaborated with the staff and have been conscious of the establishment of many warm friendships. It is of especial significance to me, as a keen proponent of collaboration between workers in mechanics and physiology, that a substantial fraction of the reconstructed laboratories is devoted to biofluid mechanics.

There are several reasons why I have chosen to talk about the mass transport of the arterial wall. Firstly, this is because of its potential involvement in the development of the highly important disease arteriosclerosis, responsible for example for coronary thrombosis and stroke. Secondly, it presents a variety of interesting problems in fluid and solid mechanics and their interaction. Thirdly, it is a field of major interest to my group in London. It will clearly not be possible within the short time I have available to cover the field exhaustively. Rather will I try to present an outline and to deal with a few topics in some detail.

In order to convey the essence of the problem to you, I shall rely heavily on a diagrammatic representation of a longitudinal section of the arterial wall (Fig. 1). Note that there is considerable specialisation visible even at a fairly gross level. The wall of a medium-to large-sized human artery might be 0.1 cm in thickness and consists of three concentric layers. These are (i) the intima which is composed of the endothelial cells and the internal elastic lamina, (ii) the media which is composed of alternative lamellae of connective tissue (elastin and collagen fibrils) and smooth muscle cells; the cells are connected to one another by various means, including elastic fibres, membrane fusion, structural glycoproteins and (iii) the adventitia which is composed of loose connective tissue (collagen and elastin fibrils) and the cells which synthesize them, the fibroblasts.

The artery has no distinct outer border, but merges into surrounding tissue. This is an important statement both in terms of its mechanics and its mass transport. Next, note other gross structural features of the wall. Arteries give off branches of different sizes. The larger ones - not as shown in the schematic diagram - also possess the three layers, the intima, media, and adventitia. They, or arteries which lie nearby, supply the vasa vasorum, the vessels of the vessel (Fig. 2). In all large arteries there are vasa vasorum in the adventitia [1]. In the arteries of large animals only -pig, horse, cow, for example, but not man, the vasa vasorum are also present in the outer media. The vasa consist of very small arteries, or arterioles, a capillary network (vessels about the size of red cells) and small veins or venules which return the blood into the general circulation.

In the adventitia are also found lymphatics and nerves. I shall only discuss the former now. Many authors have reported a dense lymphatic network to be present in the adventitia [2, 3] Lymphatics in other tissues conduct away that fraction of the water which filters out of the arterial end of the capillaries and is not returned into the venous end. The lymphatics drain, moreover, some solutes, from the extracellular space, the space surrounding the cells (Fig. 1).

Let us look briefly at some of the fluid mechanical problems relevant to arteries. Blood flow, as we normally conceive it, occurs within the main lumen of an artery (Fig. 1). The flow is pulsatile and exceedingly complicated (Fig. 3). The time-average Reynolds number (for man) is in the range a few hundred to one to two thousand. The Womersely frequency parameter α [4] takes a value in the range 1 - 20 for the fundamental frequency. The flow may become unstable and turbulent in large arteries. The luminal pressure varies during the pulse cycle from about 80 to 120 mm Hg.

I know of no measurements of blood pressure and flow in the vasa vasorum. If, however, the fluid mechanics resemble those in the microcirculation elsewhere then the time-average pressure will range from about 50 mm Hg in the arterioles to, say, 20 mm Hg in the venules. The flow may be relatively steady, but it is possible that the cyclical variations of the luminal pressure, hydrodynamic wall shear stress and arterial geometry lead to fluctuations of pressure and flow in the vasa vasorum. The hydrostatic pressure in lymphatics is reported to be close to atmospheric pressure. The lymph collecting in the lymph capillaries flows ultimately into the veins.

These observations indicate that there is a hydrostatic pressure difference of perhaps 100 mm Hg across the arterial wall and suggest that convective transport could occur from the main lumen towards the adventitia. Thus, there could be convection into the vasa vasorum and the lymphatics. Furthermore, convection would be expected to occur between the arterial and venous ends of capillaries and between the capillaries and lymphatics. I do not plan to consider the effect of the pulsatility of the pressure.

If there is convection, is there not also diffusion in the wall? There is a large concentration difference particularly for macromolecules, between the plasma and lymph. And if there are both convection and diffusion, are they not coupled?

I believe it will help in the assessment of the mass transport of the wall to say a few further words about wall structure and composition. The endothelial cells form a continuous lining layer over the surfaces of the wall exposed to blood, the intimal and vasa vasoral surfaces (Fig. 1). There has been quite extensive study of the endothelial cells of the intima but little is known about those of the vasa vasorum. The former, as shown on the slide, are about 1μm in thickness, $10\,\mu$m in width and $30\,\mu$m in length. They are joined to one another by intercell junction, which are apparently approximately 200 Å in width. It is undoubtedly possible for material, including macromolecules, to be transported across the endothelium normally. For example, if horseradish peroxidase (m.w. 40,000) is injected into the circulation in vivo, it can be shown by electron microscopy to enter both the inter-cellular junctions and pinocytic vesicles; the latter with an internal diameter of about 600 Å, have been conceived of as ferry boats, which attach to the cell membrane at the luminal or abluminal surface, open, come into equilibrium in respect of their contents with nearby fluid, seal, detach, and diffuse across the cell to reattach and eventually to come into equilibrium again with nearby fluid.

Some larger materials enter the wall, but not the intercellular junctions and the hypothesis has been advanced that size is a factor determining the route taken by material traversing the endothelium.

I would add that there would appear to be every opportunity for material to traverse the wall. Thus, the internal elastic lamina (Fig. 1) is usually fenestrated and the extracellular material is composed of collagen, elastin and a mucopolysaccharide ground substance. The ultrastructure of the lymphatic capillaries of the adventitia seems unremarkable (Fig. 4). The endothelial cells may be separated from their neighbours by distances as large as 0.5 μm. It would, accordingly, be expected that the lymphatic capillary would offer a small barrier to transport compared to the intimal endothelium.

I will now turn to my main theme, proposing it, for preference, in the form of a question - is there net transport of material across the arterial wall? It has been widely held for many years that the answer is affirmative [5, 6, 7, 2, 8, 9]. The evidence advanced includes the demonstration that labelled macromolecules can enter the normal arterial wall in vivo both from the intima and the adventitia, the finding of plasma constituents, including plasma protein, within the wall and the observation that the wall possesses lymphatics. In addition, it has been shown, using excised arteries [10] that water and certain solutes can traverse the wall and that the rate of filtration (or ultrafiltration) is pressure dependent.

While these various observations are consistent with net transport, they do not confirm its existence. Thus, the endothelium is exceedingly fragile; simply allowing the luminal pressure of the rabbit aorta in situ to fall to atmospheric results in severe corrugation of the inner lamellae and damage to and dislodgement of some endothelial cells (Fig. 5, 6) [11]. Moreover, it is difficult to perfuse an excised artery at normal luminal pressure, because the perfusate leaks from the cut adventitial vessels and contaminates the wall [12, 13]. If the adventitial vessels are ligated to prevent leakage, as has been done, then with flow through them arrested, their luminal pressure will come to equal that in the main lumen, and transudation from them can be expected [14].

The question might be asked why arterial studies should be undertaken in vitro rather than in vivo, where arteries would be undamaged. The answer lies in the greater ease of controlling such experimental variables as luminal blood pressure and flow rate and the luminal concentration of tracer. Since, as noted, the in vitro preparation tends to be unsatisfactory, we have developed a different preparation, that of the artery perfused in situ in the living animal [15] . This preparation possesses many of the advantages of the in vitro preparation and at the present level of our understanding, appears free of most of the disadvantages, including a damaged endothelium and a disrupted adventitial circulation.

The preparation is shown diagrammatically in Fig. 7. The perfusing plasma flows from the upstream reservoir via a heat exchanger to the cannulated artery and thence to the downstream reservoir. The mid-luminal pressure is about $100 \, cm \, H_2O$ and steady, i.e. it takes a normal time-average value. The flow is fully developed ($Re \approx 100$). The wall shear rate, about 100, is low for the artery. It was kept low because wall shear stress can apparently influence the uptake of macromolecules via the intima [16, 12] and it was desirable to work at low levels where the dependence is weak.

Note that the animal is anaesthetised and alive throughout the procedure and that the vessels arising from the perfused artery, except locally at the sites of cannulation, are undamaged, patent and connected with the animal's own circulation. We have found that blood flows from the animal's circulation into the perfused artery if the luminal pressure is lower than the animal's systemic arterial pressure implying inter-arterial connections. It enters the animal's circulation via these and conventional connections if the sign of the pressure difference is reversed. The addition of Evans blue dye to the perfusate and the separate injection of the dye into the animal's circulation showed the perfused artery to be effectively an "island" surrounded by the rest of the animal, i.e. by a far larger region perfused by the animal's circulation. The location of the interface between these two regions appears to depend on the relative levels of the luminal pressure and the animal's systemic arterial blood pressure.

We have found, with this preparation, that the concentration of labelled albumin in the perfusate does not change detectably during the course of an experiment, reflecting its slow rate of uptake by the artery. We have found, furthermore, that because of the small rate of leakage of label into the animal's circulation, the concentration of label in its blood and central (thoracic duct) lymph is always far below that in the perfusate. It would seem, therefore, that with a large apparently well-mixed region at low concentration present just beyond the perfused artery, conditions would obtain which would favour diffusional flux across the wall. Moreover, the possibility would exist for the establishment of an essentially steady-state transport across the wall. This situation should be contrasted with that following a single injection of labelled albumin in vivo. The plasma concentration then falls with time, there is partial transfer of the label to other plasma constituents [17] and there is a progressive rise of label concentration in extracellular fluid and lymph.

We now consider some results. We used purified radioactively labelled albumin as the tracer. The relative inertness of this material and the fact that it is confined to the extracellular space, introduced major simplifications into the experimental conditions. At the end of predetermined periods of perfusion, arteries were excised, rinsed and their radioactivety was determined and related to that of the perfusate; uptake was expressed as a tissue plasma concentration ratio C_T/C_P, cpm gm^{-1} wet tissue/cpm gm^{-1} plasma, assuming unit density. In some experiments whole wall activity was assessed while in others the wall was rapidly frozen and sectioned parallel to the intima [7, 9] in order to establish the concentration profile across the wall.

Fig. 8 shows the average profiles for four different perfusion times. The adventitial scatter is probably due to the geometric complexity of the interface. Fig. 9 shows that there is a gradual rise of the mean medial value, which becomes approximately steady at 0.0095 SEM 0.0013 (n = 11) at 30 min. The mean adventitial value, though showing large scatter initially, apparently becomes steady at about 0.04 after one hour.

A crucial test for the existence of net flux (assuming diffusional transport and an inert tracer) is whether when wall tracer concentration has become steady the wall is saturated. We sought the answer to this question in two experiments. In the first we incubated segments of the artery in vitro in plasma containing label. The average C_T/C_P profiles for 10 arteries incubated for periods ranging from 10 min - 3 hours are shown in Fig. 10. The mean time-dependent behaviour of the medial and adventitial C_T/C_P is shown in Fig. 11. In the adventitia an apparently steady value of 0.35 is reached within 10 min. In the media there is a relatively rapid rise to a value of 0.041 SEM 0.007 (n = 6) at 40 min and a slower rise thereafter to a value of 0.072 SEM 0.007 (n = 5) at 3 hours.

In the second experiment we perfused arteries in situ with label in the normal way while simultaneously applying label to their outer aspect at the same concentration as in the lumen. An artificial lymph, rather than plasma, was used as the outer irrigating fluid in order to minimise disturbance of local colloid osmotic pressure. The experiments were of 30 and 60 min duration and there was no significant difference between the medial or adventitial values at these times. The average medial C_T/C_P was 0.034 SEM 0.0032 (n = 9) and the average adventitial value was 0.22 SEM 0.034 (n = 9).

We attempt to interpret these various findings in terms of a simple model of arterial wall mass transport which assumes that albumin transport is a diffusional process. The resistance to the diffusion of macromolecules in the concentration boundary layer in blood in arteries is negligible compared to the resistance of the blood: wall interface [12] . The effective source concentration would therefore be the bulk concentration in the plasma. The effective sink concentration in the in situ perfusion experiments is presumed to have been a composite of that in the plasma, extracellular fluid and lymph in the region perfused by the animal's own circulation. We do not know the effective sink concentration but can conceive that it did not exceed the concentration of label in the animal's plasma, which was generally about 5 % of that in the perfusate.

We observed essentially steady in situ values for C_T/C_P which were much lower than the "saturation" values obtained in vitro. This finding implies steady net flux across the wall. It also implies that the inner part of the wall provides a higher resistance than the outer part; with a homogeneous wall zero sink concentration, the steady in situ value of C_T/C_P averaged across the wall, would be half the saturation value.

Our in vitro measurements show clearly that the wall is inhomogeneous. The average C_T/C_P for the media rose to 0.04 at 40 min and slowly thereafter. Evidence we mention below is consistent with the later rise being due to changes in the properties of the wall. In contrast the average C_T/C_P for the adventitia became steady in 10 min, taking a value of 0.35. Therefore, the distribution volume for albumin appears to be substantially greater in the adventitia than in the media. The time course observations are, because of the complexity of the system, less easy to interpret. They suggest, however, that in vitro, i.e. even in the absence of an adventitial circulation, the adventitia provides a lower resistance to diffusion than does the media.

We now consider results which bear on the assumption that albumin transport is diffusional. The frozen sections showed label to be distributed across the wall, a prerequisite for net flux. In the in situ perfused arteries, in which label was applied solely via the perfusate, the average steady medial C_T/C_P was 0.0095 SEM 0.0013 (n = 11). However, in the in situ experiments in which label at the same concentration was applied simultaneously to the lumen and outer aspect of the artery, the steady medial C_T/C_P was 0.032 SEM 0.0032 (n = 8). The increase is highly significant (p < 0.001) and medial C_T/C_P evidently depends markedly on external concentration approaching in these studies the in vitro saturation value.

The detailed mass transport of the artery wall is presumably highly complicated and influenced by the vasa vasorum and lymphatics. It seems, however, virtually certain, because of luminal hydrostatic pressure, that convection across the media, an avascular tissue in this vessel [1], will be from its inner surface towards its outer surface. The dependence of the medial concentration of albumin on the external concentration therefore implies that convection did not play a major role compared to diffusion in the transport of the material. It must however be appreciated (see also below) that a comparison is being made between the properties of in situ perfused arteries and arteries incubated in vitro, which had been permitted to shorten and were at zero transmural pressure.

It would be of interest to determine the distribution of resistance across the wall for albumin transport. This evidently depends on the diffusion gradient, but this is not simply related to C_T/C_P in an inhomogeneous wall. It can be calculated by scaling local steady state C_T/C_P by the local distribution volume for albumin. We have done this, using 90 min in vitro data and the average profiles for 11 arteries are shown (Fig. 12). The values in the media and adventitia are less than 0.22, indicating a steep gradient across the intima. There is an almost linear fall across the media from 0.22 near to the lumen to 0.08 near the media/adventitia boundary, while in the adventitia the value is essentially constant. These values are consistent with the media providing a considerably larger resistance than the adventitia and a not insignificant resistance compared to the intima.

In the light of comments above, concerning the validity of comparing results from stressed and unstressed vessels, these conclusions must, however, be regarded as tentative. Indeed, the results of [18] and of [19] suggest that mechanical stresses within the physiological range are capable of compressing "transport channels" within the artery wall. The former workers found that the rate of flow of water through the rabbit thoracic aorta, for a given transmural pressure (100 mmHg), was about ten-fold lower if the tissue was supported against a porous disc, when it might be expected to become compacted. Harrison and Massaro [19] found with pig aorta supported against a porous disc, that the rate of flow of water for a given transmural pressure (110 mmHg), fell by a factor of about four to a stable value during a six hour period (Fig. 13).

Results we have obtained recently suggest that physiological mechanical stress can alter the transport properties of the wall for albumin. We observed, in in vitro experiments, that the distribution volume for albumin in the media increased gradually with time and considered that this change might reflect a change in the properties of the wall. Consistent with that view, albumin uptake at 20 min was lower for freshly excised in vitro segments than for segments incubated for 180 min in unlabelled plasma prior to being incubated with label.

These changes could have been associated with relaxation of arterial smooth muscle, since excised artery segments gradually dilate. We have therefore tested the effect of some vasoactive materials on the uptake of albumin by unstressed segments incubated in vitro [20]. Noradrenaline, a constrictor of smooth muscle, was added to the incubating plasma in concentrations ranging from 10^{-11} M to 10^{-5} M. After 20 min (Fig. 13) there was no significant departure from the control value, however after 90 min and 180 min, concentrations exceeding 10^{-9} M depressed uptake. At 90 min average medial uptake was 0.072 SEM 0.006 (n = 9) in control experiments and 0.046 SEM 0.006 (n = 9) with 10^{-5} M noradrenaline (p < 0.01) (Fig. 14). At 180 min the control value was 0.081 SEM 0.004 (n = 8) (p < 0.01). In another series of experiments, also involving in vitro incuabtion for 90 and 180 min sodium nitrite, a relaxant of smooth muscle, was added to the plasma in concentrations ranging from 10^{-10} M to 10^{-2} M. At 90 min concentrations exceeding 10^{-6} M increased albumin uptake. The control C_T/C_P value at 90 min was 0.072 SEM 0.006 (n = 9). With 10^{-4} M sodium nitrite the value was 0.1 SEM 0.006 (n = 9). These results are consistent with changes of medial smooth muscle tone affecting the uptake of albumin by the media. It is therefore possible that medial smooth muscle tone and, indeed, medial mechanics, influences arterial wall mass transport. At the same time, the results reinforce the view that all studies of arterial wall mass transport should be made under closely comparable mechanical conditions.

It is appropriate in closing to consider the possible relevance of these observations to the development of arterisclerosis. A widely held theory for the condition assumes a net arterial mass transport and the accumulation of material within the wall as a causative factor [5, 8].

The lesions develop between the endothelium and the inner media [21] and the mechanisms implicated in their genesis include increased endothelial permeability, increased binding of transport materials to wall components and impaired outflow due to intimal thickening, abnormality of the internal elastic lamina, medial fibrosis and disturbance of the vasa vasorum and lymphatics [2, 8, 10, 21, 22].

The spatial resolution of our observations is inadequate to allow us to assess the resistance of the internal elastic lamina. Our investigations on the net transport of the wall are still at an early stage, however it seems reasonable to postulate that the media acts as a barrier to the outflow of material from the subendothelial region of the wall and that mechanical stress within it, including that due to smooth muscle tone, influences this property. Arterial smooth muscle tone is regulated by circulating vasoactive materials and the activity of the nerves supplying the media [23].

Our discussion thus far has been confined to inert materials which enter the arterial wall from the blood. However, many materials entering the wall react chemically within it. If the transport of a reaction product is dominantly diffusional and its concentration is higher in the wall than in the blood and lymph, then it will "drain" not only towards the vasa vasorum and lymphatics, but also towards the main lumen. For such a material, therefore, both the endothelium and the media may act as barriers to efflux from the subendothelial region.

Acknowledgements

I have pleasure in acknowledging the contributions to the work I have reported of many colleagues, both in the Physiological Flow Studies Unit and elsewhere.

References

[1] Wolinsky, H. and Glagov, S.: Nature of species differences in the medial distribution of aortic vasa vasorum in mammals. Circulation Res. 20 (1967), pp. 409-421.

[2] Jellinek, H., Veress, B., Balint, A. and Nagy, Z.: Lymph vessels of rat aorta and their changes in experimental atherosclerosis: an electron microscopic study. Exp. Mol. Pathol. 13 (1970), pp. 370-376.

[3] Hoggan, G. and Hoggan, E.: The lymphatics of the walls of the larger blood vessels and lymphatics. J. Anat. (London) 17 (1883), pp. 1-21.

[4] Caro, C.G., Pedley, T.J., Schroter, R.C. and Seed, W.A.: Mechanics of the Circulation, Oxford University Press (1978).

[5] Anitschkow, N.: Experimental arteriosclerosis in animals. In: Arteriosclerosis, ed. E.V. Cowdry, Macmillan, New York, (1933).

[6] Page, I.H.: Atherosclerosis: An introduction. Circulation 10 (1954), p.1.

[7] Adams, C.W.M.: Tissues changes and lipid entry in developing atheroma. In: Atherogenesis initiating factors, Ciba Foundation Symposium 12 (new series), eds. Ruth Porter and Julie Knight, (1973), pp. 5-37. Elsevier Associated Scientific Publishers, Amsterdam.

[8] Walton, K.W.: Pathogenetic mechanisms in atherosclerosis. Amer. J. Cardiol. 35, (1975), pp. 542-558.

[9] B r a t z l e r , R.L., C h i s h o l m , G.M., C o l t o n , C.K., S m i t h , K.A., Z i l v e r - s m i t , D.B. and L e e s , R.S.: The distribution of labelled albumin across the rabbit thoracic aorta in vivo. Circulation Res. 40 (1977), pp. 182-190.

[10] W i l e n s , S.L, and M c C l u s k e y , R.T.: The comparative filtration properties of excised arteries and veins. Am. J. Med. Sci. 224 (1952), pp. 540-547.

[11] B a l d w i n , A., C a r o , C.G., D a v i s , J. and W i n l o v e , C.P.: Structural and mechanical effects of vascular collapse. J. Physiol. 278 (1978), pp. 14-15.

[12] C a r o , C.G. and N e r e m , R.M.: Transport of ^{14}C-4 cholesterol between serum and wall in perfused dog common carotid artery. Circulation Res. 32 (1973), pp. 187-205.

[13] C a r o , C.G.: Transport of material between blood and wall in arteries. In Atherogenesis: Initiating factors. Ciba Foundation Symposium 12 (new series), eds. Ruth Porter and Julie Knight, (1973), pp. 128-164. Elsevier Associated Scientific Publishers, Amsterdam.

[14] C a r o , C.G.: Mechanics and mass transport of the arterial wall. Les Colloques de l'Institut de la Sante et de la Recherche Medical Inserm 78 (1978), pp. 33-48.

[15] C a r o , C.G. P e d l e y , T.J., S c h r o t e r , R.C. and S e e d , W.A.: Mechanics of the circulation. Oxford University Press (1978).

[16] F r y , D.L.: Certain chronological considerations regarding the blood vascular interface with particular reference to coronary artery disease. Circulation 40 (Suppl.4) (1969), p.38.

[17] W i n l o v e , C.P., D a v i s , J., D e l l , R. and R u b i n , B.R.: The use of radio-iodinated human serum albumin in measurements of vascular permeability. Cardiovascular Res. 12 (1978), pp. 609-616.

[18] Y a m a r t i n o , E., B r a t z l e r , R., C o l t o n , C., S m i t h , K. and L e e s , R.: Hydraulic permeability of arterial tissue. Circulation 49 and 50 (Suppl. III), p. 273.

[19] H a r r i s o n , R.G. and M a s s a r o , T.A.: Water flux through porcine aortic tissue due to a hydrostatic pressure gradient. Atherosclerosis 24 (1976), pp. 363-367.

[20] C a r o , C.G. and L e v e r , M.J.: Effect of vasoactive agents on the albumin space of the arterial wall. Proceedings of Physiological Society, University College Meeting, March 1980.

[21] G o f m a n , J.W. and Y o u n g , W.: The filtration concept of atherosclerosis and serum lipids in the diagnosis of atherosclerosis. In: Atherosclerosis and its origin, ed. Sandler, M. and Bourne, G.H. New York Academic Press, (1963), pp. 197-229.

[22] A d a m s , C.W.M.: Vascular histochemistry in relation to the chemical and structural pathology of cardiovascular disease. London, Lloyd-Luke, 74 (1967).

[23] K e a t i n g e , W.R.: Blood vessels. In Brit. Med. Bull. 35 (1979), p. 249, ed. E. Bulbring and T.B. Bolton.

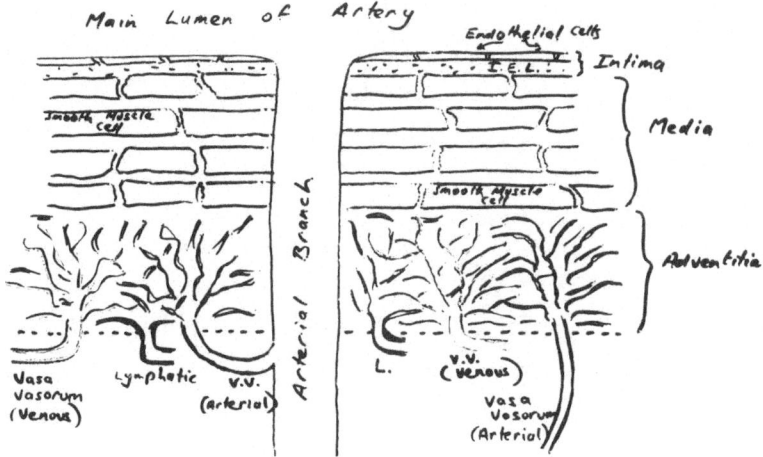

Fig. 1: Diagram of longitudinal section through arterial wall.

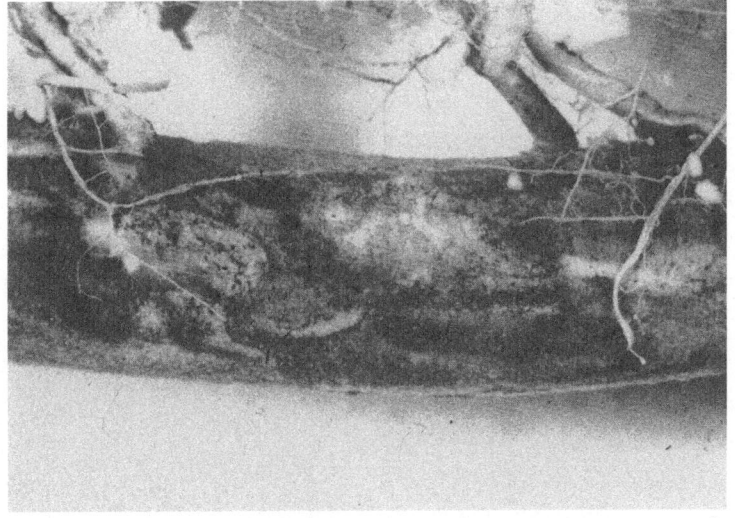

Fig. 2: Cast of rabbit aorta showing vasa vasorum.

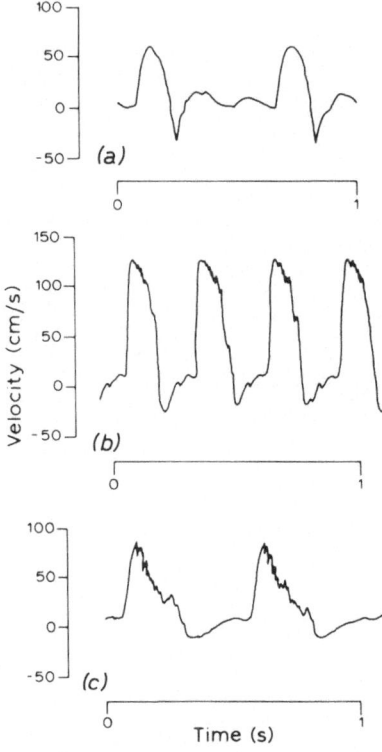

Fig. 3: Blood velocity in the dog aorta. High frequency disturbances, possibly turbulence, are seen on some traces.

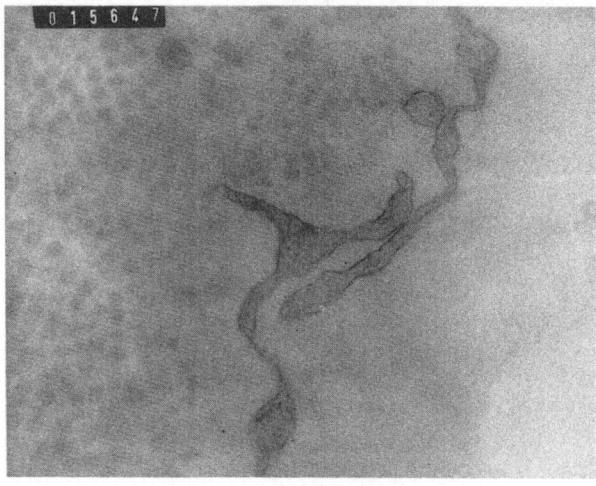

Fig. 4: Electron micrograph of a lymphatic from the adventitia of the rabbit aorta.

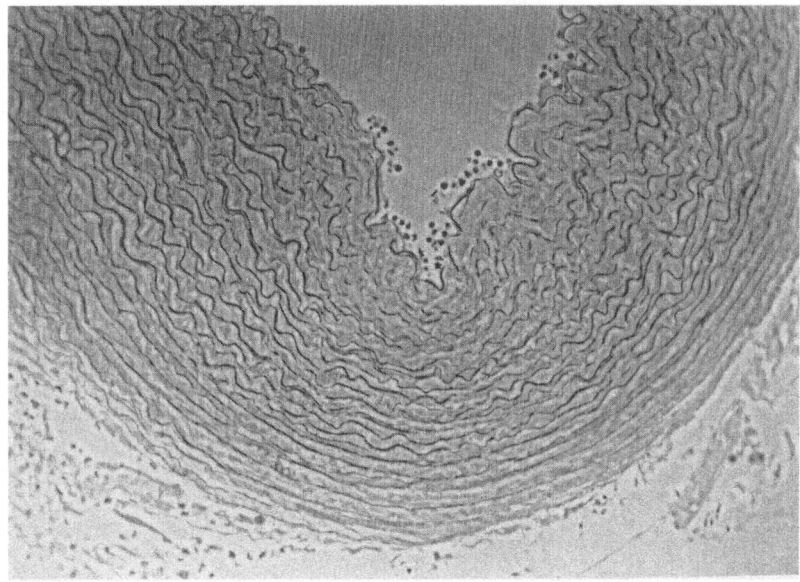

Fig. 5: Light micrograph showing a high curvature region in the rabbit thoracic aorta in situ at zero transmural pressure.

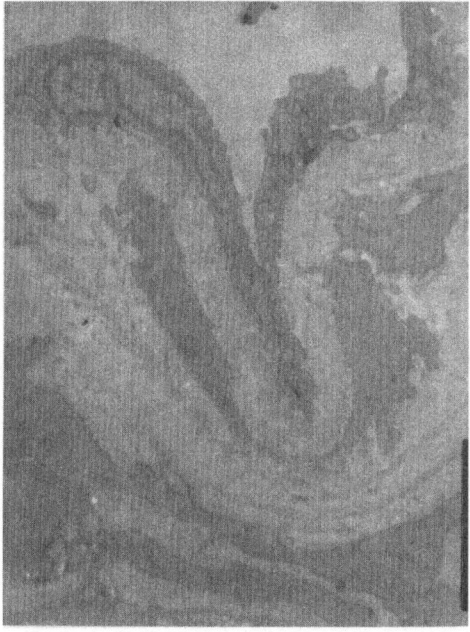

Fig. 6: Electron micrograph showing the intima of part of a high curvature region in the rabbit thoracic aorta in situ at zero transmural pressure.

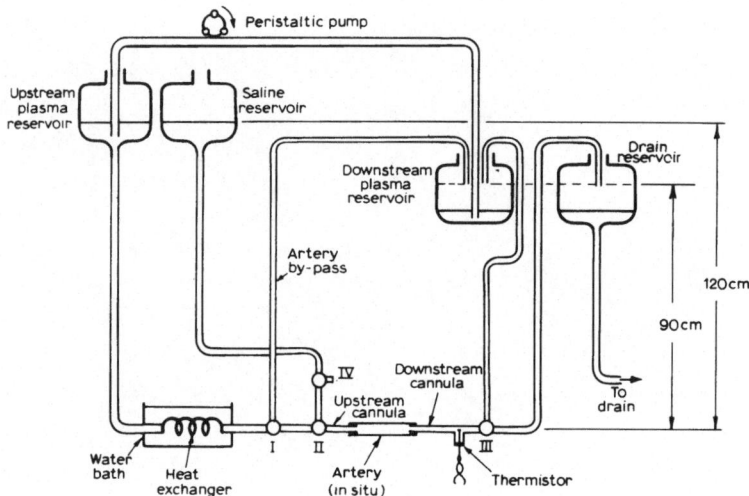

Fig. 7: Perfusion apparatus.

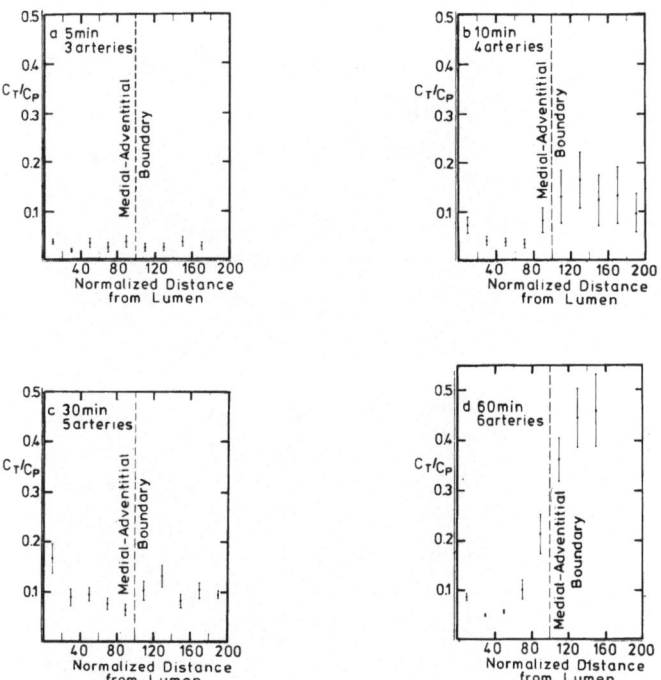

Fig. 8: C_T/C_P profiles, obtained in in situ perfusion studies of varying duration.

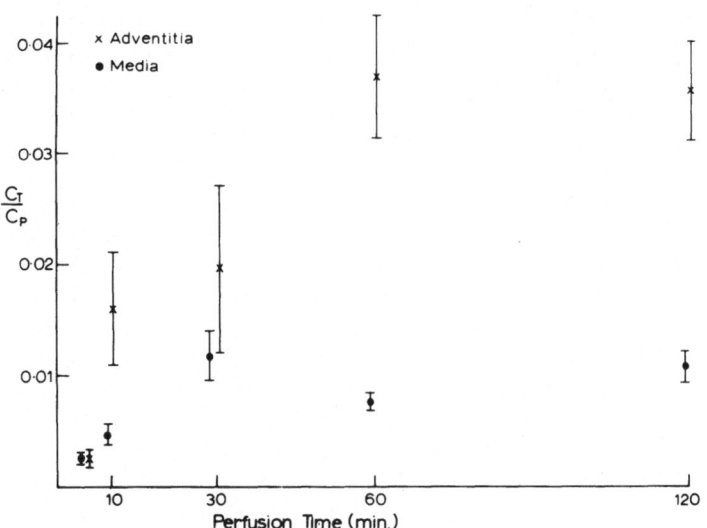

Fig. 9: Mean medial and adventitial C_T/C_P values for in situ perfusion studies of varying duration.

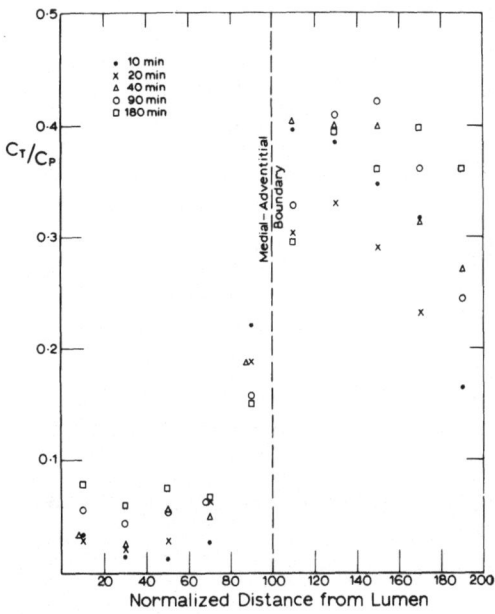

Fig. 10: C_T/C_P profiles obtained in in vitro studies of varying duration.

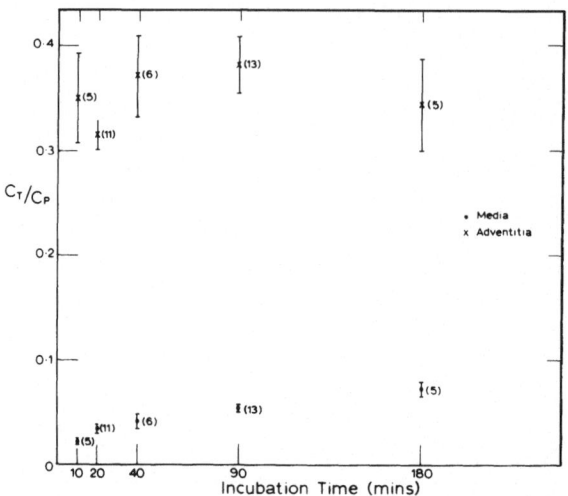

Fig. 11: Mean medial and adventitial C_T/C_P values for in vitro studies of varying duration.

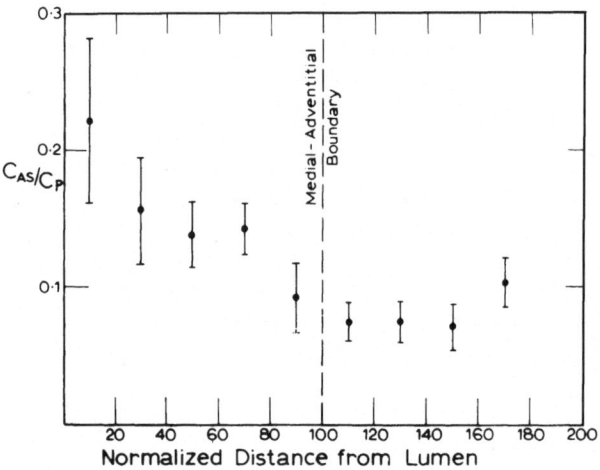

Fig. 12: The diffusion gradient obtained by scaling steady state in situ perfusion values by 90 min in vitro incubation values.

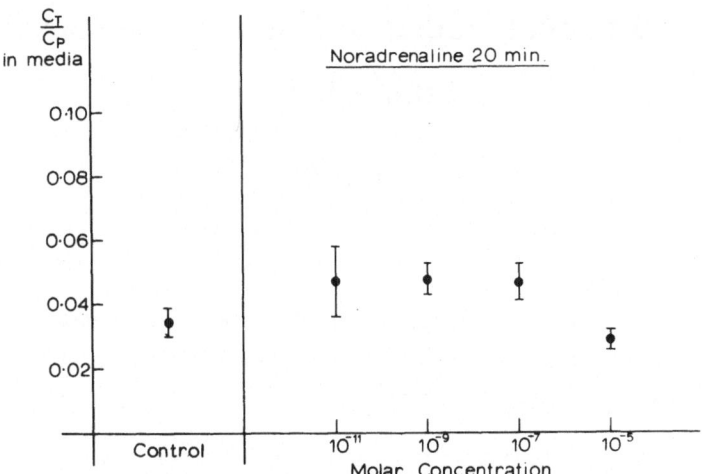

Fig. 13: The influence of varying concentrations of noradrenaline on mean medial C_T/C_P (20 min incubation in vitro).

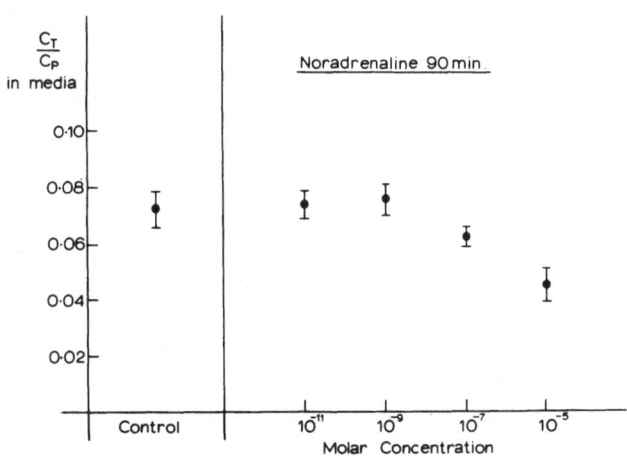

Fig. 14: The influence of varying concentrations of noradrenaline on mean medial C_T/C_P (90 min incubation in vitro).

Numerical Studies of Nonhomogeneous
Fluid Flows

R. Peyret
Nice

1. Introduction

This lecture presents some numerical studies of nonhomogeneous fluid flows which have been done during these last years in the Computational Fluid Mechanics Group of the Laboratory of Theoretical Mechanics in the University Paris VI.

These studies are based upon the numerical solution of the unsteady Navier-Stokes equations. The nonhomogeneity is described variously according to the problem considered. First, two problems of stratified flows will be studied. The first problem concerns the flow induced by the collapse of a homogeneous region inside a stably stratified fluid at rest. The fluid is assumed to be non diffusive so that the nonhomogeneity of the fluid is described by an advection equation. The second problem deals with the evolution of a jet of homogeneous fluid into a stably stratified fluid at rest; in this case, the stratification is assumed to be due to thermal effects so that the nonhomogeneity will be described by an advection - diffusion equation.

In the third problem presented here, the nonhomogeneity is due to the presence of solid particles in suspension in a fluid. This problem is related to the motion of microorganisms in liquid layers. The representation of particles is a discrete one: the motion of each particle is considered separately and it is based upon a set of rules.

The common feature of all these problems is their high degree of unsteadiness. Moreover, the motion is also highly variable in space. So, the numerical method must be sufficiently accurate in time as well as in space, generally of second order accuracy, and very stable: implicit schemes are systematically used. The first part of the present paper will be devoted to a discussion of the numerical method chosen to solve the unsteady Navier-Stokes equations in the formulation of primitive variables with a view to pointing out some of the essential features associated with this formulation.

2. The Numerical Solution of the Navier-Stokes Equations

The Navier-Stokes equations in dimensionless form are written as:

$$\frac{\partial \vec{U}}{\partial t} + A(\vec{U}) + \nabla p - \vec{F} = 0 \qquad (2.1a)$$

$$\nabla \cdot \vec{U} = 0 \qquad (2.1b)$$

where $A(U)$ is expressed by one of the following equations according to the choice of the form of the convective terms:

$$A(\vec{U}) = \nabla \cdot (\vec{U}\vec{U}) - Re^{-1}\nabla^2\vec{U}, \tag{2.2a}$$

$$A(\vec{U}) = (\vec{U}.\nabla)\vec{U} - Re^{-1}\nabla^2\vec{U}. \tag{2.2b}$$

In these equations, $\vec{U} = (u, v)$ is the velocity vector, p the pressure, t the time and $Re = U_*$ $L \rho_*/\mu$ is the Reynolds number where U_*, L, ρ_* are characteristic scales and μ is the viscosity of the fluid. The equations are of Boussinesq type and the forcing term $\vec{F}$ is related to the nonhomogeneity of the fluid. The specification of $\vec{F}$ as well as the supplementary equations describing the nonhomogeneity will be considered later for each one of the applications.

2.1 Discretization with Respect to Time

Two types of discretization with respect to time have been used to solve (2.1). The first one is a two-level scheme [1]:

$$\frac{1}{k}(\vec{U}^{n+1} - \vec{U}^n) + \Theta[A(\vec{U}^{n+1}) + \vec{F}^{n+1}] + (1-\Theta)[A(\vec{U}^n) + \vec{F}^n] + \nabla p^{n+\tau} = 0, \tag{2.3a}$$

$$\nabla \cdot \vec{U}^{n+1} = 0 \tag{2.3b}$$

where k is the time step $(t = nk; n = 0, 1,...)$, Θ is a constant $0 \le \Theta \le 1$. In the applications, the value $\Theta = 1/2$ (Crank-Nicholson scheme) has been chosen. The present gradient is evaluated at a certain time $(n + \tau)k$ with $0 < \tau \le 1$, depending on the choice of Θ.

The second scheme is a three-level scheme [2]:

$$\frac{1}{2k}(3\vec{U}^{n+1} - 4\vec{U}^n + \vec{U}^{n-1}) + A(\vec{U}^{n+1}) + \nabla p^{n+1} - \vec{F}^{n+\gamma} = 0, \tag{2.4a}$$

$$\nabla \cdot \vec{U}^{n+1} = 0. \tag{2.4b}$$

This scheme is second order accurate (if $\gamma = 1$) while the scheme (2.3) is of second order accuracy only for $\Theta = \tau = 1/2$. Due to their implicit nature, the schemes (2.3) with $\Theta \ge 1/2$ and (2.4) with $\gamma = 1$, are unconditionally stable (linear stability of von Neumann). If $\Theta < 1/2$, necessary conditions of stability for the scheme (2.3) associated with a second order accurate, centered approximation to the spatial derivatives are:

$$\frac{k}{Re\ h^2} \le \frac{1}{4(1-2\Theta)} \quad , \quad (|u| + |v|)^2 k \cdot Re \le \frac{4}{1-2\Theta} \tag{2.5}$$

Since three levels of time are involved in (2.4), the scheme does not determine the solution at the first time step, given suitable initial conditions. The first time step is treated using the Crank-Nicholson scheme. For subsequent time cycles, the use of the three-level scheme rather than the Crank-Nicholson can be justified, in certain cases, by the fact that harmonics of short wavelength are better damped.

2.2 Discretization with Respect to Space.

The staggered MAC mesh (Fig. 1) has been used in all the computations reported here, and the derivatives have been approximated with entered differences of second order accuracy. In the case (2.2a), typical examples of discretization of the nonlinear terms are

$$(\frac{\partial}{\partial x} u^2)_{\ell+1/2,m} = \frac{1}{h} \left[(u^2)_{\ell+1,m} - (u^2)_{\ell,m} \right],$$

$$(u^2)_{\ell,m} = \frac{1}{4} (u_{\ell+1/2,m} + u_{\ell-1/2,m})^2 \ ;$$

and

$$(\frac{\partial}{\partial y} uv)_{\ell+1/2,m} = \frac{1}{h} \left[(uv)_{\ell+1/2,m+1/2} - (uv)_{\ell+1/2,\ m-1/2} \right],$$

$$(uv)_{\ell+1/2,m+1/2} = \frac{1}{4} (u_{\ell+1/2,m+1} + u_{\ell+1/2,m}) (v_{\ell,m+3/2} + v_{\ell,m+1/2}).$$

A typical example in the case (2.2b) is

$$(v \ \frac{\partial u}{\partial y})_{\ell+1/2,m} = \frac{1}{8h} (v_{\ell+1,m+1/2} + v_{\ell,m+1/2} + v_{\ell,m-1/2} + v_{\ell+1,m-1/2})$$

$$(u_{\ell+3/2,m} - u_{\ell-1/2,m}).$$

In both cases, the second order derivatives are approximated with standard centered differences; moreover, the continuity equation (2.1b) is approximated at the center point of the cell, say:

$$\frac{1}{h} (u_{\ell+1/2,m} - u_{\ell-1/2,m}) + \frac{1}{h} (v_{\ell,m+1/2} - v_{\ell,m-1/2}) = 0. \tag{2.6}$$

2.3 Solution of the Finite-Difference Equations.

After discretization with respect to space as described above, the equations (2.3) or (2.4) lead, at each time step, to a nonlinear algebraic system for the unknowns which are the values of velocity and pressure at mesh points, say u_h, v_h and p_h:

$$u_h \equiv u^{n+1}_{\ell+1/2,m} \ , \quad v_h \equiv v^{n+1}_{\ell,m+1/2} \ , \quad p_h \equiv p^{n+\tau}_{\ell,m}$$

and, possibly, the value $\vec{F}_h = (0, F_h)$. The finite-difference equations are written in the symbolic form:

$$\mathcal{M}_u (u_h, v_h, p_h) = 0, \tag{2.7.a}$$

$$\mathcal{M}_v (u_h, v_h, p_h, F_h) = 0 \ , \tag{2.7.b}$$

$$\mathcal{D} (u_v, v_h) = 0, \tag{2.7c}$$

where (2.7a,b) are the momentum equations and (2.7c) is the continuity equation. These equations are solved with the iterative procedure (index: ν):

$$u^{\nu+1}_h - u^{\nu}_h + \kappa \mathcal{M}_u (u^{\nu}_h, v^{\nu}_h, p^{\nu}_h) = 0,$$

$$v_h^{\nu+1} - v_h^{\nu} + \kappa \mathcal{M}_v (u_h^{\nu}, v_h^{\nu}, p_h^{\nu}, F_h^{\nu}) = 0 ,$$

$$p_h^{\nu+1} - p_h^{\nu} + \lambda \, \mathcal{D} (u_h^{\nu+1}, v_h^{\nu+1}) = 0. \qquad (2.8)$$

The calculation of $F_h^{\nu+1}$ will be considered later. The parameters κ and λ are constants which are chosen to insure the convergence of the procedure. Necessary conditions for convergence can be obtained by carrying a von Neumann stability analysis of (2.8) considered as an approximation to a "time-dependent system" (ν being identified with an index of time). The system is simplified by neglecting convective and forcing terms. The condition that the spectral radius of the amplification matrix is not larger than one yields the criterion

$$\kappa > 0, \ \lambda > 0, \ \frac{\kappa}{h^2}(\frac{\alpha}{Re} + \beta \, \frac{h^2}{k} + 2\lambda) \leq 1,$$

where $\alpha = 4\Theta$, $\beta = 1/2$, for (2.3) and $\alpha = 4$, $\beta = 3/4$ for (2.4).

In order to improve the convergence, the iterative procedure described above can be modified by using in each of (2.8) the values of the unknowns at iteration $\nu+1$ as soon as they are computed (the Gauss-Seidel technique). It must be noticed that procedure (2.8) is the so-called "artificial compressibility method" [3], [4.1], [5] which is used here at each time step.

2.4 Remarks on the iterative Procedures

2.4.1 The Poisson Equation for the Pressure

Let us consider the Navier-Stokes equations (with $\vec{F} = 0$) discretized by the two-level scheme (2.3), say:

$$\nabla p^{n+\tau} + S_1(\vec{U}^{n+1}) + S_0(\vec{U}^n) = 0, \qquad (2.9a)$$

$$\nabla \cdot \vec{U}^{n+1} = 0, \qquad (2.9b)$$

where

$$S_1(\vec{U}^{n+1}) = \frac{1}{k} \vec{U}^{n+1} + \Theta \, A(\vec{U}^{n+1}),$$

$$S_0(\vec{U}^n) = -\frac{1}{k} \vec{U}^n + (1-\Theta) A(\vec{U}^n),$$

and the iterative procedure (2.8) is written as:

$$\vec{U}^{n+1,\nu+1} - \vec{U}^{n+1,\nu} + \kappa \, [\nabla p^{n+\tau,\nu} + S_1(\vec{U}^{n+1,\nu}) + S_0(\vec{U}^n)] = 0, \qquad (2.10a)$$

$$p^{n+\tau,\nu+1} - p^{n+\tau,\nu} + \lambda \, \nabla \cdot \vec{U}^{n+1,\nu+1} = 0 \qquad (2.10b)$$

Now, after the elimination of $\vec{U}^{n+1,\nu+1}$ using (2.10.a), the equation (2.10.b) becomes:

$$p^{n+\tau,\nu+1} - p^{n+\tau,\nu} - \kappa\lambda\nabla^2 p^{n+\tau,\nu} = -\lambda\nabla\cdot[\vec{U}^{n+1,\nu} - \kappa S_1(\vec{U}^{n+1,\nu}) - \kappa S_0(\vec{U}^n)].$$
$$(2.11)$$

This last equation is nothing else than a special iterative procedure to solve a Poisson equation for the pressure which is given by (2.11) at convergence $\nu \to \infty$:

$$\nabla^2 p^{n+\tau} = -\nabla\cdot[S_1(\vec{U}^{n+1}) + S_0(\vec{U}^n)]. \qquad (2.12)$$

In particular, in the explicit case $\Theta = 0$, we have $\nabla\cdot S_1(\vec{U}^{n+1}) = -k^{-1}\nabla\cdot\vec{U}^{n+1} = 0$;

therefore we obtain exactly the Poisson equation which is usually considered in the "MAC method" [6, 7] as well as in the "method of Projection" [8, 4.II, 5]. This shows the close relationship between these methods and the present one. We must notice that the explicit scheme $\Theta = 0$ associated with the artificial compressibility technique has been introduced in [8] and used, in particular, in [9, 10]. Note that a Poisson equation analogue to (2.12) is obtained with the Three-level scheme (2.4).

2.4.2 Spatial Mesh

In all the computations reported here we have used the staggered MAC mesh system. It seems interesting to discuss a little the effect of the mesh on the numerical solution. The first mesh (Fig. 2a) was considered in [3, 8]. Here, the velocity and the pressure are defined at the nodes of the mesh. The advantages of such a mesh is its simplicity and also the fact that the velocity is defined, in particular, on the boundary Γ where this quantity is generally prescribed. On the other hand, one of its disadvantages is the fact that the pressure is also defined on the boundary. And as it is known, there is generally no boundary condition for pressure, so that it is necessary to devise a special technique to compute the pressure on this boundary.

Hence, we have used in [5], a mesh such that the velocity is defined at nodes but the pressure at the center point of the cell, as shown in the Fig. 2b. Therefore, the pressure is no longer defined on the boundary Γ and we can use the same formulas to compute the whole pressure field.

Finally, the mesh that we are now using is the MAC mesh proposed in [6, 7]. The mesh (Fig.2c) differs from the previous one by the location of discretization points for the velocity. A small disadvantage of this mesh is that only one of the velocity components is defined on each side of the boundary Γ; so it is necessary to employ non-centered differences near the boundaries. But this inconvenience is largely balanced by the advantage of the MAC mesh concerning the computation of the pressure.

As a matter of fact it is very instructive to compare the layouts of points which are involved in the discretization of the Poisson equation obtained as a result of the use of the artificial compressibility method as explained above.

In Figs. 3 the circles represent the points where the pressure is defined and the black dots represent the points which are involved in the approximation of the Poisson equation at a given point P.

In the first two cases (Fig. 3a, 3b), we can see that there exist two uncoupled networks of pressure points. This fact leads to the existence of two solutions for the pressure; these solutions differ from each other by an arbitrary function of time and there is apparently no reason for these two arbitrary functions to be identical. So, the existence of the two pressure fields could lead to oscillations in the computed pressure; but the pressure gradient itself is not oscillatory, the same is true for the computed veloctiy. However, we must notice that it could be possible to couple (weakly) the two fields by way of the special technique needed to define the pressure on Γ in the case of mesh (a) or by a modification of the approximation of $\nabla \cdot \vec{U}$ near Γ in the case of mesh (b).

Finally, it is seen in Fig. 3c, that such a phenomenon of uncoupling does not appear in the case of the MAC mesh; this feature makes this mesh very convenient.

2.4.3 Treatment of the Pressure Near the Boundaries

The last remark concerns the treatment of boundaries when using primitive variables. Let us assume the solution of the Navier-Stokes equations is computed in a domain with the velocity $\vec{U}$ given on the boundary Γ, say $\vec{U}_{\Gamma}^{n+1}|_{\Gamma} = \vec{U}_{\Gamma}^{n+1}$. There is no boundary condition for the pressure p. When using iterative procedure (2.8) and when the MAC staggered mesh is utilized, we can obtain the pressure in the whole field without taking special care of the boundaries: the values of the pressure near the boundaries are computed with the same current algorithm. Therefore, we do not explicitly prescribe boundary conditions for the pressure.

But we have previously noticed the very close analogy which exists between the Projection of MAC methods and the present one in its explicit version $\theta = 0$. However, as is known, in the methods which make explicitly use of a Poisson equation for the computation of the pressure, it is necessary to consider boundary conditions for this quantity. In the following, we make more precise the relationship between these various methods regarding the treatment of the boundaries.

First, let us consider the method of Projection, as introduced in [5]. It is a fractional step method. At the first step, we begin to compute explicitely a provisional value $\vec{U}^*$

$$\frac{1}{k}(\vec{U}^* - \vec{U}^n) + (\vec{U}^n \cdot \nabla)\vec{U}^n - \frac{1}{Re}\nabla^2\vec{U}^n = 0, \tag{2.13}$$

which is the momentum equation without pressure gradient. Then, at the second step, we correct $\vec{U}^*$ by considering the equations :

$$\frac{1}{k}(\vec{U}^{n+1} - \vec{U}^*) + \nabla \cdot p^{n+1} = 0, \tag{2.14a}$$

$$\nabla \cdot \vec{U}^{n+1} = 0. \tag{2.14b}$$

By taking the divergence of (2.14.a) and by making use of (2.14.b) which states that $\vec{U}^{n+1}$ must be a divergence free vector, we get the Poisson equation

$$\nabla^2 p^{n+1} = \frac{1}{k} \nabla \cdot \vec{U}^* . \tag{2.15a}$$

This last equation is nothing else than the Poisson equation (2.12) with $\Theta = 0$. More precisely $\vec{U}^*$ is nothing else than $-k \, S_0(\vec{U}^n)$ with $\Theta = 0$.

The boundary condition for p is obtained by projecting the vector equation (2.14.a) on the normal unit $\vec{v}$ to Γ. Thus, we obtain the Neumann condition

$$(\frac{\partial p}{\partial v})_\Gamma^{n+1} = - \frac{1}{k} (\vec{U}_\Gamma^{n+1} - \vec{U}_\Gamma^*) \cdot \vec{v} . \tag{2.15b}$$

where $\vec{U}_\Gamma^*$ is the (not yet defined) value of $\vec{U}^*$ on Γ.

To sum up, (2.13) gives $\vec{U}^*$, then the solution of the Neumann problem (2.15) gives p^{n+1} and, finally, equation (2.14.b) gives $\vec{U}^{n+1}$.

The essential feature of the method is that the numerical solution is independent of the value $\vec{U}_\Gamma^*$. This assertion [5] is based upon the two following points: (i) $\vec{U}^*$ at inner points is independent of $\vec{U}_\Gamma^*$ because it is calculated with the scheme (2.13) which is explicit, (ii) the value $\vec{U}_\Gamma^*$ appears in the Neumann problem (2.15) simultaneously in the right-hand side of the Poisson equation (2.15.a) and in the Neumann condition (2.15.b) and it cancels identically. In order to prove this assertion it is sufficient to analyze the discretization of (2.15) for the points located near the boundary Γ only, since $\vec{U}_\Gamma^*$ appears in the problem only for these points. Let us consider the point P (Fig. 4); the approximation of (2.15.a) at this point can be written as:

$$\frac{1}{h^2} \left[(p_{2,m}^{n+1} - p_{1,m}^{n+1}) - (p_{1,m}^{n+1} - p_{0,m}^{n+1}) + (p_{1,m+1}^{n+1} - p_{1,m}^{n+1}) \right.$$
$$\left. - (p_{1,m}^{n+1} - p_{1,m-1}^{n+1}) \right] = \frac{1}{kh} \left[(u_{3/2,m}^* - u_\Gamma^*) + (v_{1,m+1/2}^* - v_{1,m-1/2}^*) \right], \tag{2.16a}$$

and the Neumann condition (2.15.b) is approximated by

$$\frac{1}{h} (p_{1,m}^{n+1} - p_{0,m}^{n+1}) = -\frac{1}{k} (u_\Gamma^{n+1} - u_\Gamma^*) . \tag{2.16b}$$

Now, it is easy to see that, when the value of $p_{1,m}^{n+1} - p_{0,m}^{n+1}$ given by (2.16.b) is brought into (2.16.a), the unknown quantity u_Γ^* cancels from both sides of the equation (2.16.a). From this, one concludes that the solution is independent of this value u_Γ^*. In particular, we can choose u_Γ^* = u_Γ^{n+1} and we get a zero normal derivative for the pressure on Γ. This result is purely numerical and does not imply that the real normal pressure gradient is zero.

This method of projection was first applied in [5, 11] with the mesh of Fig. 3b. When it is associated with the MAC mesh, the method becomes identical with the MAC method [6, 7] as long as the boundary conditions are not concerned. But it is possible, as was recognized in [12], to devise an analogous treatment for these conditions. The MAC method, as described in [7], considers the scheme (2.3) in its explicit version $\Theta = 0$. The discretized approximation of (2.3.a) is written successively at points $(\ell+1/2, m)$ and $\ell - 1/2, m)$ for u^{n+1} and at points $(\ell, m + 1/2)$, $(\ell, m -1/2)$ for v^{n+1}; then, bringing these quantities into the approximation of (2.3.b) given in (2.6) and stating that the discretized divergence of $\vec{U}^{n+1}$ is zero, we get the Poisson equation satisfied by the pressure. At point P (Fig. 4), this equation is:

$$\frac{1}{h^2}[(p_{2,m}^{n+1} - p_{1,m}^{n+1}) - (p_{1,m}^{n+1} - p_{0,m}^{n+1}) + (p_{1,m+1}^{n+1} - p_{1,m}^{n+1}) - (p_{1,m}^{n+1} - p_{1,m-1}^{n+1})]$$

$$= \frac{1}{kh}[(u_{3/2,m}^{n} - u_{1/2,m}^{n}) + (v_{1,m+1/2}^{n} - v_{1,m-1/2}^{n})] \qquad (2.17a)$$

$$- \frac{1}{h}[(a_{3/2,m}^{n} - a_{1/2,m}^{n}) + (b_{1,m+1/2}^{n} - b_{1,m-1/2}^{n})]$$

where a and b are the two components of $A(\vec{U})$. At the same time, we have to consider the boundary condition for the pressure given by the projection of the momentum equation (2.2.a) on the normal $\vec{v}$ to the boundary Γ, say:

$$(\frac{\partial p}{\partial \kappa})_{1/2,m}^{n+1} = -(\frac{\partial u}{\partial t} + a)_{1/2,m}^{n+1}$$

which is discretized as

$$\frac{1}{h}(p_{1,m}^{n+1} - p_{0,m}^{n+1}) = -\frac{1}{k}(u_{1/2,m}^{n+1} - u_{1/2,m}^{n}) - a_{1/2,m}', \qquad (u_{1/2,m} = u_\Gamma).(2.17b)$$

At this point, it is easy to see that (2.17) are the same as (2.16): the only change lies in the manner of writing them, but this change involves a difference in the interpretation of the right-hand-sides: in (2.16), u_Γ^* is considered as an unknown quantity to be defined, while in (2.17) the analogue of u_Γ^*, which is $(u_{1/2,m}^{n} + k\, a_{1/2,m}^{n})$, needs a spatial approximation for $a_{1/2,m}^{n}$ which involves the undefined quantities $u_{-1/2,m}^{n}$, $v_{0,m+1/2}^{n}$. Generally, the definition of these quantities corresponds to a special noncentered discretization of $a_{1/2,m}^{n}$; moreover, a simplified form of (2.17.b) if often used.

However, considering equations (2.17) we can bring the expression of $p_{1,m}^{n+1} - p_{0,m}^{n+1}$, as given by (2.17.a) and we ascertain that $a_{1/2,m}^{n}$ disappears from both sides of the equations. Therefore, we can conclude that the quantities $u_{-1/2,m}^{n}$, $v_{0,m+1/2}^{n}$ can be arbitrarily chosen in the current centered discretization of $a_{1/2,m}^{n}$. On the other hand, $v_{0,m+1/2}^{n}$ appears also in $b_{1,m+1/2}^{n}$, but here the problem is that which was mentioned, in the above section, about the approximation of derivatives near the boundary when using the MAC mesh.

In conclusion, we can say that the numerical solution given by the MAC method is independent of the evaluation of the normal pressure gradient on Γ (by the way of the evaluation of $a_{1/2,m}^{n}$) provided it is compatible with the approximation of the R.H.S. of the Poisson equation. This compatibility is essential in order to have the exact analogy of the cancelling of u_Γ^* in (2.16). Final y, if $a_{1/2,m}^{n}$ is taken equal to $-(u_{1/2,m}^{n+1} - u_{1/2,m}^{n})/k$ which corresponds to a special choice of

$u^n_{-1/2,m}$, we again get a zero normal pressure gradient on Γ. Note that homogeneous conditions are used in [13] in a different context.

Finally, in the present method, explicit or not, based upon the artificial compressibility iterative procedure, the compatibility is automatically satisfied.

3. Applications

3.1 Homogeneous Region Collapse in a Stratified Fluid [14]

This application concerns the flow induced by the collapse of a region Ω_o of homogeneous fluid into a domain Ω_1 filled with a stratified fluid at rest (Fig.5). The stratification is assumed to be obtained by salinity , so that the nonhomogeneous nature of the fluid is characterized by the concentration * $\tilde{c}$ and the density * $\tilde{\rho}$ is connected to the concentration through the sate law

$$\tilde{\rho} = \tilde{\rho}_o [1 + \alpha \ (\tilde{c} - \tilde{c}_o)] .$$

Initially in Ω_1 the fluid is stably stratified, linearly with respect to the ordinate * $\tilde{y}$, say $\tilde{c}_s(\tilde{y})$; so that

$$\tilde{c}_s \ (H) \ = \tilde{c}_2, \qquad \tilde{c}_s(-H) \ = \tilde{c}_1, \qquad \tilde{c}_1 > \tilde{c}_2.$$

The pressure* $\tilde{p}_s(\tilde{y})$ corresponding to $\tilde{c}_s(\tilde{y})$ is then hydrostatic. In Ω_o, the concentration is uniform and equal to $\tilde{c}_o = (\tilde{c}_1 + \tilde{c}_2)/2$; in other words the density $\tilde{\rho}_o$ in Ω_o is equal to the mean density in Ω_1. By using the following the characteristic scales: Length $= 2\ell$, Velocity $= \nu/(2\ell)$, Time $= (2\ell)^2/\nu$, pressure $= \tilde{\rho}_o \nu^2/(2\ell)^2$, concentration $= H (\tilde{c}_1 - \tilde{c}_2)/\ell$; where ν is the kinematic viscosity , the Reynolds number in the Navier-Stokes equations (2.1) becomes equal to 1. In these equations, p is the pertubation of pressure with respect to the hydrostatic pressure $p_s(y)$ and the forcing term $\vec{F} = (0, - Gr \sigma)$ where σ is the pertubation of concentration with respect to $c_s(y)$ and

$$Gr = \alpha g \ (2\ell)^4 \ \frac{\tilde{c}_1 - \tilde{c}_2}{2H\nu^2}$$

is the Grashof number. The fluid is assumed to be nondiffusive so that σ satisfies the advection equation (in dimensionless form) :

$$\frac{\partial \sigma}{\partial t} + u \frac{\partial \sigma}{\partial \kappa} + v \frac{\partial \sigma}{\partial y} - v = 0. \tag{3.1}$$

The initial conditions at t = 0 are:

$$u = v = 0 \qquad \text{and} \quad \sigma = \begin{cases} y \text{ in } \Omega_o \ , \\ 0 \text{ in } \Omega_1 \ , \end{cases}$$

and the boundary conditions on AA'C'C are

$$u = v = 0.$$

----

* The quantity is with dimensions.

On this boundary, (3.1) yields $\partial\sigma/\partial t = 0$; then we have $\sigma = 0$ on AA'C'C at any time.

The Navier-Stokes equations (2.1) with $A(\vec{U})$ given by (2.2.a) are discretized with the three-level scheme (2.4) and $\gamma = 1$. The space derivatives are approximated with second-o der accuracy, centered differences as explained in section 2.2. Equation (3.1) is discretized with the same mplicit three-level scheme but the spatial derivatives are approximated with up-wind, non-centered 3-point differences of second order accuracy.

The corresponding finite-difference equation is written in the symbolic form:

$$\mathcal{M}_\sigma (\sigma_h , \, u_h , \, v_h) = 0$$

where

$$\sigma_h \equiv \sigma_{\ell,m+1/2}^{n+1} \quad , \quad u_h \equiv u_{\ell+1/2,m}^{n+1} \quad , \quad v_h \equiv v_{\ell,m+1/2}^{n+1} \ .$$

The resulting system is iteratively solved by being inserted into the procedure (2.8) , say

$$\sigma_h^{\nu+1} - \sigma_h^\nu + \mu \, \mathcal{M}_\sigma (\sigma_h^\nu , \, u_h^{\nu+1} , \, v_h^{\nu+1}) = 0$$

where the parameter μ must satisfy the condition of convergence

$$0 < \frac{\mu}{24k} \left[18 + 12\,C + (5 + 3\sqrt{2})\,C^2 \right] \leq 1 \quad \text{where } C = k(|u| + |v|)/h.$$

The illustrative results presented in Figs. 6 to 8 correspond to Gr = 24.5, $L/(2\ell) = 10$, $H/(2\ell) =$ 2.5, h = 0.1, k = 0.05, $\varkappa = 0.002$, $\lambda = 0.8$, $\mu = 0.025$. Fig. 6 gives a sketch of the initial state. Due to the Boussinesq approximation the flow is symmetrical with respect to the horizontal axis OO'. Consequently, only results in the lower half of the domain are shown. Fig. 7 shows the profiles of velocity: the main feature is the tendency for the homogeneous fluid to move upward and laterally. This is better seen in Fig. 8 which shows instantaneous streamlines at various times. As time increases, the first two vortices spread laterally while their centers remain practically fixed and their magnitude decreases continously. Moreover, at time 0.7 we can notice the birth of new eddies near the bottom of the domain. These eddies go upward and, at a later time – which is not shown here – a third pair of vortices will be created at the bottom, while the others will be largely damped.

3.2 Homogeneous Jet into a Stratified Fluid [1, 15]

This application concerns the intrusion of a homogeneous fluid into a stratified fluid at rest (Fig. 9). Here, the stratification as assumed due to the thermal effects: the density $\tilde{\rho}$ is connected to the temperature $\tilde{T}$ through the state law:

$$\tilde{\rho} = \tilde{\rho}_0 \left[1 - \beta\,(\tilde{T} - \tilde{T}_0) \right] \ .$$

Initially, the fluid at rest is stably stratified linearly with respect to $\tilde{y}$, say $\tilde{T}_s(\tilde{y})$ so that

$$\tilde{T}_s(H) = \tilde{T}_2 , \qquad \tilde{T}_s(-H) = \tilde{T}_1 , \qquad \tilde{T}_1 < \tilde{T}_2 ,$$

and the corresponding pressure $\tilde{p}_s(\tilde{y})$ is hydrostatic. The fluid emitted through the aperture BB' has a uniform temperature $\tilde{T}_o = (\tilde{T}_1 + \tilde{T}_2)/2$, equal to the mean temperature of the fluid at rest. The entry profile is parabolic, U_o being its maximum value. The dimensionless variables are defined by the characteristic scales: length = 2d, velocity = U_o, pressure = $\tilde{\rho}_o U_o^2$, time = $2d/U_o$, and temperature = $H(\tilde{T}_2 - \tilde{T}_1)/d$. In the Navier-Stokes equations (2.1), p is again the pressure pertubation with respect to the hydrostatic pressure and the forcing term is $\vec{F} = (0, Ri, \Theta)$ where Θ is the temperature pertubation and

$$Ri = \beta g \left(\frac{2d}{U_o}\right)^2 \frac{\tilde{T}_2 - \tilde{T}_1}{2H}$$

is the Richardson number. The temperature pertubation Θ is the solution of the equation

$$\frac{\partial \Theta}{\partial t} + u \frac{\partial \Theta}{\partial \kappa} + v \frac{\partial \Theta}{\partial y} + v = \frac{1}{Re\ Pr} \nabla^2 \Theta \tag{3.2}$$

where Pr is the Prandtl number.

At the initial time t = 0, the fluid is at rest:

$$u = v = 0 \quad \text{and } \Theta = 0$$

Concerning the boundary conditions, two cases have been considered:

In case I, [1], the fluid is emitted into as basin of finite length equal to $L/(2d)$ in dimensionless variables, and it is forced to leave the domain through the horizontal sides $y = \pm H/(2d)$ where its velocity and temperature are prescribed. So the boundary conditions are written:

On BB'   : $u = \psi(t)(1 - 16 y^2)$, $v = 0$, $\Theta = -y$,

on AB, A'B', CC'   : $u = v = 0$, $\Theta = 0$,

on AC   : $u = 0$, $v = -\psi(t)U(\kappa)$, $\Theta = 0$,

on A'C'   : $u = 0$, $v = \psi(t)U(\kappa)$, $\Theta = 0$.

The entry velocity on BB' is progressively established by the way of the function $\psi(t)$ (Fig. 10). The expression for ψ as well as that for U are given in [1] (U = const. except near the end points).

In case II, [15], the fluid is emitted into a channel of infinite length and it is allowed to leave the domain of computation freely through the downstream vertical side CC'. The difficult problem is to devise artificial boundary conditions on CC' which would permit the fluid to leave the domain freely without pertubing the upstream region nor creating numerical instabilities. The problem, which is already delicate because of the velocity-pressure formulation, is enhanced here by the stratification and the propagation of internal waves which is associated with it. In such flows, the disturbances are felt at a very large distance from their source.

After several tests, the following conditions on CC' have been found to be the most satisfactory:

(a) $\qquad \dfrac{\partial v}{\partial \kappa} = 0$,

(b) $\qquad \dfrac{\partial u}{\partial \kappa} + \dfrac{\partial v}{\partial y} = 0$,

(c) $\qquad \dfrac{\partial^2 \theta}{\partial \kappa^2} = 0$,

(d) $\qquad \dfrac{\partial p}{\partial y} = -\dfrac{\partial v}{\partial t} - v\dfrac{\partial v}{\partial y} + \dfrac{1}{Re}\dfrac{\partial^2 v}{\partial y^2} + Ri\,\theta$.

The velocity v is computed from (a), then the continuity equation (b) gives u and the condition (c) gives θ ; finally, the pressure p is computed from the approximate momentum equation (d). On the other boundaries, the conditions are:

on BB' $\qquad$: $u = \psi(t)(1-16y^2)$, $\qquad v = 0$, $\theta = -y$,

on AB, A'B' $\quad$: $u = v = 0$, $\dfrac{\partial\theta}{\partial\kappa} = 0$,

on AC, A'C' $\quad$: $u = v = 0$, $\theta = 0$.

The Navier-Stokes equations (2.1) with (2.2b) and the equation (3.2) are approximated with the Crank-Nicholson scheme (equation (2.3) with $\theta = \tau = 1/2$).

Second-order accurate centered differences are used in the Navier-Stokes equations but centered, fourth-order accurate differences are used to approximate the convective term in (3.2) in order to minimize the truncation error associated with this term compared with the diffusion term. The discretized version of (3.2) is

$$\mathcal{M}_\theta\,(\theta_h,\,u_h,\,v_h) = 0$$

where $\theta_h = \theta_{\ell+1/2,m+1/2}^{n+1}$, and it is inserted into the general procedure (2.8) by

$$\theta_h^{\nu+1} - \theta_h^{\nu} + \chi\,\mathcal{M}_\theta(\theta_h^{\nu}\,,\,u_h^{\nu+1},\,v_h^{\nu+1}) = 0,$$

with the condition of convergences:

$$0 < \frac{2\chi}{h^2}\left(\frac{1}{Re\,Pr} + \frac{h^2}{4k}\right) \le 1.$$

The results presented here correspond to : $H/(2d) = 3.5625$, $L/(2d) = 5$ (case I) or 6 (case II), $Pr = 10$ and

Re	$\kappa = \chi$	λ
10	0.195×10^{-2}	0.875
100	0.293×10^{-2}	0.593
250	0.320×10^{-2}	0.544

Fig. 11 presents horizontal velocity profils in some sections (case I). Here again, the flow is symmetrical with respect to the y-axis and we show only the profiles in the lower half of the domain. This assumption of symmetry would be questioned and we might think that, for some Reynolds numbers considered here (Re = 100 or 250), the jet would not be stable and would not remain symmetrical. However, experimental [16] and numerical [17] studies carried out with homogeneous fluid have shown that a jet in an infinite environment remains symmetrical for higher Reynolds numbers. Moreover, several computations [15] have been done in the case II without imposing symmetry and, even more, by prescribing non symmetrical oscillatory profiles for the entrance velocity. The results showed that for Re = 250, even without stratification, the flow has a tendency ultimately to recover the symmetry at some distance of the entry. Finally, it must be noticed that stratification has a stabilizing effect, so it seems quite reasonable to assume here the symmetry of the flow.

Two main features, characteristic of stratified flows, are visible in Fig. 11. First, the wavy shape of the profiles which corresponds to the presence of eddies and reverse flow; secondly, the larger the Richardson number the more the profiles exhibit high maxima at a large distance from the entry: the disturbances created by the intrusion are felt at a large distance ahead of it ("upstream influence"). Figs. 12 show the evolution of the horizontal velocity u along axis y = 0. We can observe the existence of maxima much larger than the entry velocity when stratification is present and the Reynolds number is sufficiently large (Fig. 12b - d). The location of these overshoots corresponds to local constrictions of the jet due to the presence of eddies (see Figs. 13, 14). Both cases I and II are compared in Fig. 12d : up to some distance, depending on the stratification the two graphs are very close to each other. The following Figs. 13, 14 present some streamlines patterns. The differences between flows with or without stratification is very pronounced. In the latter case (Fig. 13c, Fig. 14b) there exists only one eddy and it remains attached to the wall near the entrance, while vortices are continously created and convected with the fluid when stratification is present (Figs. 13a, b, 14a). Besides, when Ri = 0, the lateral expansion of the jet in the case II (Fig. 14b) is prevented only by the presence of the horizontal walls; on the other hand when stratification is present (Fig. 14a) the buoyant force constricts the motion of the fluid which is then channeled in the neighborhood of the axis, and we can observe the wavy shape of the streamlines in the core region. Figs.15 illustrate the evolution of the motion in time: the birth of the first eddy near the entrance (Fig. 15a), then the convection of this eddy and the creation of new ones (Fig. 15b) while the fluid becomes constricted near the axis ; then (Figs. 15c, d), the phenomenon continues and the fluid is more constricted near the axis with streamlines nearly horizontal.

The streamline patterns do not make visible the motion of the homogeneous intrusion into the stratified fluid. The advance of the tongue of homogeneous fluid is shown in Fig. 16. The black region at the left is a picture of the homogeneous fluid obtained by an emission of markers at the entry. Moreover, in order to get an idea of the motion of the fluid ahead of the tongue, some markers have been set out in the whole domain. Again, the striking effect of stratification is evident. In the non stratified case (Fig. 16a), the particles are largely scattered by viscous effects, while in the stratified case (Fig. 16b) the homogeneous intrusion enters inside the channel while remaining concentrated near the axis and putting the fluid in motion well ahead of it. A precise analysis of the law of evolution in time of the tongue is carried out in [15].

3.3 Convection by Motile Particles [2]

This example deals with a problem of convection induced by motile solid particles in suspension in a fluid of constant density initially at rest. From a physical point of view, this problem called "bioconvection" is connected to the motion of microorganisms such as protozoa in a liquid layer. Protozoologists have observed that cultures of certain microorganisms exhibit striking patterns reminiscent of thermal convection. The mechanism creating the patterns is a result of two facts: (1) the mean density of the organisms exceeds that the ambient fluid and (2) the swimming response of the organism includes a "negative geotaxy", that is a tendency to swim upward (represented here by a velocity $U_o \vec{j}$) in response to the gravitational field. Taken together, these two facts imply that a potentially unstable density stratification would be established when a sufficiently large number of organisms accumulate near the free surface A'C' (Fig. 17), and a convective motion will result.

The characteristic scales are : length = H, velocity = U_o, time = H/U_o, pressure = $\tilde{\rho}_o U_o^2$. The forcing term $\vec{F}$ in the Navier-Stokes equation (2.1) is

$$\vec{F} = [0, \ - \ \frac{R^*}{Re} \ \sum_{i=1}^{\mathcal{N}} \ \delta \, (\vec{r} - \vec{r}_i)]$$

where $\vec{r} = (x, y)$, $\vec{r}_i = (x_i, y_i)$ is the location of the ith particle which simulates one organism δ = Dirac function, $\mathcal{N}$ = total number of particles

$$R^* \ = \ \frac{\tilde{\rho}_o - \tilde{\rho}_p}{\tilde{\rho}_o HU_o \nu} \ g v_p$$

where $\tilde{\rho}_o$= density of the fluid, $\tilde{\rho}_p$= density of a particle, v_p = volume of a particle. The parameter R^* measures the magnitude of the Archimedean force acting on a simple particle.

The equations determing the motion of particles will incorporate three effects: (i) the advection of particles by the fluid with the velocity $\vec{U}$, (ii) a negative geotaxy (uniform drift upward $\vec{j}$) and (iii) a random component $\vec{A}$ representing the variability of motility with time and from particle to particle , let

$$\frac{d\vec{r}_i}{dt} = \vec{v}_i = \vec{U}_i + \vec{j} + \vec{A}_i \ , \qquad i = 1, \ldots, \mathcal{N} \ . \tag{3.3}$$

The various physical assumptions concerning the motion of particles and their effect on the fluid are discussed in [18] and briefly mentioned in [2]. Various models for the random components $\vec{A}_i$ are described and tested in [2]; their description will not be reproduced here.

The discretization of the Navier-Stokes equations in conservative form (2.2a) is made by means of the three-level scheme (2.4) with $\gamma = 0$. The forcing term $\vec{F}$, evaluated therefore at time nk, is discretized as $\vec{F}_h = (0, F_h)$ with

$$F_h = - \frac{\beta}{Re} \, N^n_{\ell,m+1/2}$$

where $\beta = R^*/h^2$ and $N_{\ell,m+1/2}$ is the number of particles within the square PQRS of Fig. 18. Equation (3.3) is discretized as follows:

$$\vec{r}^{n+1}_i = \vec{r}^n_i + k \, [\vec{U}^{n+1}(\vec{r}^n_i) + \vec{j} + \vec{A}^{n+1}_i] \, , \quad i = 1,..,\mathcal{N} \, , \quad (3.4)$$

with the condition

$$|\vec{U}_i + \vec{j} + \vec{A}_i| \, k < h \, , \qquad i = 1,..,\mathcal{N} \, ,$$

which expresses the fact that no particles cross two adjacent horizontal of vertical lines of the mesh in a single time step. Note that $\vec{U}^{n+1}(\vec{r}^n_i)$ in (3.4) is defined by interpolation from the four nearest neighbours.

At the initial time $t = 0$, the fluid is at rest : $u = v = 0$, and the initial configuration of particles is given. The boundary conditions are

on A'ACC' : $u = v = 0$,

on A'C'   : $\frac{\partial u}{\partial y} = v = 0$.

Rules for the treatment of particles near the boundaries (wall and free surface) are described in [2].

Fig. 19 shows the initial configuration of particles. In the computations reported here, $\mathcal{N} = 756$, L/H = 2, Re = 5, h = 0.1, k = 0.02, $\varkappa = 0.345 \times 10^{-2}$ and $\lambda = 0.828$. Fig. 20 shows the result of a computation without motility, that is to say $\vec{v}_i = \vec{U}_i$ in (3.3) : the particles fall down and the ultimate state is the rest. Figs. 21 and 22 give some results with geotaxy but without random motion ($\vec{A}_i = 0$). Fig. 21 shows patterns obtained for $\beta = 5$ and we desplay the vertical velocity component $v \, (\varkappa, 1/2)$: this provides a good indication of the overall vortical structure of the flow field. In Fig. 22 we give results for $\beta = 15$ with the corresponding instanteneous streamlines of the flow superimposed on the particle pattern. We now describe the sequence of events (up to $t = 3$) revealed by a film from which the figures are extracted. We shall refer below to the particles initially contained in certain of the 36 equal squares labeled as Aa, Ab, Ba, etc., in Fig. 19.

$\underline{\beta = 5}$ (see Fig. 21). The initial motion of the particles in the column Bc, Bd through Ec, Ed is downward. Particles along the top of the pattern are carried toward top center by the vortical motion of the fluid, while particles Fc, Fd are moving upward. The lower edge of the pattern, well away from the center, begins to move outward and upward. As these particles fan out the column Cc - Ee, for example, has moved to the right and has rotated through an angle of about 45° by $t = 0.28$, bringing its midpoint to about the line $y = 1/2$. The pattern at $t = 0.5$ (Fig. 21a) is thus obtained as the particles Ee have begun to move not only upward but also inward toward the centerline x = 1, a result again of the vortical fluid flow. The dense band at the top of the pattern has resulted from the compression of the top rows of particles by this flow.

As t increases from 0.5 to 1.0, the change in the pattern occurs mainly in the central column, which is gradually emptied as the two arms move upward and inward (Fig. 21b). By t = 2 the arms have met and particles have begun to collect along the upper edge. By t = 2.08 a plume has begun to fall, reaching the bottom by t = 2.4 and leaving behind it a "hole" devoid of particles. The inertia of this falling swarm of particles then carries the lower edge of the pattern down with it, until by t = 2.8 the distance between the edge and the boundary is about half that of Fig. 21c. By t = 3.0 the accumulation of particles along the lower edge has moved outward and upward as in the initial stages, and begun its ascent to the top.

While the distribution of particles at t = 3 is not close to the initial one, the area of the pattern is roughly the same, the morphological movements are similar, and it seems likely that as particles ascend they will accumulate near top center and will eventually again form a plume. There is thus evidence suggesting that the pattern remains time dependent and is conceivable periodic with period of about 3 or 4.

β = 15. (See Fig. 22). The initial motion downward and outward is more vigorous and there is little tendency for the extremities of the arms to rise. By t = 0.5 the pattern is roughly horizontal. The main new feature of the subsequent morphology is the formation of two new fluid vortices, resulting in two distinct particle aggregates. This is found to occur as particles near the center (that is, near the line x = 1) are able to ascend, with particles along the upper edge moving to the left and right. New vortices appear by t = 0.84 and now grow as the central column is cleared of particles. By t = 1.4 a dense accumulation occurs at the upper edge of each of the two aggregates, leading to simultaneous plume formation. The plumes reach the bottom of the aggregates at t = 1.9. The dense bands at the bottom of the aggregates in Fig. 22c are the remnants of the plumes. As t increases from 2 to 3, the motion of the particles within each aggregate is down over a roughly central column but up along the aggregate periphery on either side of the column. The particle trajectories within each aggregate are therefore found to be roughly like two adjacent vortical eddies of opposite sense, superimposed on vortical streamlines which thread through the aggregate.

The mechanism by which the value of β changes the number of aggregates thus appears to be essentially dynamic. For β = 15 the fluid domain cannot apparently support a single aggregate (such as that of the initial pattern); in effect by increasing β the weight of the particles has increased without any change of motility. Finally, Figs. 23 and 24 show patterns obtained in the case where a random motion is introduced ($\vec{A}_i \neq 0$), Figs. 23 and 24 correspond, respectively to β = 5 and β = 15. The effect of the stochastic component in the motility appears to produce a somewhat diffuse version of the pattern for simple geotaxy, at least in the parameter range shown here; however, we must note that for β = 25, these can emerge, according to the model of random motion, two of three aggregates (Fig. 25).

4. References

[1] P e y r e t , R.: Unsteady evolution of a horizontal jet in a stratified fluid. J. Fluid Mech., 78 (1976), pp. 49-63.

[2] C h i l d r e s s , S. and P e y r e t , R.: A numerical study of two-dimensional convection by motile particles. J. Méca. 15 (1976), pp. 753-779, Erratum 16 (1977), p. 803.

[3] C h o r i n , A.J.: A numerical method for solving incompressible viscous flow problems. J. Comp. Phys. 2 (1967), pp. 12-26.

[4] T e m a m, R.: Sur l'approximation de la solution des équations de Navier-Stokes par la méthode des pas fractionnaires. I. Archiv rat. Mech. Anal., 32 (1969), pp. 135-153. II. - ibid. - pp. 377-385.

[5] F o r t i n, M., P e y r e t , R. et T e m a n, R.: Résolution numérique des équations de Navier-Stokes pour un fluide incompressible. J. Méca. 10 (1971), pp. 357-390. (See also Lecture Notes in Physics 8 (1971), pp. 337342, Springer Verlag.

[6] H a r l o w , F.H. and W e l c h , J.E.: Numerical calculation of time dependent viscous incompressible flow. Phys. Fluids, 8 (1965), pp.2182-2189.

[7] H a r l o w, F.H., W e l c h , J.E., S h a n n o n , J.P. and D a l y ,B.J.: The MAC method. Los Alamos Scientific Lab. Report LA-3425, (March 1966).

[8] C h o r i n , A.J.: Numerical solution of the Navier-Stokes equations. Math. Comp. 22 (1968), pp. 745-762.

[9] V i e c e l l i , J.A.: A computing method for incompressible flows bounded by moving walls. J. Comp. Phys. 8 (1971), pp. 119-143.

[10]L i u , N.: Finite difference solution of the Navier-Stokes equations for incompressible three-dimensional internal flows. Lecture Notes in Physics 59 (1976), pp. 300305, Springer-Verlag.

[11]L a D e v e z e ,J. et P e y r e t ,R.: Calcul numérique d'une solution avec singularité des équations de Navier'Stokes : écoulement dans un canal avec variation brusque de section. J. Méca 13 (1974), pp. 367-396.

[12]E a s t o n , C.R.: Homogeneous boundary conditions for pressure in the MAC method. J. Comp. Phys. 9 (1972), pp. 375-379.

[13]A m s d e n .A.A. and H a r o w ,F.H.: The SMAC method. Los Alamos Scientific Lab. Report No. LA-4370 (1970).

[14]M a n t e l, B.: Etude numérique du comportement d'une zone de fluide homogène au sein d'un fluide stratifié. Thèse 3e cycle, Mécanique Théorique, Université Paris VI, (1979).

[15]R e b o u r c e t ,B.: Etude numérique de jets instationnaires en fluides homogènes et stratifiès. Thèse 3e cycle, Mécanique Théorique, Université Paris VI, 1980.

[16]B e a v e r s , G.S. and W i l s o n ,T.A.: Vortex growth in jets. J. Fluid Mech. 44 (1970), pp. 97-112.

[17]G r a n t , A.J.: Numerical model for instability in axisymmetric jets. J. Fluid Mech. 66 (1974), pp. 707-724.

[18]L e v a n d o w s k y , M., C h i l d r e s s ,S., S p i e g e l , E.A. and H u n t e r , S.H.: A mathematical model for pattern formation by swimming micro-organisms. J. Protozool., 22 (1975), pp. 296-306.

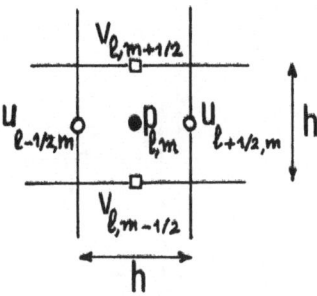

Figure 1

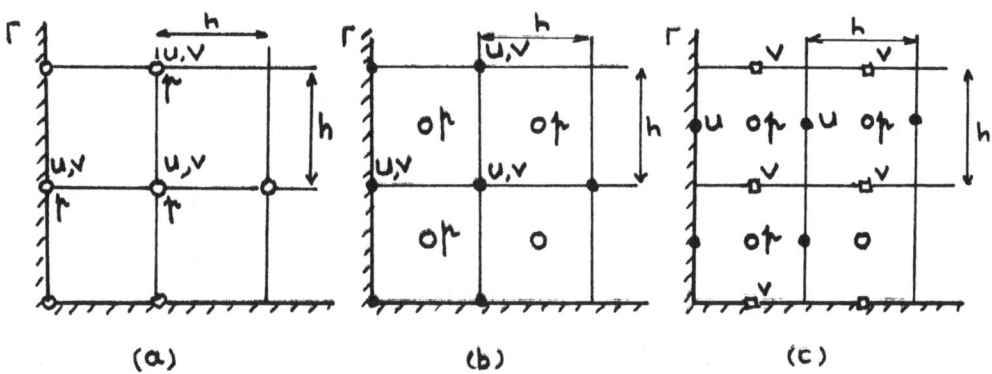

Fig. 2: Discretization mesh.

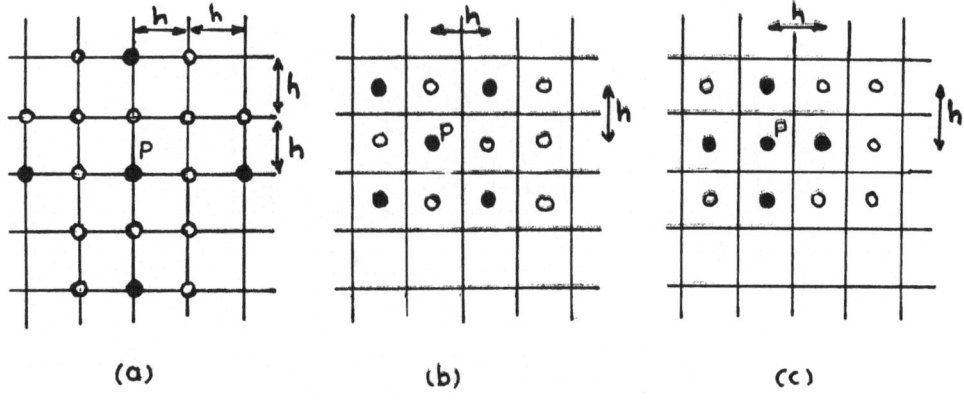

Fig. 3: Layout of points involved in the discretization of the Poisson equation for the pressure.

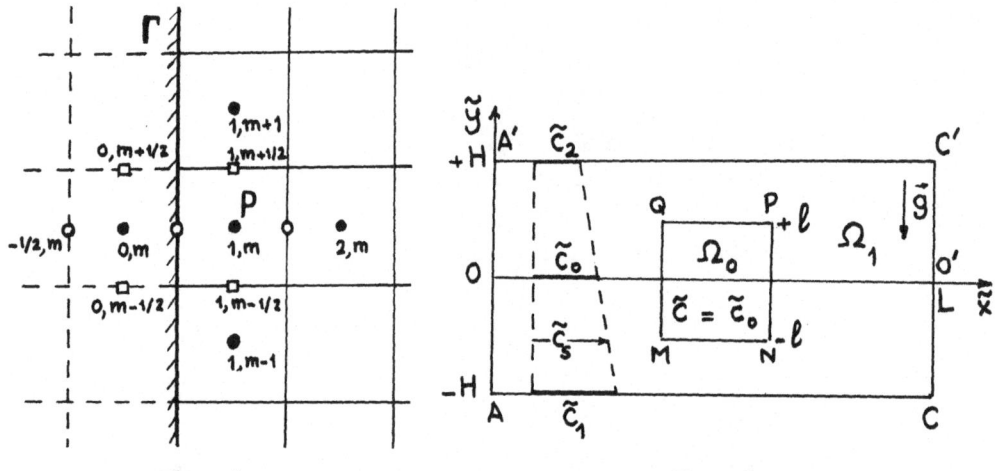

Figure 4   Figure 5

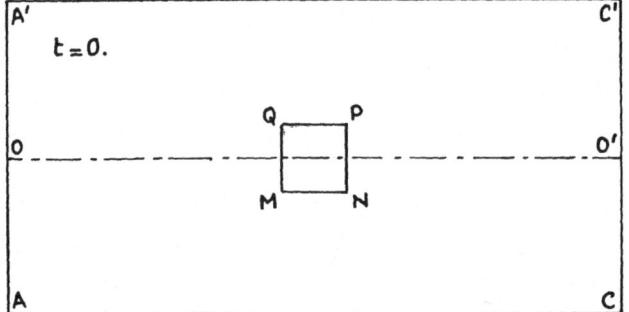

Figure 6: Initial configuration

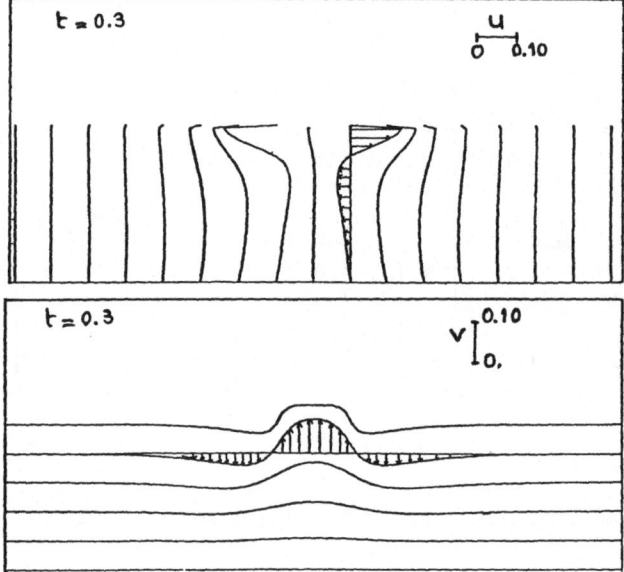

Fig. 7: Velocity profiles, Gr = 24.5 (lower half domain).

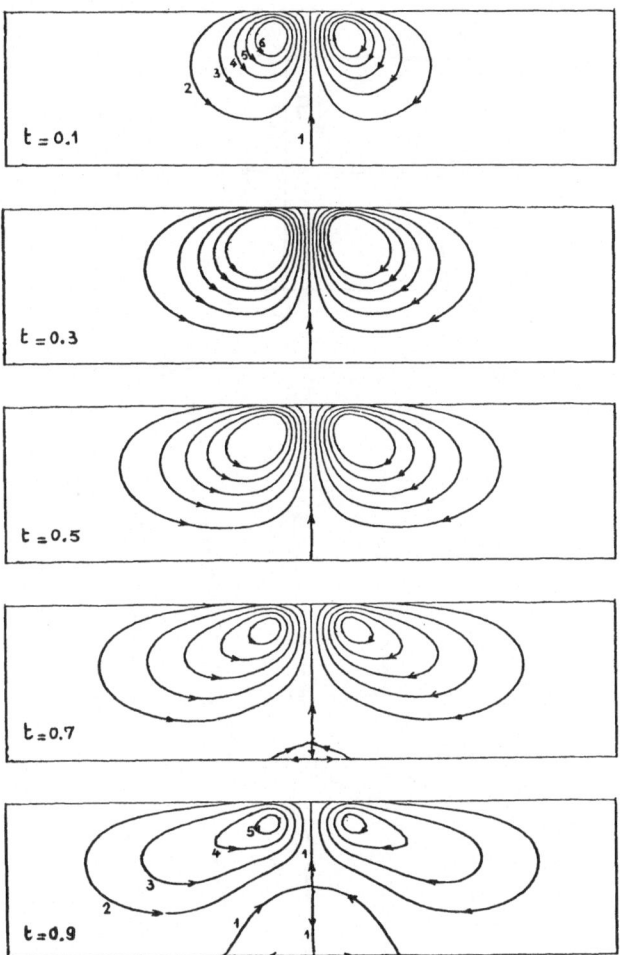

Fig. 8: Instantaneous streamlines, Gr = 24.5 (lower half domain),
curve (1) : $\psi = 0$ (2) : $\psi = -0.0051$
(3) : $\psi = -0.0103$ (4) : $\psi = -0.0155$
(5) : $\psi = -0.0206$ (6) : $\psi = -0.0258$

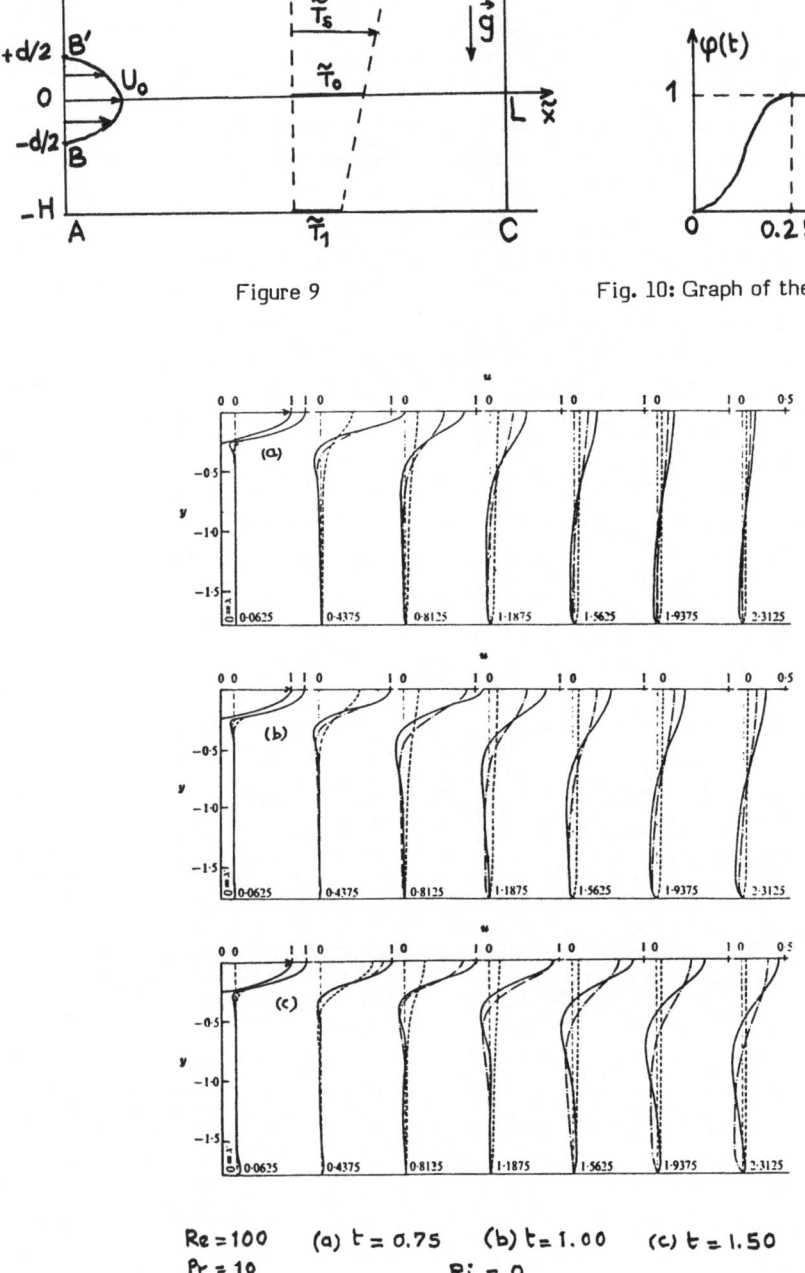

Figure 9

Fig. 10: Graph of the function $\psi(t)$.

Re = 100   (a) $t = 0.75$ (b) $t = 1.00$ (c) $t = 1.50$
Pr = 10   - - - - - -   Ri = 0
            — · — ·   Ri = 32
            ————   Ri = 64

Fig. 11: Case I: Finite basin-horizontal velocity profiles u(x,y,t) (lower half domain).

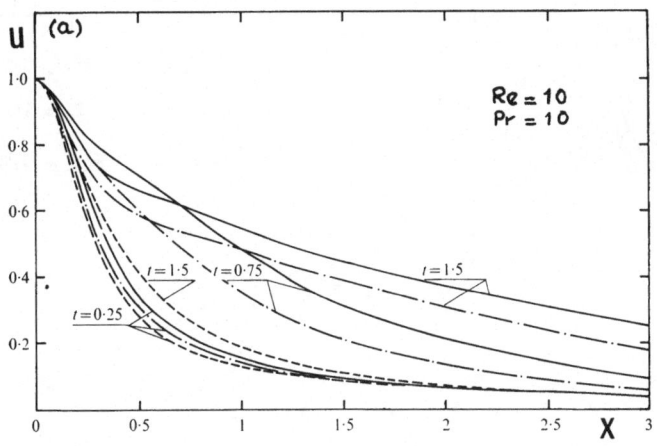

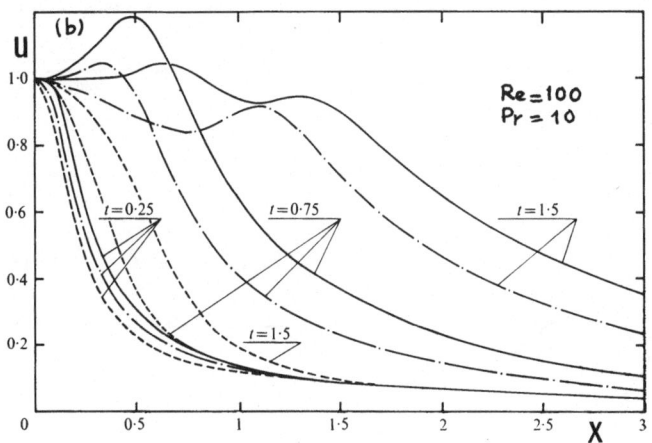

Fig. 12: Velocity u on the axis y = 0.
(a), (b): Case I - finite basin.
--- Ri = 0
-·- Ri = 32
——— Ri = 64

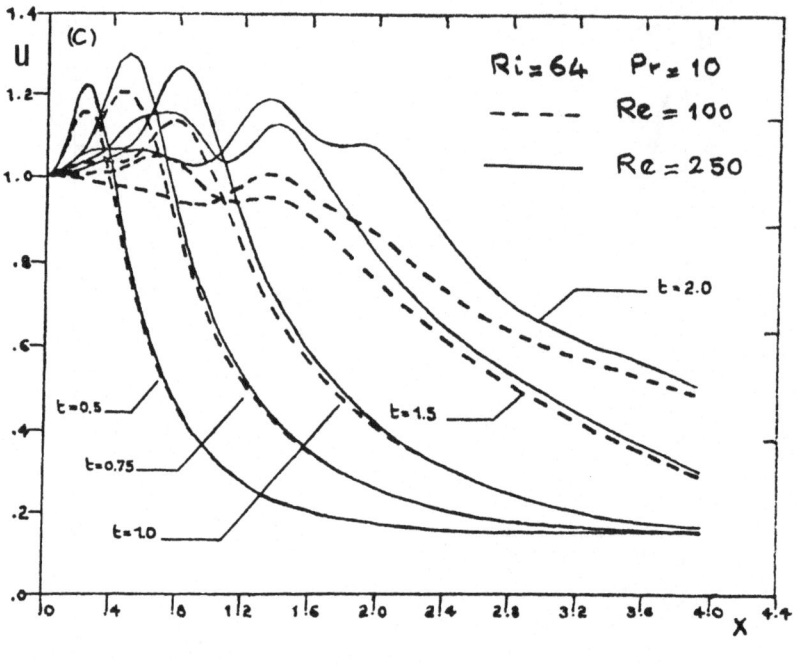

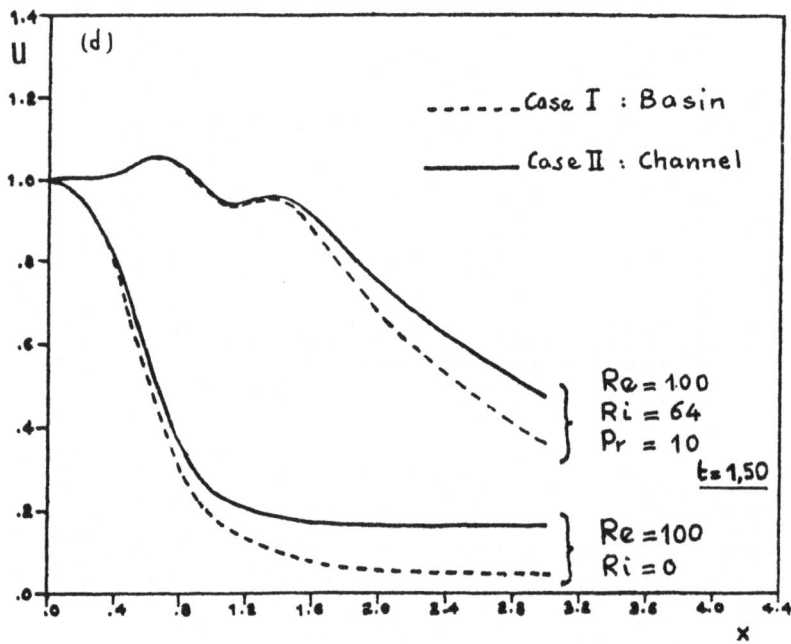

Fig. 12: Velocity u on the axis y = 0.

(c): Case II. Semi-infinite channel.

(d): Comparison between case I and case II.

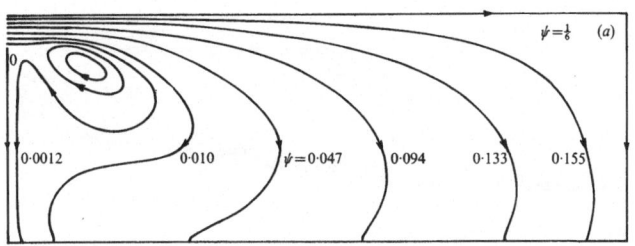

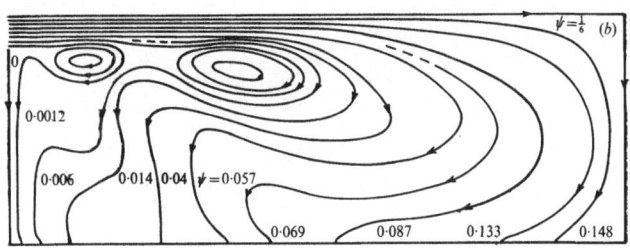

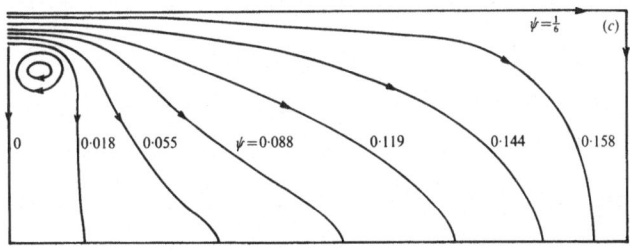

Fig. 13: Case I: Finite basin-instantaneous streamlines (lower half domain).

(a) t = 0.75	Re = 100	Ri = 64	Pr = 10
(b) t = 1.50	Re = 100	Ri = 64	Pr = 10
(c) t = 1.50	Re = 100	Ri = 0	

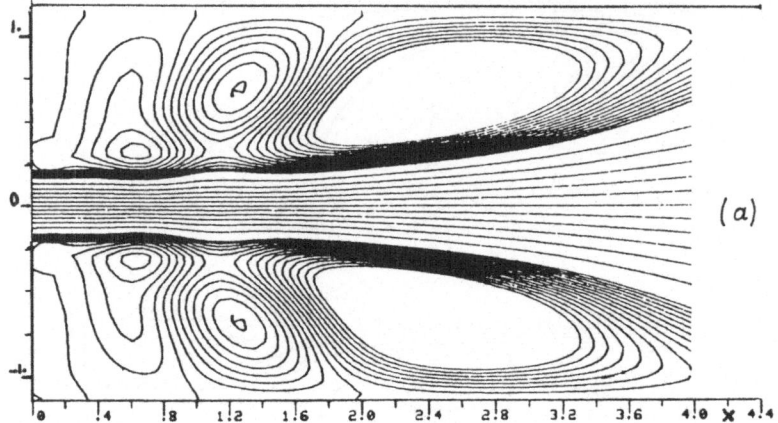

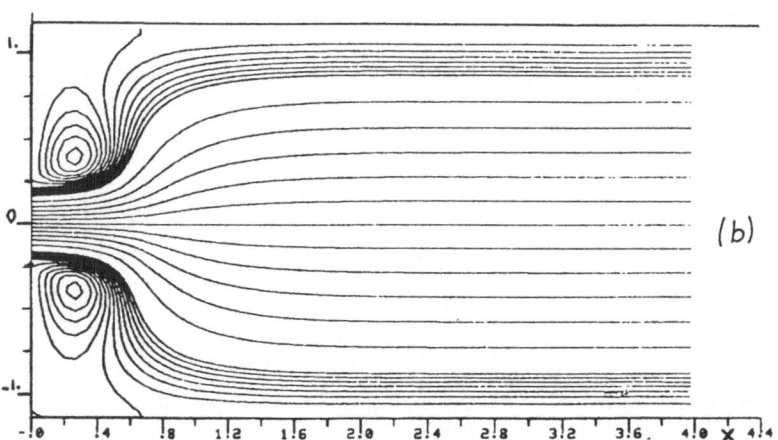

Fig. 14: Case II: Semi-infinite channel - instantaneous streamlines.
      (a) t = 1.50   Re = 100   Ri = 64   Pr = 10
      (b) t = 1.50   Re = 100   Ri = 0

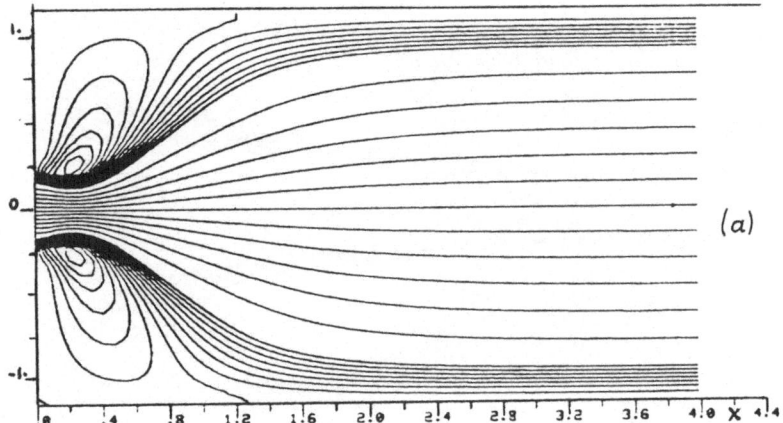

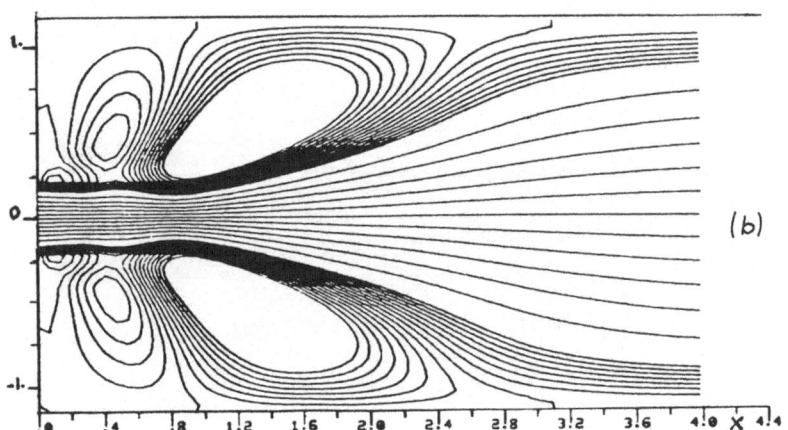

Fig. 15: Case II. Semi-infinite channel. Instantaneous streamlines.
Re = 250, Ri = 64, Pr = 10.
(a) t = 0.5   (b) t = 1.00

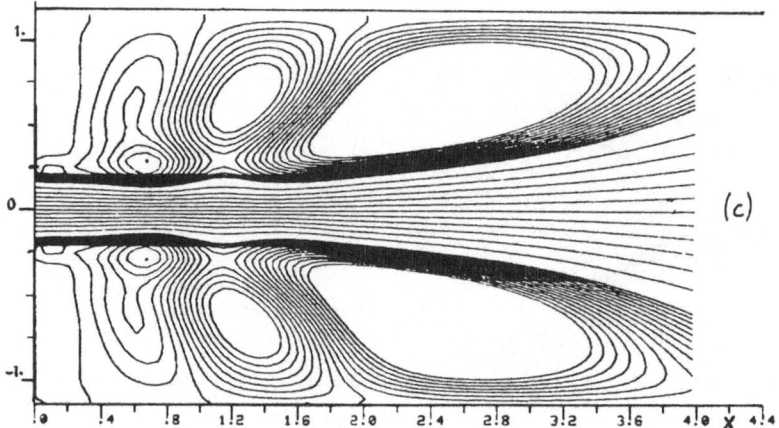

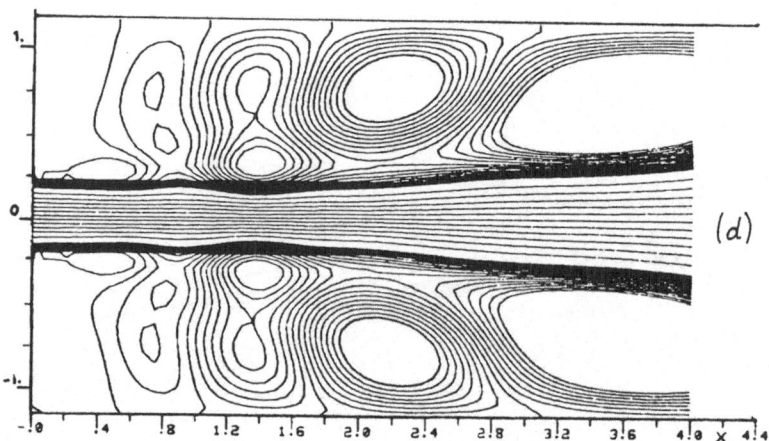

Fig. 15: Case II. Semi-infinite channel. Instantaneous streamlines.
Re = 250, Ri = 64, Pr = 10.
(c) t = 1.50   (d) t = 2.00

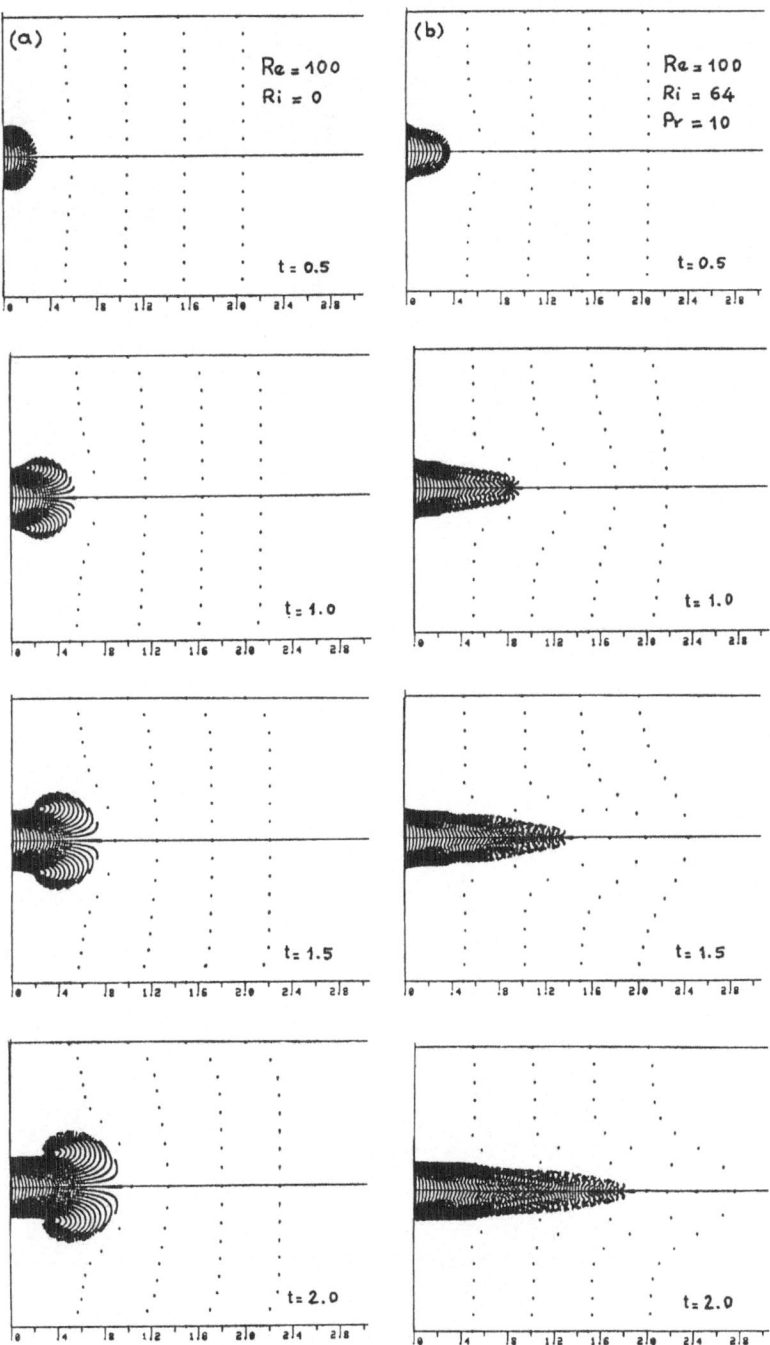

Fig. 16: Case II: Semi-infinite channel - motion of the intension.
(a) Re = 100   Ri = 0
(b) Re = 100   Ri = 64   Pr = 10

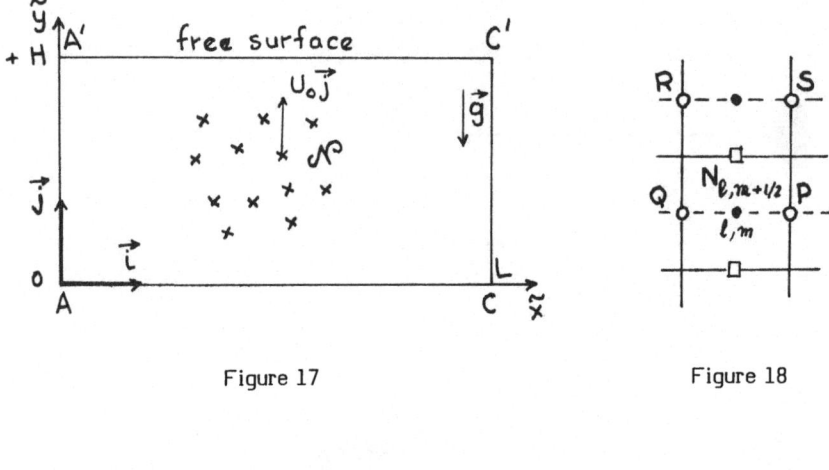

Figure 17   Figure 18

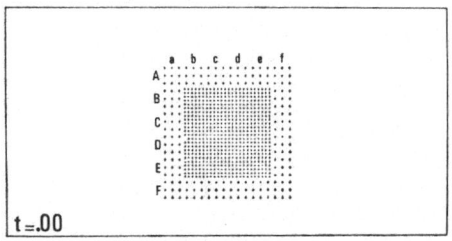

Figure 19: Initial configuration of the particles with $\mathcal{N}$= 756. Letters along edge refer to
36 equal squares containing 6 or 36 particles.

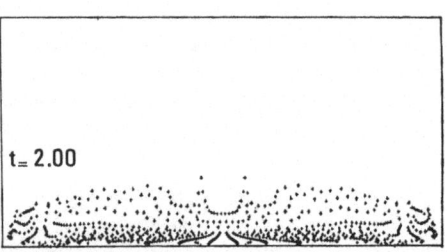

Fig. 20: Particle pattern: Case without motility $\vec{v}_i = \vec{U}_i$.

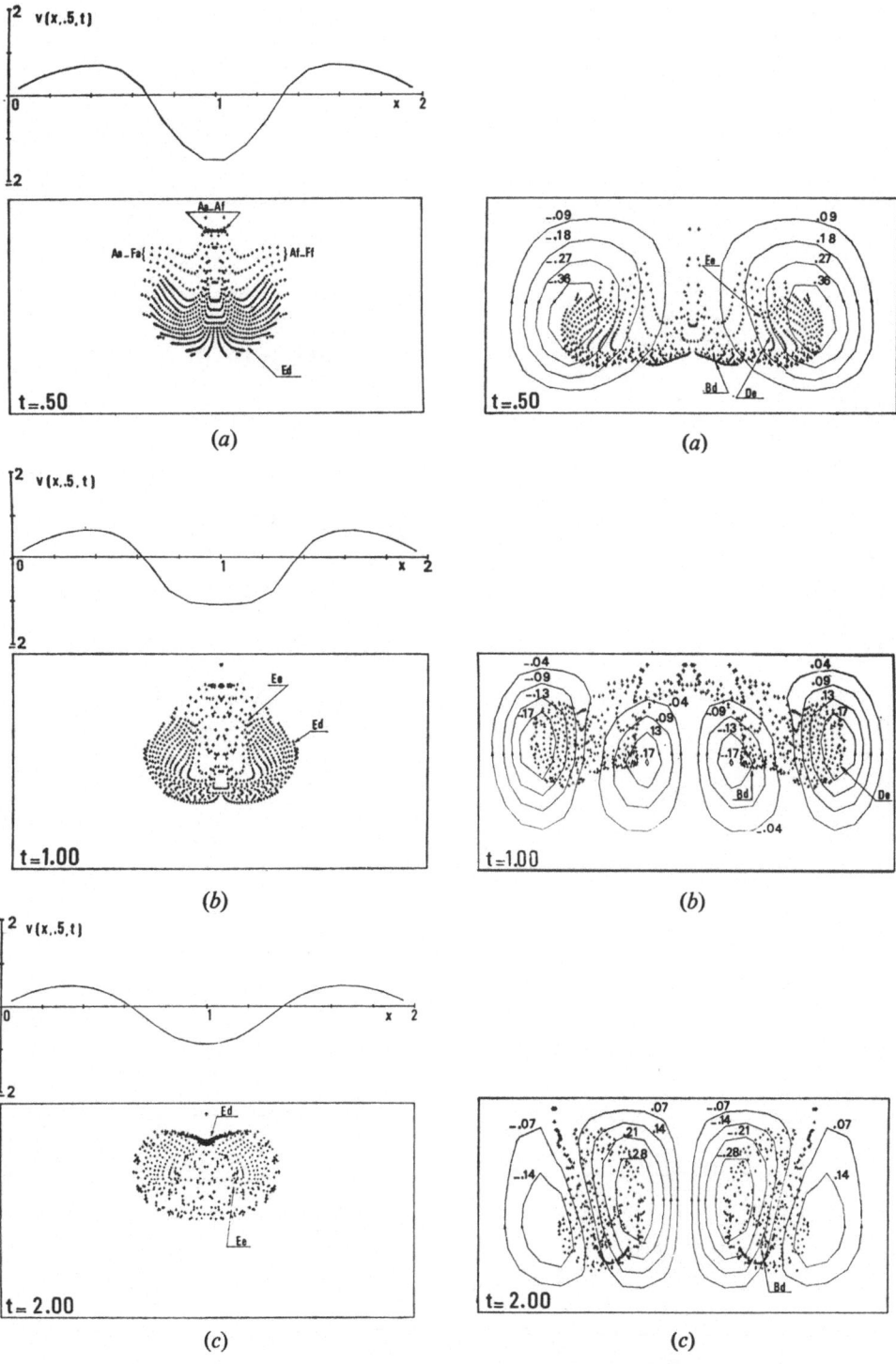

Fig. 21: Particle patterns and centerline values of v, $\beta = 5$, $\vec{A}_i = 0$.

Fig. 22: Particle patterns with superimposed instantaneous streamlines, $\beta = 15$, $\vec{A}_i = 0$.

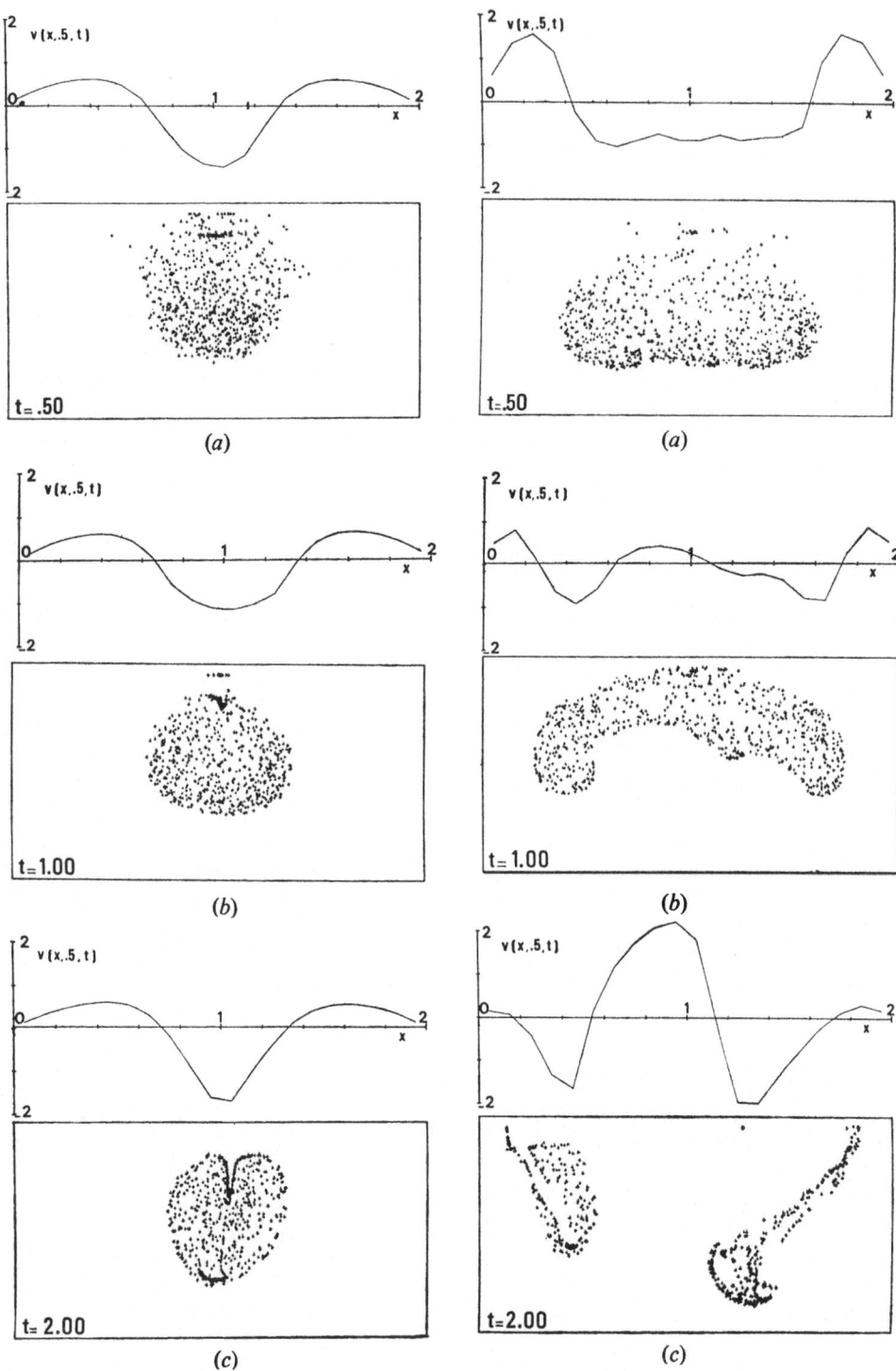

Fig. 23: Particle patterns and centerline values of v, $\beta = 5$, $\vec{A_i} \neq 0$.

Fig. 24: Particle patterns and centerline values of v, $\beta = 15$, $\vec{A_i} \neq 0$.

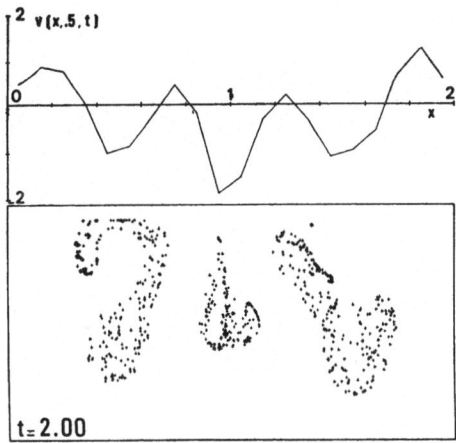

Fig. 25: Particle patterns and centerline values of v, $\beta = 25$, $\vec{A}_i \neq 0$.

Finite-Difference Techniques for Vectorized Fluid Dynamics Calculations

Editor: D.L. Book
1981. 60 figures. VIII, 226 pages
(Springer Series in Computational Physics)
ISBN 3-540-10482-8

Contents: Introduction. – Computational Techniques for Solution of Convective Equations. – Flux-Corrected Transport. – Efficient Time Integration Schemes for Atmosphere and Ocean Models. – A One-Dimensional Lagrangian Code for Nearly Incompressible Flow. – Two-Dimensional Lagrangian Fluid Dynamics Using Triangular Grids. – Solution of Elliptic Equations. – Vectorization of Fluid Codes. – Appendices A-E. – References.

R. Peyret, T. D. Taylor

Computational Methods for Fluid Flow

1982. Approx. 126 figures. Approx. 320 pages
(Springer Series in Computational Physics)
ISBN 3-540-11147-6

Contents: Numerical Approaches: Introduction and General Equations. Finite Difference Methods. Integral and Spectral Methods. Relationship Between Integral, Spectral and Finite Difference Methods. Specialized Methods. – Incompressible Flows: Finite Difference Solutions of the Navier-Stokes Equations. – Finite Element Methods Applied to Incompressible Flow. Spectral Method Solutions for Incompressible Flows. Turbulent Flow Models and Calculations. – Compressible Flows: Inviscid Compressible Flow. Viscous Compressible Flows. – Concluding Remarks.

Symposium of Numerical and Physical Aspects of Aerodynamic Flows

19–21 January 1981
California State University, Long Beach, California
Editor: T. Cebeci
1982. Approx. 346 figures. Approx. pages
ISBN 3-540-11044-5

This symposium treated numerical and physical aspects of aerodynamical flows. The contributions have been updated for publication and cover Numerical Fluid Dynamics, Interactive Steady Boundary Layers, Singularities in Unsteady Boundary Layers, Transonic Flows and Experimental Fluid Dynamics. Each section begins with a critical review and introduces the reader to the papers which follow. It is hoped that this volume provides a good overview of current knowledge, helps to set priorities for future developments and will due to the careful editing keep its value over the years.

D. P. Telionis

Unsteady Viscous Flows

1981. 132 figures. XXIII, 408 pages
(Springer Series in Computational Physics)
ISBN 3-540-10481-X

Contents: Basic Concepts. – Numerical Analysis. – Impulsive Motion. – Oscillations with Zero Mean. – Oscillating Flows with Non-Vanishing Mean. – Unsteady Turbulent Flows. – Unsteady Separation. – References.

F. Thomasset

Implementation of Finite Element Methods for Navier-Stokes Equations

1981. 86 figures. VI, 162 pages
(Springer Series in Computational Physics)
ISBN 3-540-10771-1

Contents: Introduction. Notations. – Elliptic Equations of Order 2: Some Standard Finite Element Methods. – Upwind Finite Element Schemes. – Numerical Solution of Stokes Equations. – Navier-Stokes Equations: Accuracy Assessments and Numerical Results. – Computational Problems and Bookkeeping. – Appendix 1: The Patch Test of the $P1$ Nonconforming Triangle: Sketchy Proof of Convergence. – Appendix 2: Numerical Illustration. – Appendix 3: The Zero divergence Basis for 2-D $P1$ Nonconforming Elements. – Three Dimensional Case. – References.

Springer-Verlag
Berlin
Heidelberg
NewYork

Lecture Notes in Physics

Selected Issues from
Lecture Notes in Mathematics